WORKING RIVER

HAER#ID-2-1

WORKING RIVER

Agriculture, Hydropower, and the Reshaping of Idaho's Snake River

Volume 1

Office of the Idaho Attorney General

Stephen Wee and Scott Miltenberger
foreword by Clive J. Strong

Ridenbaugh Press
Carlton, Oregon

WORKING RIVER: Agriculture, Hydropower, and the Reshaping of Idaho's Snake River

For more information, contact Ridenbaugh Press, P.O. Box 834, Carlton OR 97111.

Printed and bound in the United States of America.

First edition December 2022

10 9 8 7 6 5 4 3 2 1

Front cover image: American Falls Dam. Bureau of Reclamation.
https://www.usbr.gov/projects/index.php?id=24

Back cover image: Construction at Milner Dam. Library of Congress.
https://www.loc.gov/resource/hhh.id0139.photos/?sp=2

Library of Congress Cataloging in Publication Data

Stephen Wee and Scott Miltenberger

Strong, Clive (foreword)

Working River

Bibliography

1. History. 2. Idaho Law. 3. Idaho Economy. 4. Idaho National Resources.

I. (author last, first). II. Title.

ISBN 978-0-945648-54-3 (softbound)

Ridenbaugh Press

P.O. Box 834, Carlton OR 97111

Phone (503) 852-0010

www.ridenbaugh.com

stapilus@ridenbaugh.com

Contents

Foreword

by Clive J. Strong, former deputy attorney general

The Snake River in Idaho is a perfectly plain and obvious feature of the state, whether at ground level or from outer space. The uses to which it has been put, and the human conflict over what should be done with it, however, are not nearly so obvious. For more than a century there have been many heated controversies over who should have the better right to use the waters of the Snake River – and how and for what purposes the water should be used. The most significant of these controversies was whether hydropower generation should be subordinate to upstream development. *Working River*, a phrase coined by former Governor and U.S. Senator Len B. Jordan to describe the Snake River, describes the origin and evolution of the subordination policy, how it led to the "Swan Falls Controversy" in the late 1970s, and ultimately the 1984 Swan Falls Settlement, which for the first time, established "a proper balance among competing demands" for the limited water supply of the Snake River."

From the days of the first settlers in Idaho in the mid-1800s, the waters of the Snake River were diverted to irrigate the rich soils of the Snake River plain to create what Mark Fiege identified as an "Irrigated Eden," in his book of the same name. The federal government, the State, and private irrigators collaborated on water projects so extraordinary that by 1980 more than 3.1 million acres of the Snake River Plain were irrigated, about two-thirds of those acres using surface waters of the Snake River system.

Almost concurrent with agricultural development, Idaho Power Company ("Idaho Power" or "the Company") began acquiring and constructing dams on the Snake River to harness the hydropower potential of the river. With the construction of its Hells Canyon Complex in the mid-1900s, Idaho Power operated eleven dams on the Snake River extending from American Falls to Hells Canyon. The dams were designed to use essentially the entire flow of the Snake River to generate electricity.

Under a strict application of Idaho's prior appropriation doctrine – *which grants priority of use of the water to whoever first puts it to beneficial use* – Idaho Power's hydropower water rights could have blocked upstream diversions of water for irrigation and precluded much of the irrigation development on the Snake River Plain. Despite their seeming incompatibility, however, hydropower and irrigation development proceeded together, decade after decade, based on the widely held understanding among irrigators, the State, and the United States that in times of shortage, hydropower generation would be considered subordinate or secondary to the use of water for irrigation.

The potential for conflict between irrigation and hydropower development was identified as early as Idaho's State Constitutional Convention in 1889. Delegate Isaac N. Coston recognized that the then-nascent hydropower industry could pre-empt all future irrigation development within the river basin., "If the waterpower of this country can be used to prevent irrigation of the country, it can be held by virtue of a prior right, good-bye to all the prosperity we expect to come from the use of the water in irrigating our plains and developing this country." Despite Coston's warning, the State Constitution ratified in 1890 did not expressly address the potential for conflict between hydropower generation and agricultural development.

A formal policy subordinating the use of Snake River water for hydropower purposes to the needs of irrigated agriculture was not articulated until 1920. After completion of the Minidoka dam in 1906, the U.S. Reclamation Service (after 1923, the Bureau of Reclamation, or "Reclamation") and the State began to look for additional opportunities to reclaim lands in the Upper Snake River Basin. Reclamation identified American Falls as an ideal site for a large storage reservoir. In March of 1920, United States Director of Reclamation A.P. Davis and Idaho Governor D.W. Davis designated a joint federal-state Board of Engineers to consider the fundamental issues surrounding a reservoir at American Falls: the available water supply; the size of the reservoir necessary to meet existing and contemplated irrigation requirements; and the irrigation and power interests to be affected by the project. Idaho Power's existing hydroelectric facility was identified as one of the obstacles.

At the Board's first meeting, Idaho Commissioner of Reclamation W.G. Swendsen asserted the state's interest in "the formation of a definite plan for the development of the remaining resources of the Snake River water supply on a broad and comprehensive basis which would insure to the state the maximum utility of the possibilities of the stream." In the April 1920 report that followed, the Board concluded that the "greatest use of the water of Snake River will be found in the dedication of the entire flow of the stream to irrigation in so far as the water can be economically appropriated."

With "[t]he lowest point on the river at which diversion is practicable in large amounts [being] Milner Dam," the Board contemplated the building of a reservoir at American Falls with the requisite capacity to make use of "a large percentage of all the waters originating in the watershed above Milner Dam" for irrigation purposes. "The net effect of this condition," wrote the Board, "will be to dry up the river below Milner Dam during the irrigation season, also to as great an extent as possible below American Falls Reservoir, during the non-irrigation season."

"The principle involved therefore," concluded the Board, "is to secure as nearly as possible a total use of the waters for irrigation above Milner Dam, and to secure the greatest possible use of power below Milner Dam."

This became what is now commonly referred as a "Milner Zero flow policy."

To effectuate this policy, the Board concluded: "In granting power rights in the future the Federal Government and the State should so far as possible provide restrictions requiring its eventual surrender when and as the waters are required for application to land."

With a definite plan in place for the development of the Snake River above Milner, Reclamation initiated discussions with Idaho Power to acquire its American Falls hydropower facility. The negotiations between Reclamation and the Company were long and involved; but ultimately, the parties agreed on a contract in 1923 by which the Federal government acquired most of the Company's American Falls project. While Idaho Power retained the right to use flows Reclamation released from the reservoir to generate hydropower, it had no right to demand the release of water from reservoir except for limited storage rights the

Company obtained under the contract. Idaho Reclamation Commissioner W. G. Swendsen characterized the contract as protecting Snake River irrigators and being consistent with the state constitutional provisions that declared the rights of agriculturalists in water to be superior to other interests.

While the 1923 contract established the priority of use of water for irrigation over hydropower in the Snake River Basin above Milner Dam, it marked the beginning of a long and contentious relationship between Reclamation and Idaho Power over how and to what extent the Company's hydropower water rights would be subordinated to future agricultural development. This conflict played out initially in proceedings before the Federal Power Commission (FPC), which under the 1920 Federal Power Act has authority over the licensing of private hydropower projects on navigable rivers.

Reclamation's policy, consistent with the 1920 Report, was to seek inclusion of subordination clauses in the FPC licenses for private hydropower facilities on the Snake River that provided all rights to the use of water for power purposes be considered subordinate to all rights acquired by the United States or other parties for irrigation. The 1934 FPC License for the Twin Falls project, for example, included Article 14, which reads:

> As a condition of this license, the Licensee, for itself, its successors and assigns, hereby stipulates and agrees that all rights to the use of water for power purposes heretofore or hereafter required for the development of power at the site of this project shall be held and considered at all times to be subject, inferior, and subordinate to all rights heretofore or hereafter acquired by the United States or other parties for irrigation purposes, [with two limited exceptions].

Although Idaho Power consented to inclusion of Article 14 in its FPC License for the Twin Falls project, in subsequent FPC licensing proceedings the Company resisted inclusion of subordination clauses in its licenses. The Company argued State law governs whether and to what extent its use of water for power is subordinate to agricultural development. Additionally, the Company argued for compensation for the impact of agricultural development on its facilities.

As downstream out-of-state interests contemplated the construction of new dams downstream from Milner in the 1920s, Idaho amended Article 15, Section 3 of the Idaho Constitution in 1927 to expressly empower the state to subordinate future hydropower water rights. The State, however, for unknown reasons did not enact legislation to confirm that use of water for hydropower is subordinate agricultural development until 1984. Instead, the State implemented the policy on a project-by-project basis.

When Reclamation announced its intention to construct a high Hells Canyon Dam, Idaho Power publicly embraced Idaho's hydropower subordination policy to garner State and irrigators support for its own Hells Canyon project. The Company argued that if Reclamation's Hells Canyon project was constructed the federal government would take the irrigators' water rights and interfere with agriculture in southern Idaho. Idaho Power stated that if it were allowed to construct its Hells Canyon hydroelectric project there would be no threat to the irrigators' water rights. Thomas Roach, Idaho Power's president, testified before the FPC on the Company's applications for its three proposed hydroelectric power dams at Oxbow, Brownlee, and Hells Canyon that:

> Our company for a period of 37 years or more has had a very firm and fixed policy of complete coordination of the use of the Snake River waters for the development of hydroelectric power with the needs of that water for irrigation and has followed the policy of always placing the use of that water for irrigation in a prior position to the use of the water for hydroelectric development.
>
> As far back formally as 1947, in our hearing, our initial hearing, before the Oregon Hydroelectric Commission, that policy was stated and made a formal part of our application to the Oregon Hydroelectric Commission, and currently all of our State permits in the State of Idaho carry in them a specific provision which preserves for irrigation not only now but in the future a prior claim on the water, with the claim for hydroelectric energy being secondary to that of the irrigator or the farmer.

Idaho Power agreed to the inclusion of the following subordination clause in the FPC license for the project:

> The project shall be operated in such manner as will not conflict with the future depletion in flow of the waters of Snake River and its

> tributaries, or prevent or interfere with the future upstream diversion and use of such water above the backwater created by the project, for the irrigation of lands and other beneficial consumptive uses in the Snake River watershed.

At the request of Governor Len B. Jordan Idaho Power agreed to the inclusion of a subordination clause in its hydropower water right for the C.J. Strike, which was under development at the time of the Hells Canyon controversy. Additionally, the Company agreed to the subordination of hydropower water rights for the Hells Canyon project, but for unknown reasons the State did not include subordination clauses in most of the project water rights.

With assurances from Idaho Power, the State and irrigators supported the Company's project over Reclamation's. After a long and acrimonious fight, detailed in Karl Brooks' *Public Power, Private Dams: The Hell's Canyon High Dam Controversy* book, Idaho Power won the right to build three hydropower dams in Hells Canyon. Agricultural development in the Snake River Basin above Hells Canyon proceeded for over two decades with the settled expectation that Idaho Power's hydropower water rights were subordinated to upstream development.

Thus, it came as a surprise to all when in 1983, the Idaho Supreme Court, in response to legal proceeding initiated by a group of conservation activists and ratepayers before the Idaho Public Utilities Commission, issued a decision putting into question the long-held understanding that Idaho Power's hydropower water rights were subordinated to upstream irrigation and other consumptive uses of the waters of the Snake River.

While the legal proceeding centered on Idaho Power's Swan Falls hydropower water rights the question of water rights was not their immediate concern – the Company's plan to build the Pioneer coal fired-plant was. Idaho Power held water rights to 8,400 cfs (cubic feet per second) for its Swan Falls facility with priority dates *from 1900 to 1919*. None of the water rights for the facility contained a subordination clause. Conservationists and ratepayers, attacking the Company's inclusion of the full value of the Swan Fall facilities in its rate base, implicitly argued that if the Company had protected its water rights, Pioneer would not be necessary.

In response, Idaho Power sued a number of upstream water users and the State, asserting that its Swan Falls hydropower water rights were not subordinated. The Company's lawsuit provoked outrage. The State and irrigators successfully argued before the Fourth Judicial District Court of Ada County that Idaho Power's FPC license for the Hells Canyon Complex subordinated *all* the Company's hydropower water rights to upstream development.

Unexpectedly, however, the Supreme Court reversed the District Court and held that the subordination clause in the Hells Canyon license only applied to the Company's Hells Canyon Complex. The Court's decision, however, left open the question of whether the Company had subordinated its hydropower water rights through other means, such as waiver, forfeiture, or abandonment.

After the Supreme Court's decision, Idaho Power initiated a second lawsuit asserting its hydropower water rights had priority over approximately 7,500 upstream water rights. The court's decision and Idaho Power's second lawsuit fractured the alliance between the irrigators, the State, and Company. Idaho Power's defense of the Supreme Court decision was viewed by the State and irrigators as a betrayal of the promises the Company made during the Hells Canyon licensing proceeding. They saw the Supreme Court's decision as a threat to the foundation of Idaho water law and if allowed to stand it would install Idaho Power as the "water master of the Snake River." Thus, began a two-year acrimonious battle in the Fourth Judicial District Court for Ada County and the Idaho Legislature. The intensity of the conflict is reflected in then-Idaho Attorney General Jim Jones' book *A Little Dam Problem*.

After fighting to a standstill in the 1984 legislative session, both sides recognized litigation was not likely to lead to a resolution of the century old conflict over whether use of water for hydropower should be subordinate to agricultural development. Thus, the State and Idaho Power over the course of six months negotiated the 1984 Swan Falls Agreement. As more fully described in the 2016 Idaho Law Review article *Understanding the 1984 Swan Falls Settlement*, "[t]he use of the term "Agreement" was somewhat of a misnomer in that it did not resolve the lawsuits or the overall controversy. Rather, the Agreement set forth a process for implementing "a series of judicial, legislative, and administrative actions" set forth in the parties' October 1, 1984,

Framework for Final Resolution of Snake River Water Rights Controversy. The parties agreed the actions set forth in the Agreement were in the public interest and if implemented would resolve the controversy to the mutual satisfaction of the parties.

As initially drafted, the Agreement proposed to resolve the dispute over the subordination of Idaho Power's water rights by: 1) increasing the Swan Falls minimum stream flow in the State Water Plan to 3,900 cfs in the summer and 5,600 cfs in the winter, 2) recognizing the Company is entitled to unsubordinated water right equivalent to the Swan Falls minimum flow, and 3) making the Company's hydropower water rights for its facilities below Milner in excess of the minimum flow below subordinatable to future upstream development approved under proposed criteria for the reallocation of water available under the subordinatable rights. As explained in *Understanding the Swan Falls Agreement*, the Agreement was amended at the eleventh-hour to ensure the State not Idaho Power would have the authority to determine whether to reallocate water available under the subordinatable water rights. The parties proposed legislation by which the State would exercise its constitutional authority to regulate and limit the use of water for hydropower to create a trust for the benefit of Idaho Power and the people of Idaho. The Agreement as finally drafted provided that the State "would assert the existence of water rights held in trust by the State and that the Snake River is fully appropriated as needed to enforce the State Water Plan." The method for settling the trust, however, was not addressed in the Agreement, and ultimately led in 2006 to a three-year long battle between Idaho Power and the State that threatened to undo the Swan Falls Settlement.

Anticipating the need to relitigate the question of whether Idaho Power's hydropower water rights were subordinated, the State retained JRP Historical Consulting, LLC in 2006 to conduct an historical study of the origin and evolution of Idaho's hydropower subordination policy. While JRP report was conducting it historical review, the State prevailed on its motion for summary judgment before the SRBA Court asserting that the Governor holds in trust the subordinatable hydropower water rights for the Idaho Power hydroelectric facilities below Milner Dam for the benefit of Idaho Power and the people of Idaho. Following the April 18, 2008 SRBA Court decision, the State and Idaho Power Company entered into settlement discussions that led to the *2009 Swan Falls Reaffirmation Agreement* in which the parties' reached agreement on the ownership and elements of the hydropower water rights for Idaho Power's facilities.

Although JRP's report was never submitted to the SRBA Court, Attorney General Lawrence Wasden determined that publication of JRP's historical study is in the public interest as it provides context for the policies implemented as part of the Swan Falls Settlement. Working River helps to inform how policies such as the Zero Flow at Milner, and the subordination of Idaho Power's hydropower water rights reshaped the Snake River.

The silver lining from the Swan Falls controversy is out of the controversy Idaho not only memorialized its hydropower subordination policy, but in the process of doing so formally adjudicated all rights to the use of water in the Snake River Basin and led to amendment of the Idaho State Water Plan to established "a proper balance among competing demands" for the limited water supply of the Snake River." With these tools, the State is now able to address looming water management issues such as the conjunctive management of the Snake River Basin.

Clive J. Strong

Working River represents research commissioned by the Idaho Attorney General's Office to assist Attorney General staff in understanding and responding to issues arising in the SRBA regarding the subordination of Idaho Power's water rights. JRP Historical Consulting, LLC was retained by the Attorney General in 2006 to conduct a comprehensive historical review of the relationship between irrigation and hydropower development in the Snake River Basin. Although ultimately the SRBA issues were settled by the 2009 Reaffirmation Agreement, the Attorney General determined that it is in the public interest to publish JRP's research to help inform the public of the origin and reasons for Idaho's hydropower subordination policy.

For its research, JRP reviewed historical records not only in Idaho but also across the United States, documents that captured the events sketched out in the book from Idaho's statehood up to the 1984 Swan Falls Settlement. JRP historians Stephen Wee and Scott Miltenberger, together with staff historians and research assistants examined materials in the National Archives in Washington, D.C., Denver, Co., College Park, Md., Seattle, Wa., as well records in the Library of Congress, the California State Library in Sacramento, Ca., University of California in Davis, Ca., Idaho State Historical Society Archives, the University of Idaho, Idaho State University, and Boise State University. JRP also examined records at the Harry S. Truman and Dwight D. Eisenhower Presidential Libraries, Federal Energy Regulatory Commission in Washington, D.C., the U.S. Bureau of Reclamation in Denver, Co. and the Idaho Department of Water Resources. In all, JRP collected and reviewed nearly 10,000 documents. Copies of these documents will be deposited at the Idaho State Historical Society Archives.

Working River takes this story up to the Swan Falls Settlement. Readers interested in following the story of the Swan Falls Controversy in the years since – and more detail about subordination – may want to review these law review articles:

The Origin and Evolution of Hydropower Subordination Policy on the Snake River: A Century of Conflict and Cooperation, 46 Idaho L. Rev. 119 (2009);

Understanding the 1984 Swam Falls Settlement, 52 Idaho L. Rev. 223 (2016); and *Understanding the Snake River Basin Adjudication*, 52 Idaho L. Rev. 53 (2016).

The law review articles carry the development of policy forward from the 1984 Swan Falls Settlement through the Snake River Basin Adjudication, which was largely completed in 2014. Readers may also be interested in an oral history of the adjudication, *Through the Waters*, from the Idaho State Bar, which described the origins and development of the SRBA through the viewpoints of many of its participants.

The authors wish to recognize the contributions of the Office of the Idaho Attorney General in making this book possible. Attorney General Wasden was a steadfast supporter of this project. The following staff of the Attorney General's Office assisted the authors by providing documents and review of the book: Clive Strong, David Barber, Michael Orr, Ann Vonde, David Stanish, Pat Boehm, Shasta Kilminster-Hadley and Victoria Wigle.

The authors further wish to express their gratitude to Renee Jaussaud, Marene Baker, and the staff at the National Archives in College Park, Maryland, Denver, Colorado, and Seattle, Washington, as well as the countless librarians and archivists at the Truman and Eisenhower presidential libraries, Boise State University Special Collections, and the Idaho State Historical Society – most especially former Idaho State Archivist Rod House.

Stephen Wee and Scott Miltenberger

PART ONE

FROM SAGEBRUSH DESERT TO IRRIGATED LANDSCAPE, 1860-1927

Sec. 3 *Priority of appropriation shall give the better right as between those using the water; but when the waters of any natural stream are not sufficient for the service of all those desiring the use of the same,...those using the water for agricultural purposes shall have preference over those using the same for manufacturing purposes....*

Article XV, Sections 3 and 5, of the Idaho State Constitution, 1890

Without irrigation, agriculture, and hence, permanent settlement, could not be thought of. The farmer then must select a location near a stream whose waters can be turned from its course and spread over the soil. With the stream diverted and flowing through his ditches and laterals, he can laugh at the cloudless skies; for now he has the much-needed moisture under perfect control, the only essential agency originally lacking, but which, placed at his disposal in this manner, makes the conditions for the highest degree of agricultural prosperity simply ideal.

D. W. Ross, Idaho State Engineer, 1900

A river is more than an amenity, it is a treasure. It offers a necessity of life that must be rationed among those who have power over it.

Justice O. W. Holmes, *New Jersey v. New York, 1931*

1

The Irrigation Preference

When Idaho became a territory in 1863 its people saw the vast sagebrush desert of the Snake River Plain as an obstacle on their journey to the rich farm lands west of the Cascade Mountains. It was a dry and dangerous wasteland, with a large and treacherous river running through and serving as a barrier to the poor trails they used. But, a generation later, when Idaho territory was transformed into a state, they saw this vast and mostly unpopulated region differently. They now saw it with ambition: They sought to transform the forbidding desert into bountiful agricultural land. This was no frontier con artist's fantasy Geography not only had made southern Idaho a desert, but also had furnished the means to remake that desert into rich farmlands. That means was the Snake River.

The earliest American settlers, glimpsing the potential, diverted water from the Snake River and its tributaries with primitive canals and ditches diverting from easily tapped watercourses.

As the most accessible lands were developed, as agriculture expanded and matured, and as natural conditions complicated further irrigation, Idahoans applied increasing administrative and technological sophistication to their quest of turning the high mountain desert into an agricultural oasis. They experimented with an array of mechanisms for moving water where they wanted it. They formed canal companies and irrigation districts, and soon after statehood they began to partner with the federal government and its agencies. They constructed dams high in the Upper Snake River Basin to develop carryover storage as an antidote to drought, conserve winter flood flows, and provide electrical energy for cities, towns, rural families, and to run irrigation pumping plants.

But the focus of this water use was on irrigation.

The water supply of the Upper Snake River Basin became so thoroughly devoted to irrigation that in the summer the river literally stopped flowing at Milner Dam where the last of the major irrigation canals extracted the remaining water in the Snake River Basin, leaving its bed dry. They could do that because hidden beneath the desert was a huge aquifer which stored vastly more water than the man-made reservoirs, which rehydrated the river below Milner. Since the Second World War, Idaho water users developed this groundwater supply directly by drilling wells, and thereby brought more than a million additional acres of desert into cultivation.

Despite periodic disputes over water, Idahoans have demonstrated a long-standing commitment to irrigation as the primary use of the waters of the Snake River Basin – a commitment forged early in the state's history and enshrined as a central tenet of the state's constitution.

The lay of the land

In the arid environment of southern Idaho and the Snake River Basin, water played an important role in building and reshaping the landscape. In modern times its diversion by man for irrigation brought fundamental and widespread changes in land use to the plains of Southern Idaho.

The source of much of that water is the rain and snow falling on the mountains around the Snake River Basin, an area that encompasses nearly 109,000 square miles of present-day Wyoming, Utah, Nevada, Idaho, Oregon, and Washington. The Snake River Basin encompasses about 87 percent of Idaho, the whole of the state except its northern panhandle and its extreme southeastern corner.

The Snake River originates on the west side of the Continental Divide in northwest Wyoming, in Yellowstone National Park. Streams pour from these mountains and converge on the main branch of the river, its south fork, which furnishes the outlet for Shoshone and Lewis lakes and flows generally south through Jackson Lake, then westerly through Wyoming's Snake River Canyon to Alpine Junction at the Idaho-Wyoming state line.

The Snake River enters Idaho at the eastern side of today's Palisades Reservoir. Below Palisades Dam, which was built by the Bureau of

Reclamation in 1957, the river bends northwestward, enters Swan Valley, and flows 60 miles through a wide canyon in lava country before opening onto the arid eastern Snake River Plain near the small town of Heise. From Heise the river swings generally south and west traversing a roughly crescent-shaped plain across south-central Idaho to Milner, a distance of more than 200 miles. In this section, the Snake flows first northwesterly to its confluence with Henrys Fork (also known as the Snake River's north fork) at Rigby, then begins its southwesterly and westerly course across the gently rolling lava plateau that slopes toward the river.[1]

The river passes through the cities of Idaho Falls and Blackfoot before it is impounded by American Falls Dam, a storage reservoir built by the Bureau of Reclamation in 1927 as part of the Minidoka Reclamation Project. About fifty miles downstream, after receiving the waters of Raft River, the Snake River enters another Bureau irrigation reservoir, Lake Walcott, which is impounded by Minidoka Dam, constructed in 1906. About fifteen miles downriver from Minidoka Dam at a place known historically as "the Cedars," another irrigation dam, Milner Dam, forms a small reservoir.

In this reach from Rigby to Milner, the river is lies close to the elevation of the plains around it. For that reason, the largest diversions for irrigation purposes are made in this area, since the water is relatively easy to divert. Indeed, the diversions in this reach of the river are so great that immediately below Milner, the Snake River goes essentially dry during the irrigation season.

Milner is not the end of the Snake River, however, since the river is replenished by irrigation-return flows, and by groundwater. The Snake River Plain Aquifer, one of the most productive aquifers in the world, underlies about 11,000 square miles of the Snake River Plain, and it spouts water into the Snake River from springs which emerge along the river banks below Milner. The most dramatic such display is at Thousand Springs in the Hagerman Valley where water from the aquifer

---

1 Henrys Fork is sometimes referred to as the North Fork of the Snake River in early documents while the main river above Henrys Fork is sometimes designated as the South Fork. The more common usage is to refer to the north branch as "Henrys Fork" and the southern branch above Henry's Fork as "the Snake River." Henry's Fork rises in Henry's Lake in the northwestern corner of Idaho and is joined in its southerly course by Warm, Fall and Teton rivers, which drain the west slope of the Teton Range. W. G. Hoyt, *Water Utilization in the Snake River Basin*, United States Department of the Interior, USGS Water Supply Paper 657 (Washington DC: United States Printing Office, 1935), 1.

gushes from beneath the rim rock on a steep canyon wall and cascades down into the river. The springs are thought to be the reappearance of Idaho's great Lost River which sinks into the desert some 90 miles to the north and emerges as hundreds of crystal clear 60° springs in the Hagerman Valley.

West of Milner, the river has an average fall of nearly 20 feet to the mile for the next 80 miles. The river cuts its channel deep into the lava sheet and sedimentary beds, to depths as great as 700 feet below the level of the valley plain, as it continues its journey westward to King Hill. In this Milner-to-King Hill reach the river pours down several magnificent falls, the most spectacular being the great Shoshone Falls (210 feet) and Twin Falls (180 feet). Other notable vertical drops range from 20 to 50 feet.

In addition to groundwater springs and irrigation return flow, several streams in this area contribute direct surface flow from the south: Rock Creek, Salmon Falls Creek, Bruneau River, and Castle Creek. Most of the streams flowing south out of the mountains of Central Idaho, such as the Lost River, disappear into the ground before reaching the Snake River; however, Big Wood River and Little Wood River contribute direct flow at their confluence with the Snake near Hagerman. At the far western end of the plain the volume of the Snake River is greatly augmented by the discharge of the Boise, Payette, and Weiser rivers from the north, and on the south the Owyhee and Malheur rivers; each contribute significant tributary flows before the Snake plunges into its Hells Canyon reach.

The Snake River Plain stretches west to the Owyhee Mountains near the Oregon border where the river flows through a wide valley. On the far western side of the Snake River Plain, in the area anchored by the Boise metro area, the altitude of the open plain approaches that of the river channel once again. At Payette and Weiser the streambed is only slightly entrenched, and the river runs only a few feet below the city elevations.

After passing Weiser, the river veers north, entering a gradually deepening canyon lined by the Wallowa and Seven Devils Mountains. Here begins the astonishingly narrow and deep Hells Canyon, a 150 mile stretch of the river running from Weiser to the mouth of the Salmon River. More than 40 miles of this reach has an average canyon depth of more than 5,500 feet. In its Hells Canyon reach the Snake

River drops in elevation over 1,200 feet and is impounded by three hydroelectric power dams built and operated by Idaho Power Company: Brownlee (1958), Oxbow (1961), and Hells Canyon (1967). Shortly after emerging from the Hells Canyon Wilderness (created in 1975), the Snake River is joined by the Salmon River, one of its largest tributaries, and flows northerly to the river cities of Lewiston, Idaho and Clarkston, Washington. There it receives the flow of the Clearwater River, its largest tributary.

The Snake River then enters the state of Washington and flows 140 twisting miles through the Palouse region to its confluence with the Columbia River at Pasco, Washington.

In all, the Snake River flows through its basin 1,050 miles, dropping from an elevation of 9,840 feet on the western slope of the Rocky Mountains in Yellowstone National Park, Wyoming to 340 feet where it joins the mighty Columbia River in Washington. It is the Columbia's principal tributary and contributes about one-fourth of the river's total flow.[2]

Under natural conditions, upper reaches of the Snake River fluctuated greatly in volume from season to season each year. In the spring, as snow melted in the mountains of Wyoming and eastern Idaho, the river rose in its bed and sometimes spilled over its banks as it emerged onto the Snake River Plain. As summer approaches and the snow disappears, the volume of water carried in the river's channel falls. By early autumn, the river's flow is a fraction of its spring flood. Yearly variations in precipitation influence the annual cycle as the amount of rain and winter snowfall determine the magnitude and duration of spring and summer flows. Groundwater levels further affect percolation rates of underground water passing through gravels and into streams, and consequently, the volume of water carried in the Snake River. Groundwater contributions to the total flow of the river are significant. The river lost or gained flow in relation to its surrounding water tables, so surface flow in the river has varied widely, over time or over season or both, in particular reaches of its channel.[3]

---

2 Hoyt, *Water Utilization*, 1-2, 6-7, 22-23, and 227; and Harold T. Stearns, Lynn Crandall, and Willard G. Steward *Geology and Ground-Water Resources of the Snake River Plain in Southeastern Idaho*, United States Department of the Interior, USGS Water Supply Paper 774 (Washington, DC: Government Printing Office (GPO), 1938), 1-2 and 5-6.

3 F. H. Newell, "Idaho," in Department of the Interior, Census Office, *Report on Agriculture By Irrigation in the Western Part of the United States at the Eleventh Census: 1890* (GPO, 1894), 136-138; U. S. Department of Commerce, Bureau of the Census, *Fourteenth Census of the United*

The Snake River Basin has been subject to wetter and dryer cycles of precipitation and river run-off. Extreme drought years such as 1905, 1910, and 1919 brought catastrophe for agriculture and stimulated talk of developing storage facilities on the river, providing backup water during droughts. From 1929 to 1942 farmers endured an unusually long period of low run-off, the most severe period of water shortage for basin irrigators in the twentieth century. On the other hand, a cycle of very high runoff occurred between 1950 and 1957 and from 1965 through 1976 – although this second generally wet period was punctuated by sub-normal run-off in 1968 and 1973.

Historically, geological and hydraulic studies have divided the Snake River into three or more sections based upon key topographical, hydrological, and geological characteristics associated with each river section. Commonly, the term "Upper Snake River" refers to the section of the Snake River from its headwaters in Wyoming to Milner Dam. This stretch of the river includes its tributary sources in the mountains and the vast agricultural region downstream where the river opens onto the eastern Snake River Plain down to Milner Dam, near the city of Twin Falls, where the stream begins its descent into a deep canyon. The "Upper Snake River" is further subdivided into "upper" and "lower" (lower case) sub-regions at American Falls. The river's canyon stretch from Milner downstream to Lewiston is of an entirely different nature than the river above Milner and is referred to as the "Middle Snake River" (or the Canyon of the Snake River). At Lewiston the Snake River joins the Clearwater River, leaves Idaho, and flows 140 miles west to its confluence with the Columbia River; this reach of the river is referred to as the "Lower Snake River."[4]

Above Milner Dam, the Snake River, running as it does barely below the plains, is well-suited to irrigated agriculture. Below Milner, the cascading river in its canyon reach is better adapted to hydroelectric power development. The contrast is so stark that historical references to these sections of the river often treat the two reaches as two different

---

States Taken in the Year 1920, Volume VII: Irrigation and Drainage (GPO, 1922), 163-166.

4 See W. G. Hoyt, "Report Relative to Application of Idaho Power Company for a Preliminary Permit to Develop Upper Salmon Falls Site on Snake River, Idaho: Permit No. 19, Federal Power Commission, Idaho 26643," July 1921. Folder: D-100, 12-G-44; and "Report Relative to Water Power Resources of Snake River and Status of Public Lands between Milner and Weiser, Idaho," August 1922. Folder: D-100, 12-G-51, Water and Power Branch, Records Concerning Stream Classification, 1894-1961 (WPB, 1894-1961), Records of the United States Geological Survey, Record Group 57 (RG 57), National Archives at College Park, MD (NARA II); and Hoyt, *Water Utilization*; and Stearns, Crandall, and Steward, *Geology and Groundwater*.

or independent river systems. U.S. Geological Survey hydraulic engineer W. G. Hoyt, for example, noted the striking dissimilarity in a 1921 report to the Federal Power Commission:

> In general the river may be divided into two main sections. First, that portion upstream from Milner, and second, the portion downstream from Milner. In the portion of the river upstream from Milner the use of water for irrigation will predominate over the power use, while in the section downstream the use of water for power purposes will undoubtedly predominate, since the larger portion of the land adjacent to the river is at such an elevation that it cannot feasibly be irrigated by water pumped or diverted below Milner dam.[5]

Above Milner, and particularly between American Falls and Milner, the Snake River very slowly recedes between lava ridges and gradually retreats from the level of the plain. The river remains accessible in this reach above Milner and surface diversions and pump irrigation can be readily accomplished.[6] By contrast, the river's deep and rapid descent into its canyon below Milner makes irrigation diversions more difficult.

The vast Eastern Snake Plain Aquifer acts as a giant underground reservoir for a considerable part of the surface run-off and return flow from irrigated areas diverting water from the Upper Snake River. The two primary outlets for this groundwater basin are far apart and are manifested in springs emerging above American Falls and in springs and canyon side-flow below Milner. At American Falls the spring flow is captured by a storage reservoir completed in 1927. The spring flows entering the river between Milner and King Hill emerge so far below the elevation of the surrounding plain that the water supply is outside the practical reach of either direct diversions or pump irrigation systems. This increased underflow, coupled with a large number of concentrated vertical falls between Milner and King Hill – most notably, Dry Creek Falls, Twin Falls, Shoshone Falls, Auger Falls, Pillar Falls, Kanoka Rapids, and Upper and Lower Salmon Falls – made the Middle Snake River reach an excellent location for hydropower generation facilities.[7]

---

5 Hoyt, "Report Relative to Application of Idaho Power Company for a Preliminary Permit to Develop Upper Salmon Falls Site," July 1921, 5. Folder: D-100, 12-G-44, WPB, 1894-1951, RG 57, NARA II.

6 Hoyt, *Water Utilization*, 7-9, 26, 34, 120-121.

Oddly, the hydroelectric power potential of the river below Milner actually benefited from the surface water diversions of upstream irrigators that dried summer surface flow at Milner. W. G. Hoyt completed a comprehensive field examination of power sites on Snake River between Milner and Weiser for the U.S Geological Survey, the first such survey in the area, in 1921. Hoyt observed that the continuous flow from springs that replenished this reach of the river "makes almost the entire stretch an important power resource *regardless of any future irrigation diversions*." [emphasis added] In fact, he concluded: "Practically all the power can be developed without material interference with existing structures or proposed developments of any kind." Observing the power sites on Snake River below Milner Dam, Hoyt argued that the canals and irrigated tracts of the Twin Falls North Side Land & Water Company had a measurable beneficial streamflow effect in their favor.

> The main canal of the Twin Falls North Side Land & Power Co. receives water diverted at Milner and runs nearly parallel to Snake River, although at a considerable distance to the north. This canal furnishes water for irrigating a large tract between it and the river. The operation of this irrigation system has increased both the ground water and the flow of the springs....Although it is not known how much irrigation has increased the flow of these springs it has increased the flow and additional irrigation should increase it still more.[8]

Further studies of this unique aspect of the relationship between irrigation and power generation on the Upper and Middle Snake River bolstered Hoyt's early observations about the nature of the underground flows and their value to the hydroelectric power industry. Hoyt continued to monitor the Snake River for the USGS and 14 years later wrote an influential water supply paper, *Water Utilization in the Snake River Basin* (1935). In it, he again noted that conditions at Milner split the Snake River into two separate rivers. A peculiar combination of geologic, hydrologic, and topographic features had facilitated nearly complete diversion and use of the river's flow for irrigation above Milner without having seriously damaged the usefulness of the river for

---

7 Hoyt, "Report Relative to Water Power Resources of Snake River and Status of Public Lands between Milner and Weiser, Idaho," August 1922, 25-26, 33, and 57-63. Folder: D-100, 12-G-44, WPB, 1894-1961, RG 57, NARA II.

8 Ibid., 3, and 26. Folder: D-100, 12-G-44, WPB, 1894-1961, RG 57, NARA II.

generation of hydroelectric power in the spring-fed canyon reaches of the river below Milner Dam.[9]

Map 1, taken from Hoyt's 1935 *Water Utilization* study vividly illustrates the contrast between the two areas, above and below Milner.

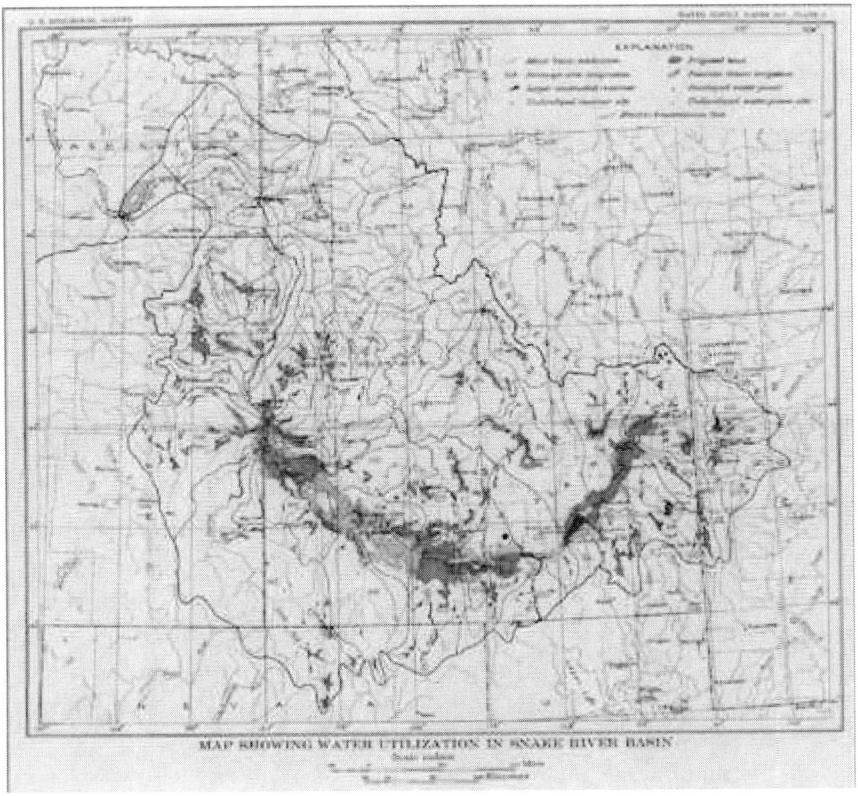

Water Utilization Map from Hoyt's 1935 USGS Study of the Snake River Basin. The areas in green are irrigated lands (the preponderance of which are above Milner or along Snake River tributaries), and the areas in yellow are possible locations of future irrigation. Red dots denote developed hydropower sites (the majority of which are located below Milner) and pink dots undeveloped power sites (the majority of which are also located below Milner). *Source*: W.G. Hoyt, *Water Utilization in the Snake River Basin*, United States Department of the Interior, USGS Water Supply Paper 657 (Washington DC: United States Printing Office, 1935), Plate 1.

9 Hoyt, *Water Utilization in the Snake River Basin*, 35-36, and 231-244.

Mining, irrigation, settlement

Mining preceded irrigation in Idaho; it was the main early industry in the territory. Elias D. Pierce discovered gold in the Clearwater drainage in 1860, and mining quickly spread to the Boise Valley. Homesteaders, farmers, and ranchers, who made a living supplying miners with food and feed for their animals, followed in the early 1860s. Outside of the mountain areas, irrigation was necessary to raise crops, and by 1863 farmers began to divert water from the Boise River and irrigate lands in the adjacent valley. Irrigation spread largely by individual enterprise into the Payette and Weiser valleys and north to Lewiston, at the confluence of the Snake and Columbia rivers.

The Snake River flowed through many inhospitable and treeless plains of sagebrush and bunch grass, but the well-drained, warm and porous volcanic ash in much of the soil in the southern part of the state was generally quite level and incredibly fertile. Combined with the ample sunshine and perennial stream flow of the snow-fed Snake, the great valley of Idaho had tremendous agricultural potential. As that opportunity became better known late in the nineteenth century, and as irrigation technology improved, thousands of irrigation farmers flocked to the barely-settled region.

Farmers first settled by tributaries of the Snake River. In the Boise River bottom lands below Boise, and in similar lands in the lower valley of the Payette and Weiser rivers, elementary agricultural development made possible the growing of hay, grain, fruit, and vegetables for local consumption in the gold and silver mining camps. In these river valleys, streams could easily be turned from their course by ditches and laterals and spread over the soil. At the end of the 1860s nearly two-thirds of the farms in Idaho still were located in Ada County (which then included more of southwestern Idaho than it does today) and most of the wheat, potatoes, and other marketable crops were grown in Boise Valley.

Railroad lines entered southern Idaho Territory and the Snake River region from the east in the early 1880s, and they too fostered irrigation. Before the railroad's arrival, the Snake River Plain from Pocatello westward was a sagebrush desert, but town and agricultural development followed the rails. Construction of the Oregon Short Line began in 1881 and was completed in 1884. The railroad connected with the main Union Pacific line in Wyoming and crossed the Idaho border passing through Montpelier, Soda Springs, and Pocatello by 1883; it continued west building freight and passenger stations at Minidoka, Shoshone, King Hill, Mountain Home, Nampa, Riverside, and Weiser as it crossed the Snake River Plain.

Between 1880 and 1890, the population of the Snake River Plain increased from about 500 to 15,000 people. Settlement and community building stimulated development of cropland and the livestock industry along railroad corridors. Cereals, hay, and potatoes, which were the bulk of agricultural production in the pioneer era, gave way to diversified higher value crops that could economically be transported by rail to mid-western and eastern centers of population. Manufacturing enterprises that used local agricultural products, like sugar beets, which were suitable to the area's soil and climate, gave further impetus to agricultural growth. Over the next two decades the Oregon Short Line, its successor, and their several branch lines would grow to serve much of the state. The railroads carried tourists to enjoy the physical beauty of the great basaltic rims and waterfalls of the Snake River region. Through popular articles, travel guides and books, the Union Pacific and Oregon Short Line executives advertised southern Idaho as "the choicest garden spot of the Rocky Mountains" and emphasized its natural advantages for irrigated agriculture, while opening the region to immigration, development, and outside markets.[10] Southern Idaho's resources were vast, and largely undeveloped. Settlers continued to pour into the lower Snake River valley around the Boise, Bruneau, and Payette rivers. In the 1880s experienced Mormon farmers, many of whom had worked on railroad construction crews and saw the potential in Idaho's fertile volcanic soils, moved onto the eastern Snake River plain. Church leaders in Salt Lake City encouraged this migration north, to expand the base of the church by colonizing the Snake River country with Latter-day Saints. Experienced from their Utah days in forming irrigation institutions, the Mormons rapidly founded canal companies and water users associations to finance, construct, operate, and maintain their irrigation works.

---

10 D. W. Ross, Biennial Report of the State Engineer to the Governor of Idaho, 1901-1902 (Boise: Capital Printing Office, 1903), 45, 68, 77-78; Leonard J. Arrington, *History of Idaho*, Volume 1 (Moscow: University of Idaho Press, 1994), 313-335, 340; George Athearn, "The Oregon Short Line," *Idaho Yesterdays* 13 (Winter 1969-1970), 2-18.

Through the 1890s and early 1900s, still more settlers, many of them non-Mormon stockmen, farmers, miners, and small-town businessmen from the Great Plains and Midwest, acquired land and founded towns in the upper valley. These non-Mormon farmers, whose crops required irrigation, availed themselves of the use of the public waters by appropriating them and building simple ditches with temporary brush dams for their diversion, often time neglecting to post and record notices of appropriation or to make public their intent other than by the actual act of construction.[11]

These small water conveyance systems were frequently owned by individual appropriators who owned or squatted on the land that was irrigated from them, but larger canals were also owned by irrigation companies or individuals who rented or sold the right to use the water flowing in them. These canal companies, organized to build the larger canal systems, did not own the land, but simply constructed the ditches, charging settlers for the water delivered and keeping the ditches maintained for an annual fee. This method of irrigation proved unpopular and in 1895 the Idaho legislature passed an act providing for organization of irrigation districts to facilitate purchase of these private canal systems and their water rights by the water users themselves. Irrigation districts proliferated in areas occupied by settlers whose lands were currently irrigated.

In the vast unsettled arid tracts of southeastern Idaho, the Carey Act provided a means for the state to promote land settlement by providing for completion of the necessary dams, reservoirs and canals.[12]

Throughout the early stages of irrigation development in the upper Snake River Basin above Idaho Falls, reservoirs were of little consequence since water was easily diverted from the Snake River flowed nearly at the level of its broad fertile flood plain.

While the quantity of land that could be cultivated by irrigation in southern Idaho stood at an estimated ten million or more acres and appeared virtually unlimited, Idaho's first state engineer, F. J. Mills, cautioned that conservation of water supply was "a matter of highest importance and interest" to the state. In his *First Biennial Report* to the

---

11 Leonard J. Arrington, "Irrigation in the Snake River Valley: An Historical Overview," *Idaho Yesterdays: The Quarterly Journal of the Idaho Historical Society* 30:1-2 (Spring/Summer 1986), 3-11. Mark, Fiege, *Irrigated Eden: The Making of an Irrigated Landscape in the American West* (Seattle: University of Washington Press, 1999), 81-95.

12 C. J. Brosnan, *History of the State of Idaho* (Chicago: Charles Scribner's Sons, 1918), 186-188.

governor of Idaho in 1896, Mills remarked that while the Snake River generally provided sufficient minimum natural flow for late irrigation, many of the smaller streams in Idaho were practically useless for irrigation purposes because all of their flow occurred in the early spring from melting snow and disappeared when most needed to water thirsty crops. Those waters, if stored, could open new areas to irrigation. Even on some of the larger streams commanding the best irrigated acreage of the state, Mills continued, utilization of the natural flow of the rivers and streams already had begun to reach its limits without the assistance of storage reservoirs. When snow that fell on the mountains melted off early, the irrigation season was too short. There was no issue of greater importance facing Idaho farmers, Mills argued, than development of storage facilities for irrigation, and nothing more urgent than for the state "to exercise full control" and initiate an ambitious dam construction program. He suggested that the governor sponsor legislation to expand the state engineer department's duties and powers to include supervision over a state-funded dam construction program.[13]

Yet, no such state program was initiated by the governor or the Idaho legislature. Judged by the small number of storage reservoirs existing in the state, little progress was made over the next decade until the rise in agricultural land values justified construction of the large Upper Snake River irrigation projects under provisions of the Carey Act and the Federal Reclamation Act.[14]

Passage of the Carey Act of 1894 and the National Reclamation Act of 1902 made possible large-scale irrigation development by common ownership and control of the land and water resources. In the vast unsettled arid tracts of southeastern Idaho, the Carey Act provided a means for the state to promote land settlement by providing for completion of the necessary dams, reservoirs and canals.

Idaho's territorial population stood at only about 15,000 people in 1870, but by 1890 at statehood it had soared to more than 88,000 inhabitants. A decade later the population nearly doubled again, to 161,000 people. Brushing aside periodic water supply problems, irrigated agriculture within the Snake River Basin rose between 1902 and 1919 from 569,286 to 2,163,892 acres, an increase of 380 percent

13 F. J. Mills, First Biennial Report of the State Engineer to the Governor of Idaho (Boise: Arch Cunningham, Printer, 1896), 6-9.
14 C. J. Brosnan, *History of the State of Idaho* (Chicago: Charles Scribner's Sons, 1918), 186-188.

in less than twenty years. In 1919, 86.9 percent of the irrigated acreage in Idaho was located within the Snake River Basin. Of these total irrigated acres within the basin, 64,832 were irrigated by direct diversion from the Snake River in 1902; by 1919 the total had risen to 716,908 acres, a more than tenfold increase. The amount of capital invested by 1920 in irrigation enterprises within the Snake River Basin was estimated at $84,317,216, representing 92.1 percent of the total irrigation investment statewide.[15]

All of this expansion occurred while little more than 1,000,000 acre-feet of storage had been developed on the Snake River above Milner Dam.

Table 1.1. Total Number of Farms and Acreage Irrigated in Idaho, 1890 to 1920[16]

Census Year	*Irrigated Farms*	*Acreage Irrigated*
1890	4,323	217,005
1900	9,188	608,718
1910	16,439	1,430,848
1920	25,283	2,488,806

Table 1.1 furnishes statistics showing the rapid increase in the number of irrigated farms and irrigated acreage in Idaho from 1890 to 1920.

On the expansive southern plains bordering the Snake River above American Falls, land with potential for irrigation by gravity flow from the Snake River and along the Henrys Fork and Teton rivers had been largely developed by the 1890s, as depicted in **Map 2**. In this upper portion of the great Snake River Basin gravity diversions were relatively easy and could be completed by tapping a stream a little above a cultivated plot with limited expense. As a result, the earliest natural flow priorities along the main Snake River and its upper tributaries are above American Falls. In contrast, below American Falls

15 A special census was taken in 1902 that presented data on irrigation by drainage basin. The 1920 census organized its data in the same manner. No other subsequent census, however, tabulated irrigation data by drainage basin. U. S. Bureau of the Census, Department of Commerce, *14th Census of the United States Taken in the Year 1920, Volume VII: Irrigation and Drainage* (GPO, 1922), 167-168.

16 U. S. Bureau of the Census, Department of Commerce, *14th Census of the United States Taken in the Year 1920, Volume VII: Irrigation and Drainage* (Washington DC: Government Printing Office, 1922), 166.

to Milner, irrigation works of great magnitude were required. Long, large capacity canals, major diversion dams, lift pumps, and power plants were required to deliver irrigation water to productive agricultural land. Consequently, these areas developed later, mostly after 1900, and were reclaimed in huge tracts served by large irrigation systems. Separate control of the land and the water that would be used for its cultivation was not feasible. Through the auspices of Carey Act of 1894 and the National Reclamation Act of 1902 irrigation development was made possible by common ownership and control of the land and water resources. Nineteen Carey Act projects were financed by private capital in Idaho between 1896 and 1918. During the same time period, the federal government built three reclamation projects in southern Idaho: the Boise Project, the Minidoka Project, and the relatively small King Hill Project. The size and cost of these large reclamation projects containing expensive reservoirs and extensive canal systems made more accurate knowledge of water supply and stream flow measurements necessary to properly plan developments. The federal government, through the U.S.G.S., took the lead in establishing stream gaging stations, collecting data, and disseminating information on stream flow measurements.[17]

The Carey Act encouraged private investment of capital and settlement of state-owned desert land by actual home-seekers. Insofar as achieving the settlement of a large body of unoccupied arid land was concerned, these projects accomplished the act's intended goal. However, the two major Carey Act projects in the vicinity of Twin Falls fell into difficulties early. While the engineering feasibility and construction of the water distribution system was carried out under supervision of the Idaho state engineer, the project was designed by overly optimistic, speculative investors whose facilities did not meet the best engineering standards of the time. Furthermore, irrigation water requirements for the climate, soil, and crop conditions present in southern Idaho was badly underestimated, water availability overstated, and project promoters had not made any provision for construction of water storage facilities as a hedge against dry years.[18]

17 C. J. Brosnan, *History of the State of Idaho* (Chicago: Charles Scribner's Sons, 1918), 187-191.
18 F. J. Mills, "Legislation Designed to Promote Irrigation Development" The Carey Act," *First Biennial Report of the State Engineer to the Governor of Idaho* (Boise: Arch Cunningham, Printer, 1896), 36-54; Leonard J. Arrington, "Irrigation in the Snake River Valley: An Historical Overview," *Idaho Yesterdays: The Quarterly Journal of the Idaho Historical Society* 30:1-2 (Spring/Summer 1986), 3-11; Mikel H. Williams, *The History of Development and Current Status*

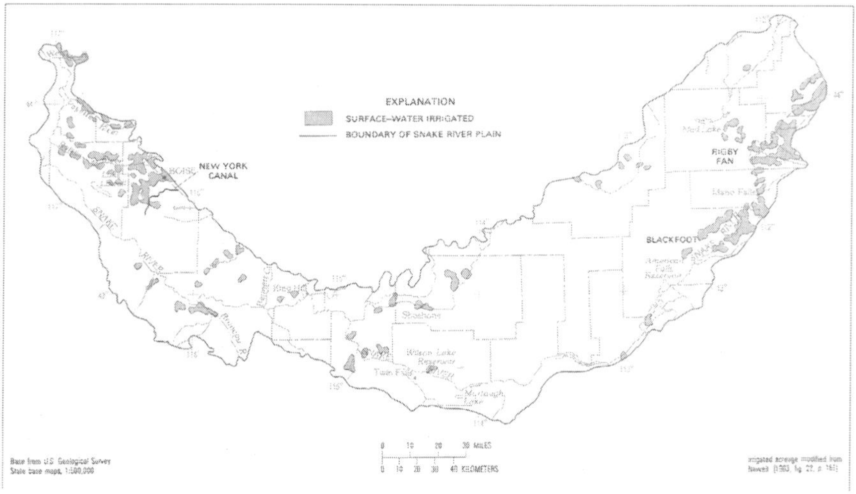

Irrigated Acreage in Southern Idaho, 1899. Source: S.A. Goodell, "Water Use on the Snake River Plain, Idaho and Eastern Oregon," Snake River Plain RASA Project, US Geological Survey Professional Paper 1408-E (GPO, 1988), Figure 7, E13.

Run-off varies widely on the Upper Snake River from year-to-year and in dry years late season irrigation water soon became insufficient to meet the water needs of would-be appropriators. Irrigators turned successively to storage reservoirs and later to groundwater pumping to supplement the natural flow of the river. Appropriative water rights to natural stream flow proved inadequate for junior irrigation developments and the relative rights of irrigators on the Upper Snake River were adjudicated by court decrees in the early decades of the twentieth century.

---

of the Carey Act in Idaho (Boise: Idaho Department of Reclamation, March 1970); Hugh T. Lovin, " A 'New West' Reclamation Tragedy: The Twin Falls Oakley Project in Idaho," *Arizona and the West* 20 (1978), 5-24; Although the majority of large private irrigation projects developed under the Carey Act in the western states were promoted by speculators with insufficient capital to carry them to completion, Idaho proved to be the most successful of the western states in terms of number of acres productively irrigated under this act. The Twin Falls South Side Project in Idaho was ultimately successful and became the largest Carey Act project in the nation. Federal assistance through co-operative development and conservation of water resources under the auspices of the Bureau and its federal reclamation projects in the Snake River Basin was a key component in its success. On Idaho's irrigation development under the Carey Act as compared to other western states, see *Statistical Appendix to the Annual Report of the Director of the Bureau of Land Management* (1958), Table 28. On the failure rate of private irrigation companies under the Carey Act generally, see Paul W. Gates, *History of Public Land Law Development* (Washington, DC: U. S Government Printing Office, November 1968), 647-652.

To acquire a reliable supply to meet project water needs during dry periods, the early twentieth century Carey Act and federal reclamation projects were required to develop holdover storage capacity as a necessary part of their agricultural operations. As a result of these historical circumstances, irrigated areas below American Falls tended to have later natural flow rights, but possessed the earlier storage priorities when compared to the areas irrigated above American Falls. These circumstances gave rise to conflicts over the delivery of water, court decrees establishing water priorities among irrigators, and to a complicated system of water exchanges between irrigators of the upper and lower segments of the Upper Snake River Basin. A special deputy appointed by the state engineer and the locally-elected watermaster coordinated and managed storage and natural flow rights on the upper Snake River above Milner Dam. Regardless of the physical location of the water accruing to each irrigator, storage or natural flow was accounted for in strict accordance with priorities as to time and amount.[19]

By the 1920s Idaho had approximately 2,500,000 acres of irrigated land and farming by irrigation was the state's greatest industry. In 1934, R. W. Faris, State Commissioner of Reclamation, estimated that fully "one-half of the people of Idaho depended upon irrigation, directly or indirectly, for their livelihood."[20] Most of the irrigated acreage before 1945 was irrigated with surface water with the centers of irrigation being located near the Snake River and its major tributaries. By this time the limits of irrigation from the natural flow of streams into ditches and canals had long since been attained, and irrigators had resorted to storage on the Upper Snake and Boise watersheds, pumping from rivers and streams to high-line canals, adopting conservation measures, and development of groundwater sources for domestic and stock watering to augment water supply for irrigation purposes. Irrigation pumping required large amounts of electrical power, which was provided by federal reclamation project power plants constructed

19 Detailed accounts of water supply and how water was measured, stored, diverted, collected and distributed according to the established Idaho State Water Right priorities on the river can be found in the district watermaster's annual reports, first prepared in 1919. See, typescript annual reports by year, "Water Distribution and Hydrometric Work, District No. 36, Snake River, Idaho," AR 20, Department of Reclamation (AR 20); unpublished reports held by the Idaho State Historical Society, Boise, ID (ISHS).

20 R. W. Faris, "Supplementary Water for Irrigation in Idaho With Particular Reference to Boise and Snake River Valleys, May 28, 1934." Folder: American Falls Unit, 1937 (1), Box 460. William E. Borah Papers (Borah Papers), Library of Congress, Manuscripts Division, Washington, D.C.

by the Bureau of Reclamation and increasingly after 1916 by Idaho Power Company (IPC).

In the second half of the twentieth century, advances in irrigation technology and favorable geologic and hydrologic conditions supported an expansion of irrigation into vast new tracts of land using the extensive Snake River Plain Aquifer as the sole source of water supply. While the Snake River itself had been the primary source of water for irrigation of croplands lying near the river, groundwater was used almost exclusively on the later-developed irrigated lands extending out into the Snake River Plain and away from areas irrigated with surface water. When demand for irrigation water increased again after World War II, irrigation-pumping technology had improved sufficiently to make groundwater pumping economically feasible. The Upper Snake River subterranean flows became a logical source of supply. More than a thousand deep groundwater wells tapped the aquifer in the 1950s and 1960s expanding the number of acres irrigated with groundwater on the Snake River Plain to some 700,000 acres by 1966.[21]

21 S. A. Goodell, *Water Use on the Snake River Plain Idaho and Eastern Oregon*, U. S Geological Survey Professional Paper 1408 – E (GPO, 1988), 11-17. Estimates of the amount of groundwater pumped for irrigation on the Snake River Plain are contained in two publications. Mundorff, Crosthwaite and Kilburn estimated that 920,000 acre-feet were pumped for irrigation needs in 1959. Norvotch, Thomas and Madison placed the estimate at 2.1 million acre-feet in 1965. See M. J. Mundorff, et al., *Groundwater for Irrigation in the Snake River Basin in Idaho*. U. S Geological Survey Professional Paper 1654 (GPO, 1964), 23; and R. F. Norvitch, et al., *Artificial Recharge to the Snake River Aquifer: An Evaluation of Potential and Effect,* Idaho Water Information Bulletin No. 12 (1969), 9.

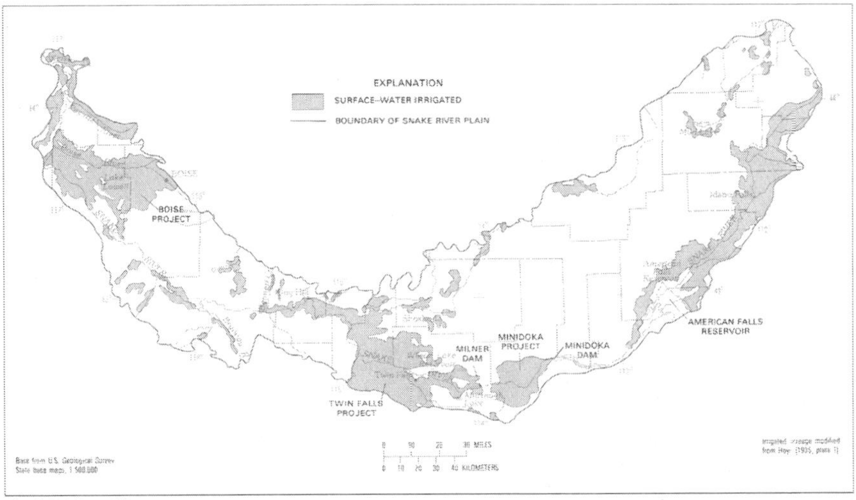

Figure 2. Irrigated Acreage, 1929
Source: S. A. Goodell, "Water Use on the Snake River Plain, Idaho and Eastern Oregon," Snake River Plain RASA Project, U. S. Geological Survey Professional Paper 1408-E (GPO, 1988), Figure 8, E14

Irrigated Acreage in Southern Idaho, 1929. *Source*: S.A. Goodell, "Water Use on the Snake River Plain, Idaho and Eastern Oregon," Snake River Plain RASA Project, US Geological Survey Professional Paper 1408-E (GPO, 1988), Figure 7, E13.

Territorial water law

In spite of the vital importance of water to Idaho even in its earliest days, there was little attempt to codify water rights in the territory prior to the 1880s.

In the wake of the discovery of gold in Idaho, miners adopted the prior appropriation doctrine – "first in time, first in right" – as the legal basis for water use on the public domain as had become local custom in the mining districts of California and across much of the American West.[22] The right to water was premised on use and was not dependent upon ownership of property, either bordering on a stream or otherwise. In practical terms, for irrigation purposes diversion and use was a necessity and the prior appropriation doctrine accommodated the

22 Robert G. Dunbar, *Forging New Rights in Western Waters* (Lincoln: University of Nebraska Press, 1983), 59-64; Wells A. Hutchins, *The California Law of Water Rights* (Sacramento: State of California, 1956), 41-51; Gordon R. Miller, "Shaping California Water Law, 1781- 1928," *Southern California Quarterly*55 (Spring 1973, 9-42; Dennis C. Colson, *Idaho's Constitution: The Tie That Binds* (Moscow: University of Idaho Press, 1991), 161-162; and Wells A. Hutchins, *Water Rights Law in the Nineteen Western States*, Volume I (Washington DC: U. S. Department of Agriculture, 1974), Chapters 7 and 8.

conditions and demands of early agricultural water users. During the territorial period when irrigation projects remained small and settlement was sparse, local customs prevailed and little formal water law developed in the territorial courts of Idaho. Only in organized mining camps and in Mormon communities did formal canons of water use prevail, but even here water law was shaped more to bring about practical results than as a considered application of abstract legal principles.[23] In the mining districts of Idaho, miners whose activities were largely confined to the public domain generally allocated water among themselves by enforcing local customs and rules embracing the familiar doctrines of prior appropriation and beneficial use. As with mining claims, water belonged to those who took first possession.[24] In a semi-arid environment, communal water laws presented a practical and feasible alternative to prior appropriation. As the Mormons expanded their irrigation frontier into southern Idaho, founders of Mormon communities observed familiar theologically-prescribed dictums for communal sharing of water resources and arbitration of disputes by local authorities – the church elders.[25]]

At the core of the appropriation doctrine is the principle that the first person to divert water from a stream and apply it to beneficial use acquires the highest priority to the use of the natural flow in times of shortage. This right is not an incident of the ownership of riparian land; instead, it is an independent right to the *use* of water.

Between the organization of Idaho Territory in 1863 and the state constitutional convention in 1889, the territorial legislature enacted only a few statutes affecting water rights and there was virtually no case law on the subject.[26]

In 1881, with the development of ditch companies (private enterprises that built canals and then sold water to individual farmers)

---

23 Donald J. Pisani, "Enterprise and Equity: A Critique of Western Water Law in the Nineteenth Century, *Western Historical Quarterly* 18 (January 1987), 15-37; and Douglas R. Littlefield, "Water Rights During the California Gold Rush: Conflicts over Economic Points of View," *Western Historical Quarterly* 14 (October 1983): 415-434.

24 The "prior appropriation" principle was reinforced by later state court rulings. In one of its earliest water-right decisions the Idaho Supreme Court in 1890 held that prior appropriators had the better title to use of water as against riparian claimants who entered land after an appropriation had been made. See *Drake v. Earhart.* 2 Idaho 750 (1890).

25 Dunbar, *Forging New Rights in Western Waters*, 9-17, 59-85; and Kate B. Carter, ed., *Pioneer Irrigation: Upper Snake River Valley* (Daughters of Utah Pioneers, 1955).

26 *Drake et al. V. Earhart*, 2 Idaho 750 (1890).

and the affiliated expansion of irrigated agriculture, the territorial legislature passed two measures that formed the basis of irrigation law in Idaho for the remainder of the territorial period. Specifically, these acts aimed to safeguard and support the substantial private investment made by those promoting development of irrigated agriculture by granting a right to acquire water by appropriation for useful purposes. The acts also provided for legal institutions to guide the administration and use of water, and to regulate canal and reservoir companies in the public interest.

The first of the territorial legislature's 1881 acts with respect to appropriation of public waters appears to have affirmed some aspects of both the English common law doctrine of riparian rights and the appropriative doctrine. In an effort to protect irrigators, the legislators enacted a water code that explicitly embraced the general principles of prior appropriation, but also provided that, "all persons, companies, and corporations owning or claiming any lands situated on the banks or in the vicinity of any stream are entitled to the use of the waters of such stream for the purpose of irrigating the land so held or claimed."

In the context of nineteenth-century irrigation, before massive reservoir projects, most of the agriculturally-productive land was located on alluvial soil adjoining streams. Protecting the water rights of persons or corporations engaged in agriculture and situated along the banks of streams seemed to be a suitable allocation of the region's water and struck a balance between riparianism and appropriation. The 1881 act favored the use of water by those needing it, but permitted only the reasonable use of water; the act granted any settler the right to use such water as "good husbandry shall require."[27] The legislature also directed the territorial courts to further acknowledge customary uses and practices that had arisen over the previous two decades; recognized the federal enactments of 1866 and 1870 providing for appropriation and use of water flowing through the public domain; and expressly adopted the appropriative doctrine as the rule of law in the territory. Thus, while the legislature did not approve the common law riparian doctrine in 1881, it embraced some of the concepts associated with it- namely that rights to water were part of the land bordering a stream and were governed by reasonable use for the common good. According to one legal scholar, the extension of these rights to landowners was a

27 Act of the Idaho Territorial Legislature, February 10, 1881.

pragmatic political compromise – a move that, despite its appearance, was more about facilitating agricultural development and expressing a "preference for irrigation," than about recognizing the common law doctrine of riparian rights in the territory.[28]

The territorial lawmakers also established procedures for water users to claim an appropriative right (for use of water) and institutions to administer the rights. The law provided that the appropriator claiming a water right was required to post a public notice of the appropriation near the point of intended diversion and to develop that claimed right through actual beneficial use in order to take possession of the right. Water users who were putting water to beneficial use could establish a right to future use and a priority date for continuing that use, superior to the rights of later (or junior) diverters. On the other hand, claimed water rights would be lost if the water was not being, or stopped being, beneficially used. This territorial water law encouraged productive use of water resources. Other territorial laws allowed water users to create water districts to administer water rights and distribute water originating from a common water source. Such districts were required to select a "watermaster," who would oversee distribution of water throughout the district under the law of prior appropriation.[29]

These territorial acts and the pending territorial Supreme Court case, *Drake v. Earhart,* provided the backdrop for debates over Idaho state water law at the 1889 Constitutional Convention. On April 2, 1889, at the urging of Republican congressional leaders who sought to have Idaho admitted to the Union, Territorial Governor Edward A. Stevenson called for a state constitutional convention to be held in Boise on July 4, 1889. On May 11, Governor George Shoup, Stevenson's replacement, subsequently re-issued this proclamation. Seventy delegates assembled in what would become the state capital. In debates lasting a little more than a month, from July 4 to August 6 when the constitution was submitted to the voters, few issues proved as contentious to these delegates as the constitutional provisions dealing with water rights.

Indeed, water was the subject of not one, but two significant debates at the convention. The first concerned what became Article I, Section 14 which dealt with the right of eminent domain and private takings of

---

28 Colson, *Idaho's Constitution,* 162-163; and Robert H. Blank, *Individualism in Idaho: The Territorial Foundations* (Pullman: Washington State University Press, 1988), 17.

29 Colson, *Idaho's Constitution*, 163.

land for water development purposes. The delegates ultimately approved of a declaration making the taking of land necessary for reservoirs, storage basins, and water conveyance systems for irrigation purposes a public use, and thereby "subject to the regulation and control of the state." Just compensation, however, was required for the taking. This type of private taking of land by one individual over another was rare in constitutional and property law but was justified as a necessity for development of irrigated agriculture on the Snake River Plain.[30]

This emphasis on irrigated agriculture continued into the second and more far-reaching debate over what became Article XV, Section 3 of the Idaho constitution. Article XV was specifically devoted to "Water Rights" and was intended to provide the framework within which Idahoans might make use of their most precious resource.[31] Six sections in the constitution elaborated on the 1881 territorial statutes regarding water,[32] but Section 3 was the most controversial. It established the general priority system – first in time, first in right – for the appropriation and use of water but specified that if water was scarce some types of water uses would be given priority over other uses. As drafted by the "Manufactures, Agriculture and Irrigation" committee, comprised entirely of delegates sympathetic to irrigation, the section stated:

> The right to divert and appropriate the unappropriated waters of any natural stream to beneficial uses, shall never be denied. Priority of appropriation shall give the better right as between those using the water; but when the waters of any natural stream are not sufficient for the service of all those desiring use of the same, those using the water for domestic purposes shall (subject to such limitations as may be prescribed by law) have preference over those claiming for any other purpose; and those using the water for agricultural purposes shall have preference over those using the same for manufacturing purposes.[33]

30 Article I: Declaration of Rights, Sec. 14 Right of Eminent Domain, Constitution of the State of Idaho, Approved July 3, 1890; Leonard J. Arrington, *History of Idaho*, volume I (Moscow: University of Idaho Press, 1994), 420-421; and Colson, *Idaho's Constitution*, 163.

31 To this day, Idaho is unique among the states for having an article of its constitution devoted exclusively to the issue of state water rights. Blank, *Individualism*, 65.

32 Section 1 recognized the doctrine of prior appropriation and declared the use of waters of the state a public use, subject to regulation and control by the state; Sections 2, 4, and 6 dealt with water sales and the right of the state to regulate such sales; and Section 5 expanded the provisions of the 1881 territorial law dealing with water districts.

33 I.W. Hart, *Proceedings & Debates of the Constitutional Convention of Idaho, 1889*, Volume II (Caldwell, Idaho: Caxton Printers, Ltd., 1912), 1117, Boise State University Special Collections (hereafter, BSU Spec. Coll.)

In proposing a preference for agriculture (outside of organized mining districts) in the use of the state's water, as law professor and Idaho constitutional historian Dennis Colson notes, the committee "was proposing a 'modified priority' or 'agricultural preference principle'" – an innovation in Western water law.[34]

Section 3 sparked considerable debate on the convention floor that crossed party lines. Delegates clashed over the merits of such a system that made first domestic and then agricultural uses superior in times of scarcity to all other uses regardless of the priority of water rights. The issues of riparian rights and just compensation for the taking of water rights of senior water right holders were also the subjects of some debate.[35] At the heart of the matter, however, was a simple question: domestic uses notwithstanding, to which enterprise should Idahoans devote their water – agriculture or manufacturing?

For supporters of Section 3, agriculture was more vital to the economic health of the state, more dependent upon an uninterrupted supply of water, and thus deserving of special constitutional protection. Citing both aridity and economic self-interest, lawyer and Ada County Republican Edgar Wilson, the youngest delegate and perhaps the most fervent supporter of agriculture, declared,

> I say that first interest of this arid country are agriculture, not manufacturing…I say that it is essential the interest and the paramount interests of this country that manufacturing be sacrificed rather than agriculture, because this is not a manufacturing country; we don't have raw materials to manufacture out of, and doubt if we ever will have.[36]

He called for "the priority of right" to be given "to those using the water for agricultural purposes as against manufacturing purposes…out of the necessity of things, for the vital interests of the arid country."[37]

Merchant, banker, and Idaho Falls Democrat Robert Anderson offered a still more practical argument. He noted that "if we stop a sawmill or a woolen mill or any other kind of a mill for the purpose of

34 Colson, *Idaho's Constitution*, 163 and 165.

35 Ibid., 165-172.

36 Hart, *Proceedings & Debates*, 1360. For a brief biography of Wilson, see Colson, *Idaho's Constitution*, 7.

37 Ibid., 1361.

distributing water for a few days to agriculturalists, we cause them to lose the profits they would make during those days." However, Anderson proceeded, "if the farmers are deprived of their water for a week or two weeks, they work for years thereafter at a sacrifice."[38] Republican lawyer John Gray, also from Ada County, justified his support for agriculture over manufacturing as a matter of convenience versus necessity: "I can't live on cotton or wool or anything of that kind; I want something to eat...."[39]

To William McConnell – an early ditch company pioneer from Latah County and future governor of the state – fell the task of being principal advocate for the proposed Section 3.[40] McConnell employed both practical and political arguments in making his case. He observed that without preference, downstream farms would be at the mercy of upstream manufacturers or speculators, or ironically enough, ditch companies. He suggested that the manufacturers "could crush those farmers by turning the water again on to his wheel and taking it away from the farms," while others "might conclude at any time after those farms were in a high state of cultivation, that he could use this water temporarily for such a time as should compel those men to abandon their properties or sell them out for a mere nothing."[41] "Let us not," McConnell called to his fellow delegates, "place anything in this constitution, which will place those agriculturalists, who are necessarily poor people, in the power of any incorporation which brings out a ditch."[42]

Delegates such as James H. Beatty and James Shoup rejected such arguments, insisting that property rights were at stake and that manufacturing needed safeguarding as well. Beatty, a Civil War veteran, lawyer (soon to be appointed chief justice of the Idaho Supreme Court), and Republican from the town of Hailey in Alturas County, did not "believe such a law as that [Section 3] would be constitutional, it would be taking away the priority right of one man and giving it to another." "I know of but one way to regulate it," he declared, "and that is that the parties first in time hold the water."[43]

---

38 Ibid., 1361-1362. For a brief biography of Anderson, Colson, *Idaho's Constitution*, 9-10

39 Ibid., 1142. For a brief biography of Gray. see Colson, *Idaho's Constitution*, 7.

40 Colson, *Idaho's Constitution*, 165-167; for a brief biography of McConnell, see ibid., 11.

41 Hart, *Proceedings & Debates*, 1137.

42 Ibid., 1324.

43 Ibid., 1117. For a brief biography of Beatty, see Colson, *Idaho's Constitution*, 10.

Beatty even went so far as to suggest that a senior water claimant "under all laws of nature would be entitled to it, because he is the first man there."[44] Both Beatty and Shoup, a Republican from Custer County and brother of the then current territorial governor, also believed that Idaho's manufacturing future would be threatened. The former claimed that if "preference" were given to "the farmer," that "[y]ou will destroy the interests and rights of a hundred families in order to benefit one agriculturalist." Similarly, Shoup argued, "manufacturing establishments give employment to everybody; they give employment to strong men, they give employment to women."[45]

In the end, although the arguments of Beatty, Shoup, and others failed to defeat Section 3's offending language regarding preferences for domestic and agricultural uses over manufacturing, the measure was amended by mining interests and proponents of private property rights. Opponents of the section were led by George Anslie, Idaho pioneer, Democrat, and Boise delegate, who was able to secure an amendment providing for compensation for condemnation, under the law of eminent domain, subject to Article I, Section 14 of the constitution, for prior non-agricultural water users supplanted by the preference for agriculture.[46] As the Idaho State Supreme Court later ruled in *Montpelier Milling Company v. City of Montpelier* (1911),

> This section clearly recognizes that the right to use water for a beneficial purpose is a property right, subject to such provisions of law regulating the taking of private property for public and private use as referred to in sec. 14, art. 1, of the constitution.
>
> It clearly was the intention of the framers of the constitution to provide that water previously appropriated for manufacturing purposes may be taken and appropriated for domestic use, upon due and fair compensation therefor. It certainly could not have been the intention of the framers of the constitution to provide that water appropriated for manufacturing purposes could thereafter arbitrarily and without compensation be appropriated for domestic purposes. This would manifestly be unjust ...[47]

---

44 Ibid., 1138.

45 Ibid., 1130. For a brief biography of Shoup, see Colson, *Idaho's Constitution*, 11.

46 Ibid., 1145 and 1163. See also, Donald Crowley and Florence Heffron, *The Idaho State Constitution: A Reference Guide*, Reference Guides to the State Constitutions of the United States, Number 19, ed. G. Alan Tarr (Westport, CT: Greenwood Press, 1994), 228 and Colson, *Idaho's Constitution*, 170. For a brief biography of Ainslie, see Colson, *Idaho's Constitution*, 11.

47 *Montpelier Milling Co. v. City of Montpelier*,19 Idaho 212, 219 (1911).

At the constitutional convention, mining interests, through the efforts of lawyer and Shoshone County delegate Weldon Heyburn, also managed to acquire a special preference for water for mining purposes, over manufacturing and agriculture "in any organized mining district."[48]

Additionally, a relatively new use for Idaho's water – hydroelectric power generation, or hydropower – entered into the convention's debates, where it was commonly viewed as a manufacturing use.[49] New electrical enterprises, based upon successful schemes for long-distance transmission of electric power, allowed for setting up water wheels, turbines and generators in remote places and providing light and powering electric motors in towns, or on irrigation projects, miles away.

The first hydroelectric plant in Idaho to serve the public was built at Hailey in 1885. It harnessed the waters of Big Wood River and supplied electric current for 250 lights at a charge of $5.00 per light per month. In Boise, the first commercial electric light plant illuminated streets and buildings in 1887, just a couple of years before the delegates assembled for the state constitutional convention in the city. Electricity was provided by a hydro plant that took water from an irrigation canal on a bench south of town and dropped it sixty-two feet through a penstock to turn two turbines linked with two 15 kW Edison direct-current generators. For the next 15 years the plant continued to illuminate streets, businesses, and residences. The *Idaho Statesman* boasted in 1890 that electric street lights in the "growing and progressive" city presented "a beautiful sight ... on a moonless night."[50]

As the constitutional convention debate over Section 3 revealed, hydroelectric power generation as a type of water use was on the mind of some delegates, but it was still a novel and poorly-understood

48 Ibid., 1156, 1164-1165. See also Crowley and Heffron, *The Idaho State Constitution*, 229 and Colson, *Idaho's Constitution*, 169-171. For a brief biography of Heyburn, see Colson, *Idaho's Constitution*, 8.

49 Because the source of hydropower is water, hydroelectric power plants also had to be located on a water source. Therefore, it was not until technology to transmit electricity over long distances was perfected that hydropower became widely used. This occurred in the years following culmination of successful tests on the feasibility of long-distance transmission of three-phase alternating current using high voltage wires at the 1891 Frankfurt Exposition. See Susan M. Stacy, *Legacy of Light: A History of Idaho Power Company* (Boise: Idaho Power Company, 199[illegible]), 11-19; James Williams, *Energy and the Making of Modern California* (Akron, Ohio: University of Akron Press, 1997), 170-177; and W. F. C. Hasson, "Electric Transmission of Power Long Distances," *Transactions of the Technical Society of the Pacific Coast*, 10 (May 1893), 49-72.

50 *Idaho Daily Statesman*, March 18, 1890, March 26, 1890, and May 10, 1897.

technology that was not easily placed within the traditional categories of water use. Was it to be included under the general rubric "manufacturing," or was it something altogether different?

McConnell himself raised the issue – prophetically as it turned out for the future development of the Snake River – in addressing the delegates in defense of the committee's proposed section:

> You have here, as I understand it, quite a large irrigation canal on this river [the Boise River]. Part of the waters of that canal are used today for manufacturing purpose, in generating electricity to light this town [Boise]. It might occur, as the science and use of electricity become more fully developed in this country, that it will pay the proprietors of the ditch better to use the water entirely for the generation of electricity, and if they should so conclude and there should be no law to prevent them diverting that which is their own to that use, they could throw that country, which is now attempting to be brought under cultivation, and some of which is already blooming as a garden, out again into a waste.[51]

Such a prospect horrified Edgar Wilson, who quickly made a motion "to insert the word 'power or motor' after the word 'manufacturing'" to ensure that this type of water use was also subordinated to irrigation uses. Wilson explained:

> The case Mr. McConnell cites is the motor power of the electric light company of this city on the other side of the ditch. Of course that is not the manufacturing purpose; they might have several establishments of that kind for the purpose of generating power and monopolizing all the water of that ditch, and this word 'manufacturing' would not include it. They don't manufacture anything; they generate electricity.

Wilson's appeal prompted one un-named member to quip, "They manufacture light" – a statement that elicited a rather abrupt denial from Wilson.[52] McConnell failed to see the dissimilarity between manufacturing and power, saying, "It is a distinction without a difference," but he subsequently seconded Wilson's motion to include "power" as a subordinate use. Wilson himself tried to clarify the issue by using the example of hydroelectric power being used to run motors

---

51 Hart, *Proceedings & Debates*, 1125.

52 Ibid., 1126.

of electric street car systems in urban areas, clearly in his mind a non-manufacturing use of water power. Few others, however, recognized the importance of Wilson's proposed amendment: only four delegates voted in favor of it.[53]

The failure of Wilson's amendment left open questions of how hydroelectric power generation was to be comprehended within the terms of Section 3, questions that lingered into the twentieth century. Was it the intent of the delegates to include hydroelectric power production under the rubric of "manufacturing" at the constitutional convention? If not, did agriculture still have priority over it? In administering water rights in Idaho, did the state, acting under the constitutional provisions of Article XV, Section 3, have the power to regulate or subordinate water uses for hydroelectric power generation to agricultural uses? These questions were answered, at least in part, by the Idaho legislature and the people of Idaho in 1928 when Section 3 of the state constitution was amended to accommodate agricultural interests in the state and allow limitations to be placed on water use for power generation.[54]

The Idaho state constitution was ratified by a margin of seven to one in a special ratification election held on November 5, 1889. The constitution recognized in Article XV, Section 3 that "priority of appropriation shall give the better right as between those using the water". As Dennis Colson noted, this was indeed an innovation in Western water law. The state constitution created a modified appropriation principle, establishing the superiority of irrigation in water use in those areas of the state were agriculture predominated and the superiority of mining where it was the central industry, and reserving for the state legislature the ability to curtail priority of rights when scarcity demanded.[55]

James H. Beatty, who participated as a delegate to the Idaho constitutional convention during the summer of 1889 and spoke passionately in defense of the doctrine of prior appropriation at the convention, was commissioned as chief justice of the Idaho territorial supreme court on November 21, 1889, beginning an eighteen-year

---

53 Colson, *Idaho's Constitution*, 173.

54 Donald Crowley and Florence Heffron, *The Idaho State Constitution: A Reference Guide* (Westpost, Conn: Greenwood Press, 1994), 228-231.

55 Colson, *Idaho's Constitution*, 163, 165, 219, and 273.

judicial career. A few months later on March 6, 1890, the Idaho Supreme Court, in an opinion written by Beatty, gave its imprimatur to the appropriation principle, rejecting the riparian doctrine and reaffirming the law of priority in the use of water as the fundamental doctrine of state water law in *Drake et al. v. Earhart*. Beatty, writing for the majority, argued that "... the maxim 'first in time, first in right,' should be the settled law here. Whether or not it is a beneficial rule, it is the lineal descendant of the law of necessity."

In rejecting the common law doctrine, Beatty concluded that it was not suited to arid and semi-arid conditions in Idaho where certainty of water supply was required to support investment in water development.[56]

The state administers

Most western states have constitutional and statutory declarations and judicial decisions reinforcing the concept that waters belong to the people or the state or, as it is commonly expressed, to the state in trust for the people. The Idaho Supreme Court ruled in two early cases that title to the public waters of the state of Idaho is vested in the state for the use and benefit of all its citizens under such laws and regulations as prescribed by the legislature from time to time. The state in its sovereign capacity as representative of the people guarantees that the common rights of all the people in water are equally protected.[57]

In years following statehood, the state built an administrative apparatus to supervise the distribution and use of the public waters of Idaho.

The water law of 1881 established a simple procedure whereby notification of the intent to appropriate waters of the state was done by posting notice in a conspicuous place at or near the point of diversion

---

56 *Drake et al. V. Earhart*, 2 Idaho 750 (March 1890), 750-758. Beatty presided over the territorial supreme court's last year of existence. He was appointed by President Harrison as a federal judge for the District of Idaho on March 7, 1891 and served on the federal bench until 1907. Monique C. Lillard, "The Federal Court in Idaho, 1889-1907: The Appointment and Tenure of James H. Beatty, Idaho's First Federal Court Judge," *Western Legal History* 2:1 (Winter/Spring 1989).

57 *Woolridge v. Robinson,* 22 Idaho 236-242 (1912); *Coulson v. Aberdeen-Springfield Canal Company,* 39 Idaho 320-324 (1924).

and recording the same within fifteen days at the local county recorder's office.

In 1895 the state legislature established the state engineer's office and appropriators from that time forward were required to record a duplicate copy of their notices, not only with the county recorder, but also in the office of the state engineer. Over the next six years, state engineer D. V. Ross noted that more than five hundred times as much water was claimed from Idaho streams as the average flow in all of the streams of the state. The fundamental principles underlying administration of the public water of the state were widely misunderstood, lamented Ross in his 1900 biennial report. "Starting out with a declaration of the right of the state to control streams, we have ended with an impression prevailing in the minds of many that they become private property as soon as the water location notice is posted on their banks. The quantity claimed depends in most cases upon the fancy of the appropriator." [58]

Under the so called "Appropriation Laws," no record was kept to monitor the progress made by the appropriator under his/her claim toward completion of the works, nor whether or not the claimant put the full measure of the claimed water right to actual beneficial use. However, these principles are what guided courts in Idaho in determining the priorities and extent of water rights.

In 1903 a new system of procedure, called the "Permit System," was established by the legislature that required a person desiring to appropriate water to file an application for a permit from the state engineer to appropriate public waters of the state of Idaho.

The issuance of a permit was only the first step in the acquisition of a water right. Only after the permit holder made proof of beneficial use did the state engineer issue to an applicant a license confirming the right to use the quantity of water which his proof had shown was actually used for the purposes for which it was appropriated.

The issuance of a "license" by the state engineer's office was the last step in perfecting a water right under the provisions of the March 11, 1903 Act.[59] As for the old possessory rights established at a time

58 D. W. Ross, state engineer, "Water Rights and their Adjudication," *Biennial Report of the State Engineer to the Governor of Idaho for the Years 1899-1900* (Boise: Capital Printing Office, 1901), 90-91.

antecedent to the 1903 act, the Act itself provided for their adjudication in the state courts.[60]

By the early years of the twentieth century, the legal foundation for distribution of consumptive use water rights had been laid. Development of storage and hydropower water rights beginning in the first half of the twentieth century, however, complicated water administration in the Upper Snake River Basin. Drought and expanding irrigation demand created water shortages that sent state and federal officials, as well as private water companies, in search of favorable sites for storage reservoirs many of which were located in the channel of the Snake River.

Administration of decreed consumptive use natural flow rights versus storage rights presented a significant challenge to administration of water rights under the appropriative doctrine. District watermasters were tasked with distribution of natural flow rights on the river in strict accordance with priorities decreed by the state district courts. On the other hand, the state engineer appointed special deputies to supervise distribution of storage water released from reservoirs. During irrigation season, natural flow and released water commingled in the river making it difficult to distribute water to its rightful owners.

By the 1920s, hydroelectric power also occupied a more significant role in the life of Idaho's inhabitants. The question of allocating scarce water resources in the public interest between competing irrigation and hydroelectric power uses in the Upper Snake River Basin could no longer be avoided. The Bureau of Reclamation and Idaho Power Company reached a historic agreement in 1923 that subordinated upper basin power rights to irrigators and established operational criteria for American Falls Reservoir that had far reaching beneficial implications for future water allocation to farmers in the Upper Snake River Basin. The terms of the agreement invoked the "irrigation preference" established in the Idaho Constitution in 1890 to protect agricultural

59 James Stephenson, Jr., State Engineer, *Irrigation in Idaho.* U. S. Department of Agriculture, Office of Irrigation Stations. Bulletin No. 216 (Washington DC: Government Printing Office, 1909), 54-57. *Nielson v. Parker*, 19 Idaho 727 held that the 1903 Act was not the exclusive method for perfecting a water right in Idaho. The case recognized that a right could still be established by beneficial use. It was not until 1971 that the permit process was mandatory for establishing a water right in Idaho. For more on the legal history and underpinnings of the permit system, see Lawrence Harvey, "A Mandatory Permit System for the Acquisition of Water Rights In Idaho," 2 *Idaho Law Review* 42 (1965).

60 Idaho State Legislature, H. B. 146, Sections 37-40 (Session Laws, 1903), 223.

interests. While reasserting that the constitutional right to divert and appropriate unused water in any stream for beneficial uses could never be denied, the Idaho state legislature in 1928 nevertheless made a specific exception to this rule that permitted the state to "regulate and limit" the use of Idaho's public waters for electric power generation.[61] The "subordination" of hydroelectric power rights on the Snake River to future agricultural uses in the basin would become a hotly contested area of water law in Idaho.

In the second half of the twentieth century, another complication arose as both hydroelectric power generators below Milner Dam and irrigation pumpers came to rely on the appropriation of waste, seepage, springs, and groundwater from the Snake River Aquifer. Collectively, these "subterranean waters" became governed by the same laws that regulated the appropriation of surface waters, but their regulation under the prior appropriation doctrine has been more difficult because the source of the supply is not easy to define by statute and because of the physical and temporal aspects of groundwater flow.

Since 1948 when the first major well-pumping projects were installed in the vicinity of Minidoka, groundwater exploitation by agricultural and hydroelectric power interests has become an increasingly important water management and conservation issue.

Groundwater development has made enormous contributions to expansion of the agricultural economy in Idaho over the last half century, especially in the Snake River Basin above Milner Dam. Groundwater levels and spring discharges were at their highest historic levels in the early 1950s, in large part because flood irrigation practices and canal seepage had artificially recharged the Snake River aquifer. In recent decades widespread adoption of winter water conservation, more efficient sprinkler irrigation, reduced diversions into porous canal delivery systems, and continuing expansion of groundwater pumping over vast expanses of desert landscape have reversed the trend. As a result, many skirmishes have been fought between senior surface appropriators and junior ground water users in the state legislature, the state courts, and in administrative hearings before the Idaho Department of Water Resources over how to manage what water users acknowledge is a depleted groundwater resource that suffers from too much pumping and reduced recharge.

61 Idaho Constitution, Article 15, Section 3, as amended in 1928.

For more than a century Idahoans have developed their water resources by relying on the prior appropriation doctrine as modified from time to time by constitutional amendment and state laws that provided special preferences for agricultural water uses.

Even the rise of new environmental considerations in the 1970s – a growing awareness of the need to set aside in-stream flows for recreation, water quality enhancement, and the benefit of fish and wildlife – did not prompt an abandonment of these central state constitutional tenets. Both of these fundamental principles of water law are rooted in the Idaho Constitution adopted in 1890 and have informed the actions of state officials ever since.

2

Fashioning a More Functional River

In 1888 Congress authorized the Secretary of the Interior to organize an "Irrigation Survey" within the Geological Survey of the Department of the Interior. It would classify and segregate irrigable lands from other public domain lands in the arid sections of the United States, to set aside suitable places for reservoirs, and to provide for rights of way for ditches and canals for the purposes of irrigation.

John Wesley Powell, Director of the Geological Survey, split the arid West into several divisions to conduct hydraulic and engineering surveys, which were undertaken under the general supervision of Captain C. E. Dutton. The Idaho Division was placed under the direct charge of A. D. Foote, whose study of the irrigation capabilities of the Snake River Valley was to include the vast four million acre plain above the confluence with the Payette River. This area was so large that his work was carried out at a reconnaissance level. The resulting report, however, concluded that irrigation prospects for the Snake River Valley were as favorable as any place surveyed in the American West. While the potential was enormous, wrote Dutton, accomplishments had been tempered by a combination of factors.

> Indeed, it may be said that the eastern portion of the Snake River plains offers about the best opportunity for a splendid irrigation development of any portion of the West whose capabilities are at present understood. Settlement, however, has until the last two or three years, almost wholly avoided it. The reason is not far to seek. There is little opportunity for the pioneer stage of development. This stage requires small streams which can be attacked by the resources of a few settlers without capital. Outside the mountains the Snake has exceedingly few small tributaries. Whatever is done must require the resources of more or less capital. With very slight qualification the development must in the ordinary course begin at the second stage, that of multiple ditches. Until very recently there has been no sufficient demand for land to cause the development to leap over the first stage and begin with the second.

The irrigation movement in southern Idaho gained momentum during the following decade. In 1888 the Geological Survey established its first gaging station on the Snake River about ten miles downstream of American Falls, at a place called Eagle Rock on the Utah & Northern Railroad line. Foote undertook the reconnaissance of several good reservoir sites on the Upper Snake River, including Jackson Lake and Swan Valley (the future sites of two U.S. Bureau of Reclamation reservoirs), and identified a 200,000-acre tract of irrigable land above Twin Falls "where there was hardly an acre of waste ground in the entire tract." With the segregation of large irrigable tracts by the state and federal government for irrigation and reclamation projects on the Snake River Plain at the turn of the century, progress became rapid and formidable.[62]

Early state and federal irrigation

Unlike most of the late nineteenth century irrigation development by individuals, communities, and small canal companies above American Falls, the explosion of irrigated acreage in the Snake River Valley in the first two decades of the twentieth century was centered in the region below American Falls, in the vicinity of Minidoka and Twin Falls, and was accomplished with government assistance.

The stream of farmers who continued to enter the state at the turn of the twentieth century took advantage of liberal state land settlement and federal reclamation programs to establish irrigated farms. It was impractical for individuals to spend many thousands, or perhaps millions, of dollars to construct irrigation works of the magnitude needed. In these cases, investors with substantial capital assets had to assist by going in advance of settlers and preparing conditions for their arrival. Those investments were not misplaced. As predicted by Captain Dutton of the Geological Survey, the first decade of the twentieth century saw remarkable development in irrigation in the major agricultural counties in southern Idaho. The direct causes were two, programs developed by the United States Congress to facilitate

62 J. W. Powell, Director, Tenth Annual Report of the United States Geological Survey to the Secretary of the Interior, 1888-89 (Washington, DC: Government Printing Office, 1890), 47, 88-89, 106-108.

reclamation and encourage settlement of the arid west: the Carey Act of 1894, designed to attract large investment of private capital to irrigate arid land ceded by the federal government to certain western states; and the National Reclamation Act of 1902, a government aid program under auspices of the Reclamation Service (after 1923, Bureau of Reclamation) to promote irrigation and settlement in the West. Once these laws were in place, financing large-scale irrigation undertakings became feasible and construction of large dams to form storage reservoirs played an increasingly important part in reclamation plans.[63]

The largest spread of irrigable land that can be watered from a single point of diversion in the Snake River Valley lies near Twin Falls: 270,000 acres, more than 20 miles long, and averages greater than ten miles in width. The superior quality of the land was well-known, but for a long- time diversion of water from the Snake River for its irrigation was thought impossible because the land lies on a high plateau far above the Snake River gorge. A. D. Foote, the irrigation engineer with the U. S. Geological Survey, investigated the area for the federal government in 1889-90 and recognized the possibility of getting water upon the plain, but never got the opportunity to test his impression of the contour of the country by conducting a survey.

Ira Burton Perrine, who resided in Blue Lakes, was familiar with Foote's work and in the late 1890s inquired of Idaho state engineer D. W. Ross whether the state might be interested in investigating the site with the idea of withdrawing the land under the Carey Act for an irrigation project. The state engineer agreed that the potential reclamation of this vast tract of land was of vital interest to the state and he assisted Perrine in completing the surveys to determine the feasibility of the project. Lines were run from a spot on the river called "Cedars" for a distance of about 65 miles on the south side and 35 miles on the north side of the river. Ross later wrote of the remarkable natural resources of this region for agricultural development:

> Snake River, in its course through these lands, makes vertical falls which aggregate more than 500 feet. With the settlement of this large area the utilization of these facilities for power purposes may be reasonably expected. I think we may reasonable claim that nowhere in the inter-mountain region are so many great natural advantages grouped

63 Stephenson, *Irrigation in Idaho*, 13-19.

> together as in this section. There is here a vast body of beautiful land which can be so easily and cheaply irrigated; a water supply that has been our boast for years; and equable climate permitting a wide range of products; a range stretching for miles in every direction; and great mineral resources, both to the north and south, while the Snake River, tumbling and roaring in its unbridled strength, can be made to furnish the power necessary for the work of a million hands.

Shortly after Perrine and Ross completed their surveys, the Twin Falls Land & Water Company was organized and the state applied for segregation of 248,000 acres under the Carey Act and state laws supplementary to the act.[64]

The Carey Act encouraged the states, private investors, and the federal government to cooperatively provide land and water supplies necessary to build large irrigation projects. The act provided for the transfer of up to a million acres of federal land to each arid western state to be patented to small farmer, owing 160 acres or less. Their land would be served by irrigation companies under arrangements approved by a state reclamation engineer as to water supply, feasibility of construction, cost, capacity of the irrigation works, and character of the land to be irrigated. The private company proposing to reclaim a tract would enter into a contract with the state, specifying at what price it would furnish water rights and covering details of the project. The reclamation companies then financed and constructed the canals, made maps of land to be segregated, and sold shares of the company's water right based upon the amount of segregated acres and the available water supply. Purchasers of company stock used their stock holdings to apply for a state land patent to the segregated acreage. Once all the contracts were satisfied (i.e., the land was sold and water rights disposed of), the whole irrigation works were to be turned over to the shareholders who were required under their contract to organize an operating company.[65] This was one of the most attractive features of the Carey Act projects, observed Idaho Attorney General Frank Martin in 1901:

---

64 D. W. Ross, Idaho state engineer, "Irrigation Surveys," *Biennial Report of the State Engineer to the Governor of Idaho for the Years 1899-1900* (Boise: Capital Printing Office, 1900), 66-69.
65 Act of August 18, 1894, Ch. 301, Sec 4, 28 Stat. 422. Initially, the State Board of Land Commissioners was the state agency in Idaho responsible for administrating the Carey Act. The Idaho State Engineer administered the appropriation of water. In 1919, as part of a general reorganization of state government, the Commissioner of Reclamation (formerly the State Engineer) in the newly named Department of Reclamation was given responsibility for overseeing both programs. See "Historical Chronology – Idaho Department of Water Resources," Idaho Department of Water Resources website: http://www.idwr.idaho.gov/about/; last accessed March 17, 2008.

> These irrigation propositions under the Carey Act are of greatest merit because they finally give the consumers of the water absolute ownership. The state stands between them and the company till ownership changes, and it is the duty of the land board to regulate the price settlers must pay.[66]

Thousands of people claimed farms in the Snake River Valley under the Carey Act, and Idaho became the most successful Carey Act state in the nation, developing some 630,000 acres. Communities such as American Falls, Twin Falls, and Jerome, owed their existence to large irrigation projects initiated under the act.

The first Carey Act project in Idaho was proposed by the American Falls Canal and Power Company above American Falls in 1895. Some 51,000 acres in Bingham and Blaine counties were segregated for the project with the company receiving a patent for a 640-acre townsite subdivided into residential, commercial, industrial, and farm lots and blocks. The company initiated construction on the Aberdeen-Springfield Canal in 1901; it was completed within five years and received a natural flow right for 1,172.1 cfs from the Snake River. The American Falls Canal and Power Company went bankrupt in 1905 and a holding company, the American Falls Canal Security Company, took over and finished the canal construction project. With the contract between the state and construction company completed, the project was turned over to shareholders of the Aberdeen-Springfield Canal Company in 1910 by which time some 50,000 acres had been patented by settlers under the Carey Act.

In 1900 the first major irrigation project below American Falls was filed under provisions of the Carey Act by I. B. Perrine for his Twin Falls Land and Water Company. Between 1895 and 1930, private land and water companies launched sixty-five Carey Act projects in Idaho, covering more than one-half of the land patented under the legislation in the entire West.[67]

---

66 "Extensive Operations Proposed by Twin Falls Company: To Work Under Carey Act," *Idaho Daily Statesman*, February 18, 1901.

67 Arrington, *History of Idaho*, vol. 1, 476-480; Mark Fiege, *Irrigated Eden: The Making of an Agricultural Landscape in the American West* (Seattle: University of Washington Press, 1999), 124-127; and Hugh T. Lovin, "The Carey Act in Idaho, 1895-1925: An Experiment in Free Enterprise Reclamation," *Pacific Northwest Quarterly* 78 (1937), 122-133.

Of all the Carey Act ventures, the Twin Falls Land & Water Company's so-called South Side Project was the largest and perhaps the most successful.

In 1884, irrigation developer I. B. Perrine moved from his native Indiana to the Snake River country of southern Idaho where he selected a well-chosen farm situated in the spring-fed Blue Lakes canyon to protect his cattle from the area's harsh winters. In this secluded canyon, Perrine planted prized orchards and grew vegetables, grains, and grasses and learned of the agricultural potential of the region. By attracting outside capital to invest in building a vast system of irrigation canals to turn the arid Snake River plain into productive farm land, Perrine earned the moniker "Father of Magic Valley Irrigation."

Perrine proposed the segregation of 244,025 acres under the Carey Act on August 15, 1900. On September 3, 1900 at Salt Lake City, Perrine and his associates incorporated the Twin Falls Land & Water Company for the purpose of entering into a contract with the state of Idaho under the Carey Act to reclaim the land and to finance development of the huge irrigation project that he and State Engineer Ross had surveyed a year earlier. The company employed A. J. Wiley, who had recently completed the power dam at Swan Falls, to serve in the capacity of chief engineer and to complete the survey of the proposed dam and canal system together with a segregation map for submission to Ross.

The irrigable land to be served was in a compact area that stretched 50 miles westward from Milner, where it was proposed to divert the waters to be used for the project. This point of diversion was located between 20 and 25 miles above the falling waters at Twin Falls and Shoshone Falls, but the irrigable lands extended on the plain far to the west of these points and paralleled the Snake River. The survey maps were presented to the Office of the State Engineer on October 6, 1900. Although the approved project was grandiose in its scope, Perrine envisioned construction of an even more expansive network of canals that would one day water as much as one-half million acres of land. Financiers for construction of the necessary dam and canal system included Stanley B. Milner, for whom Milner Dam is named, a mining acquaintance of Perrine and a successful Salt Lake City banker, and local entrepreneurs James H. Lowell and Albert K. Steunenberg, brother of Idaho governor Frank Steunenberg.[68]

Over the next couple of years, a conflict over water rights involving a proposed hydroelectric power project at Shoshone Falls, and escalating project costs hindered progress of the venture from a legal and financial perspective. Perrine launched his own effort to control the power potential of the Shoshone Falls power site by filing for a water right and undertaking minimal development work on a power tunnel. He contacted several Eastern financiers to obtain an infusion of much needed capital, but was rejected. Lowell and Steunenberg left the project in 1902 and the state land board, impatient with the lack of progress on the irrigation project, called the president and secretary of the Twin Falls Land and Water Company to appear before the board in November 1902 to discuss fulfillment of their agreement for development of the project.[69]

Milner, who had invested in several Utah mining ventures, and an associate, Walter G. Filer turned to their mining connections to solicit financial support for the Twin Falls irrigation project. Milner and Filer convinced Pennsylvania iron, steel, and mining capitalist Frank H. Buhl, who was worth a reputed $25,000,000, to join the Twin Falls irrigation venture. In January 1903, the Twin Falls Land & Water Company was re-organized and the company entered into a contract with the state land board to begin construction on the Milner Dam and South Side Canal under the Carey Act. Milner transferred 65,000 shares of stock to Buhl, making him the majority owner of the company. Perrine sold his remaining stock to Milner.[70] Buhl's business partner in his mining ventures, Peter L. Kimberly, later was induced to join in the venture. In April 1903 the Buhl-Kimberly Corporation, capitalized at $5,000,000, was formed for the purpose of carrying on business as contractors for the construction of public and private utilities, railways, and gas and electric companies. Over the next three years all of the major construction at the dam and on the South Side Canal was completed.[71]

---

68 Hiram T. French, History of Idaho, 256; Salt Lake City Herald, May 3, 1906; State of Idaho, Minutes of the State Land Board, Book 1: 243, 250, 270-71, 273.

69 State of Idaho, *Minutes of the State Land Board*, Book 1: 359.

70 Idaho State Historical Society, Idaho State Archives and Library, MSS 114, Contracts and Documents.

Acting in his capacity as vice-president of the Twin Falls Company, Perrine had filed a notice of appropriation with the Office of the State Engineer for the diversion of 3,400 second-feet of water from the Snake River at the dam site known as "The Cedars." Water rights were granted by the state engineer on October 8, 1900 for 3,000 second-feet taken out on the south side of the river and 400 second-feet on the north side. The company's plan was to irrigate a total of approximately 240,000 acres, all of it, except about 30,000 acres located on the south side of the Snake River. The state sold the segregated land at 50 cents per acre and shares of the company's water rights were valued at $25.00 per acre by the company. News of the land sale, begun in June 1903, was published throughout the Pacific Northwest, in Salt Lake City, and as far east as New York.

Construction work was initiated in March 1903, at the site of a temporary rock-fill and concrete power dam designed to impede the Snake River's flow sufficiently to raise it to the necessary height to drive a turbine wheel used in the construction project. Constructed by Faris & Kesl of Boise, Idaho, the permanent dam, named for Stanley Milner, president of the Twin Falls Company, was completed on March 1, 1905. It was the first irrigation dam of any consequence on the Snake River, preceding the completion of the Bureau of Reclamation dam at Minidoka by eighteen months. During irrigation season when the headgates were opened and water filled the big Twin Falls irrigation canals to redeem the desert, the mighty Snake River below Milner Dam was reduced to a trickle.[72]

Milner Dam, an 800-foot-wide stone and concrete structure, raised the level of the stream about 57 feet above low water, allowing water to flow into the main gravity canal on the south side of the river. The crest of the wasteway was 12 feet below the top of the dam and four feet below the flow line of the canal. During very low water, the water in the dam could be raised by the use of flashboards and the entire flow of

71 The Twin Falls Land & Water Company was a subsidiary of Buhl-Kimberly, formed to construct the irrigation project. Hiram T. French, *History of Idaho*, 256; "F. H. Buhl Died After a Brief Illness," *The Evening Record* (Greenville, PA), June 8, 1918; "Peter L. Kimberly Expires Suddenly," *New* Castle (PA), June 7, 1905; Salt *Lake City Herald, May 3, 1906*. Buhl and Kimberly sold out to the United States Steel Company in 1902 and retired from the iron and steel business. One year later, they formed the Buhl-Kimberly Corporation to promote the Twin Falls irrigation project in Idaho. The corporation was also involved in mining gold, silver, and copper throughout the American West.

72 Jessie Warrington, "The Wonderful Redemption of a Desert: The Story of the Past, Present and Future of the Famous Twin Falls Country," *See Idaho First* 5 (September 1914), 3-20.

the river could be turned into the canals. A massive conduit, measuring 80 feet wide at the bottom, 10 feet deep and 120 feet wide at the top, the main South Side Canal had a carrying capacity of over 3,500 second-feet.[73] The canal ran west from the dam, roughly paralleling Snake River for thirty-five miles before splitting into the High Line Canal and Low Line Canal. The canal network with its smaller conduits and laterals totaled almost 750 miles in length and embraced approximately 200,000 acres in Cassia County within the Twin Falls Tract on the south side of the river.

Water was first turned into the canal in 1905. In 1907, Perrine expanded his canal system to the first segregation on the north side of the river, initiating land sales on irrigable lands on his Twin Falls North Side Project in Lincoln and Jerome counties. Water deliveries to this 30,000 acre tract began in March 1908.[74]

Photograph1 (below) depicts Milner Dam, and **Photograph 2** (below) depicts the headworks of Twin Falls Company's South Side, or "Main" Canal.

73 The total storage created by Milner Dam was about 80,000 acre-feet, but 77,500 acre-feet were dead storage, as the lake needed to be maintained at a high elevation to get water into the irrigation canals. "Great Twin Falls Project," *Idaho Daily Statesman*, January 1, 1902; "Splendid Boost for Twin Falls," *Twin Falls News*, October 23, 1908.

74 Arrington, "Irrigation in the Snake River Valley," 6.

Milner Dam, Spillway Gates and Irrigation Falls, Southeast View. May 1989
Source: Library of Congress, Prints and Photographs Division, Historic American Engineering Record, HAER ID, 27-TWIF.V, 1-138

Twin Falls Main (or South Side) Canal Headworks, Downstream Looking Toward the East, May 1989
Source: Library of Congress, Prints and Photographs Division, Historic American Engineering Record, HAER ID, 27-TWIF.V,1-24

In 1909 the state of Idaho accepted the Twin Falls South Side Irrigation Project as complete under the terms of the Carey Act contract. About one year later, the project works were turned over to the Twin Falls Canal Company, the operating company formed by settlers on the project. The successful completion of such a massive undertaking, irrigating more than 190,000 acres of land, impelled the commencement of other Carey Act projects in Idaho.

A second major investment in Carey Act irrigation enterprises was made by James S. Kuhn and William S. Kuhn, financiers from Pittsburgh, Pennsylvania, who invested more than twenty million dollars to tap into the agricultural empire being built in the Twin Falls country. The two Kuhns were on the directorates of many financial and industrial institutions. James S. Kuhn served as the president or a member of the board of directors of four major Pennsylvania banks, president of the Municipal and Corporation Securities Company and a director of the Colonial Trust Company of Pittsburg and the Trust Company of America, New York. He was also president of the American Water Works and Guarantee Company, a corporation founded in 1886 whose core business was the ownership of common stock of utility companies providing water service. The company's origins can be traced back to the post-civil war era when concerned citizens became interested in bringing a critical commodity – pure, abundant water – to towns and cities across America. By the early years of the twentieth century, American Water Works and Guarantee Company had become the largest holding company for power, electric traction, and irrigation enterprises in the country.

William S. Kuhn was also on the board of directors of four major Pennsylvania and national banks. He served as president and director of the United Coal Company, owner of nine major bituminous mines in Pennsylvania and western Virginia, and of the West Pennsylvania Railway system, which supplied more than 50 cities and towns in northwestern Pennsylvania with power, electric light, and inter-urban railway services.[75]

The Twin Falls irrigation projects represented their initial foray into western irrigation and land development. The irrigation movement in Idaho was given a great boost when the Kuhn brothers placed the strength of their financial organizations and expertise in building water works behind the North Side Twin Falls Project.

The Kuhn Syndicate's diversified financial interests in southern Idaho extended beyond just irrigation. They also owned and operated hydroelectric power plants at Shoshone Falls, American Falls, and Upper and Lower Salmon Falls and built steam and electric railways to service the Twin Falls area. The tracts reclaimed in Idaho by the Kuhns and their associates included the second and third segregations of the Twin Falls North Side Land & Water Company (TFNSL&WCo.), embracing 180,000 acres and the towns of Jerome and Wendell; that of the Twin Falls Salmon River Land & Water Company (1908), containing 80,000 acres and the town of Hollister; and that of the Twin Falls Oakley Land & Water Company (1908), including 50,000 acres. The higher lands were reached by using five double suction centrifugal pumping plants covering 100,000 acres and costing in excess of

75 "Closing of the Bank," *New York Times*, July 8, 1913; "Tribute from Irrigation Age to J. S. Kuhn and W. S. Kuhn of the North Side Company," *Twin Falls News*, September 8, 1908.

$800,000. Marketing of their irrigation project was widespread and aimed at farmers and aspiring farmers in midwestern and eastern states.

The Kuhns published educational articles on irrigation in an "Irrigator's Hand Book," a compilation of practical suggestions on various agricultural subjects written in 1909 and aimed primarily at the settlers on their Carey Act project lands in Idaho. Noted agriculturalist C. J. Griffith, who was also Director of Agriculture for the Agricultural Department of TFNSL&WCo., authored the handbook. These publications covered a wide variety of topics, but emphasized the suitability of the fine, light soils of this region for cultivation of potatoes, sugar beets, and other field crops that became the major staple products of the region.[76]

The massive North Side irrigation project was developed in three successive state land segregations made under the Carey Act by the TFNSL&WCo. The first segregation, initiated by Ira Perrine in 1907, included irrigable lands on the north side of Snake River from Milner Dam to the Eden lava beds. The second segregation included land extending to the Wendell area, and the third segregation incorporated irrigated farms in the Gooding to Bliss region. TFNSL&WCo. made applications for water right permits from the state engineer's office in 1908, 1909, and 1910 covering a combined total of approximately 150,000 acres of irrigable land.[77] An irrigation canal diverting at Milner Dam to deliver water to this land on the north side of the river was completed in 1908, but the normal flow of Snake River during the summer irrigation season proved inadequate.

Photograph 3 (below) depicts the North Side Canal, in relation to Milner Dam. During short water years settlers on the first segregation had prior claim to normal flow water, and it was entirely possible that in

76 Hiram T. French, *History of Idaho*, 481-482; Eugene H. Grubb, W. S. Guilford, The Potato: A Compilation of Information from Every Available Source (New York: Doubleday, Page & Company, 1912).

77 These applications were assigned Permit Nos. 3752, 4489 and 5665 (State Water Right No. 1-02015). The permits were superceded by Right No. 5 as described in the decree entitled *Twin Falls Canal Company v. Charles N. Foster, et al.* (1913). The decreed water right was assigned State Water Right No. 1-0213. The decreed right was assigned a priority date of June 16, 1908 in the amount of 3,000 cfs, the capacity of the company's main canal. W. G. Swendsen, Commissioner of Reclamation, to Twin Falls North Side Land & Water Company, April 5 1922 and Gary D. Slette, Nelson, Rosholt et al. to Norm King, Idaho Department of Water Resources, February 28, 1989. Idaho Department of Water Resource, Water Right 01-0215 Backfile.

an extremely low water year junior water right holders on the third segregation might receive no water at all. TFNSL&WCo. contemplated construction of off-stream storage reservoirs at Wilson Lake (20,000 acre-feet) and Jerome Reservoir (150,000 acre-feet) supplied through its main canal. The plan proved impractical, however, as the porous volcanic soils of the region made the reservoir sites impossible to fill.

Milner Dam with North Side Canal in Foreground
Source: Library of Congress, Prints and Photographs Division, Historic American Engineering Record, HAER ID, 27-TWIF.V,1-1

The private irrigation company turned next to the U. S. Reclamation Service (after 1923, the U. S. Bureau of Reclamation) for water from its surplus storage in Jackson Lake Reservoir, which had been completed to a capacity of approximately 300,000 acre-feet in 1907 (later enlarged in 1911 to 438,000 acre-feet). The Twin Falls company negotiated a temporary contract to deliver stored water from Jackson Lake during the irrigation season when excess water was available.

Nevertheless, the company's water problems continued. Milner Dam was only a diversion facility and it provided no seasonal or carry-over storage for the Twin Falls irrigation projects. The quantity of surplus stored water at Jackson Lake Reservoir proved too little for full development of the irrigated acreage in the expanding Twin Falls irrigation projects. The two Twin Falls projects entered into a contract with the United States on February 13, 1913 for enlargement of Jackson Lake Reservoir to approximately 800,000 acre-feet, the irrigation companies paying for the project. Under the dam enlargement project, completed in 1914, the companies recognized the prior rights of the government, and accepted a secondary right for storage from Jackson Lake, the dividing line being elevation 6753. All stored water below this elevation belonged to the government and parties who contracted with the government for stored water, commonly referred to as the "bottom right."

The Twin Falls Canal Company and North Side Canal Company's subordinate storage right was referred to as the "top storage right."[78] When the reservoir failed to fill to capacity, it was the top storage rights that were adversely affected. Practically all of the decreed natural flow rights in the upper basin in Idaho were of earlier priority date than the right to storage in the top half of Jackson Lake Reservoir.

An adequate and assured water supply only came to the Carey Act projects at Twin Falls with construction of 1,700,000 acre-foot American Falls Reservoir in 1927, but even American Falls was not enough to cope with the protracted drought of the early 1930s.[79]

Carey Act irrigation projects also set aside land for town development. Perrine laid out the city of Twin Falls in the center of his irrigated tracts in 1904 as a planned community intended as the urban centerpiece of a growing irrigation empire on the Snake River. The proceeds from the sale of city lots went to help finance the construction

78 Wyoming State Engineer, Reservoir Permit Nos. 894 (August 23, 1906), 2185 (July 24, 1911), and 2659 (June 10, 1914). Diversions and increased irrigation on tributaries of the Snake River in Wyoming in the 1920-30s had a serious and widely-recognized detrimental effect on prior rights in Idaho. Wyoming parties asserted that some of these rights, under the doctrine of relation, had an earlier priority than the latest filing for the enlargement of Jackson Lake Reservoir in 1914. The only potential remedy to Idaho interests was by means of litigation similar to that involved in Nebraska v. Wyoming, or through negotiation of an interstate compact.

79 Ibid., 3-6; Hugh T. Lovin, "Free Enterprise and Large-scale Reclamation on the Twin Falls-North Side Tract, 1907-1930," *Idaho Yesterdays* 29:1 (1985), 2-14. Jackson Lake Reservoir as originally constructed was a temporary log crib dam that stored only 300,000 acre-feet. It was washed out in 1910 and rebuilt by the Reclamation Service to a capacity of 380,000 acre-feet. When rebuilt in 1916 with the assistance of the Twin Falls canal companies, the reservoir capacity was enlarged substantially to 780,000 acre-feet, a portion of which was dedicated to exclusive use of the private companies. Hoyt, *Water Utilization*, 40-41, and 125-129.

of his irrigation project. Twin Falls became the major regional economic center serving the agricultural industry by directing the flow of supplies, crops, investment capital, and basic services to re-sculpt a barren wilderness into what promoters of the irrigated tracts advertised as "the Magic Valley," with the Snake River cast as the life-giving "Nile of Idaho." As vitally needed water converted these dry lands into a thriving irrigated farm zone, other smaller satellite communities were founded and symbolized the prosperity that irrigation could bring to the region. Lots and blocks were laid out and towns such as Hansen, Kimberly, Filer, and Buhl sprang up on the south side project. The town of Jerome, founded in 1907 by the Kuhn brothers, was the principal town of the north side project with the smaller communities of Hazelton, Eden, and Wendell also serving farmers located on the north side of the river. Union Pacific's Oregon Short Line extended branch lines through the settlements of the irrigated district making it possible to ship hay, sugar beets, potatoes, and other crops to outside markets.[80]

Over the course of a few short years, the Twin Falls country on the Snake River had become one of the largest and most successful irrigation projects so far attempted under the Carey Act. In three years' time, a barren desert of sagebrush had undergone a miraculous transformation. More than three thousand settlers with their families had been settled on irrigated farms and a half-dozen towns had sprung up boasting populations from 1,000 to 4,000 inhabitants. Modern conveniences such as a municipal water supply, sewerage, telephone system, and electric light and power were at hand in these towns. Hotels, banks, manufactories, and mercantile houses lined the streets of their commercial districts. Industrial and agricultural growth was reflected in the presence of creameries, canning factories, sugar beet refineries, flour mills, breweries, and packing houses. The hydroelectric power plant at Shoshone Falls, developed by the Kuhn's Great Shoshone & Twin Falls Power Company, was the initial source of electric power for these towns, their industries and surrounding farms, but the town boosters and agricultural promoters would soon look to other nearby falling water sites on the Snake River as necessities arose.

80 Ibid., 3-11.

Federal Reclamation and the Minidoka Project

In the early 1900s, the federal government took a direct role in developing arid lands on the Snake River Plain above the private irrigation developments at Twin Falls. On June 17, 1902, Congress passed the National Reclamation Act to promote irrigation of public lands in sixteen western states. The Act provided for the federal government to withdraw un-entered and unoccupied public lands for the purpose of reclamation through the Secretary of the Interior and to build dams, canals, and irrigation works reserved for the benefit of homesteaders settling on irrigable land within the project area. These settlers received their irrigation water supply through government reclamation works and out of the water appropriations made under state law and held by the government for the reclamation project.

The Minidoka Project established by the United States Reclamation Service on the Snake River was breathtaking in size. It delivered water through Minidoka Dam to 120,000 acres of land on both sides of the river and to more than a thousand homesteaders, none of whom owned more than 160 acres.[81]

In November of 1902, under the recently-enacted National Reclamation Act, the Secretary of the Interior withdrew from public entry a large block of vacant land lying along the Snake River between Twin Falls and American Falls for the Minidoka Project. The first federal reclamation project established in Idaho, the Minidoka Project was only one-half the size of the Kuhn and Perrine privately-financed irrigation projects near Twin Falls, but the federal government had to move forward as soon as possible or risk losing out entirely to expanding private irrigation companies on the Snake River.

Equally important, as farmers and irrigation communities were established in southern Idaho, power companies made rival claims to the waters of the Snake River to generate electricity. Although power generation was a non-consumptive use, commercial power plants relied upon continuous, even flows to produce "firm" power supplies to their customers. Irrigators who used the same surface water flow in the river viewed this as wasteful; they benefited from uneven flows. Irrigated agriculture required conservation and saving of winter flows and release

81 National Reclamation Act of June 17, 1902 (32 Stat. 388); and Leonard J. Arrington, "Irrigation in the Snake River Valley: An Historical Overview," *Idaho Yesterdays: The Quarterly Journal of the Idaho Historical Society* 30:1-2 (Spring/Summer 1986), 6.

of this stored water to operate irrigation pumps and to mature crops during the summer months.[82]

Early Minidoka Project development, prior to completion of the American Falls Dam, consisted of three key elements:

■ a combined storage, diversion, and power dam at Minidoka rapids (Minidoka Dam), located above Milner Dam and about six miles south of Minidoka;

■ a distributing system of canals and laterals diverting water of the Snake River at Minidoka Dam and serving land on both sides of the Snake River;

■ the headwaters storage reservoir, Jackson Lake Reservoir, just outside Yellowstone Park in Wyoming.

In 1903, the USRS moved ahead with its project plans by filing a claim for an appropriative water right with the state of Idaho for more than 5,600 cfs for the project, of which 2,600 cfs were for an irrigation diversion at Minidoka and 3,000 cfs for power development at the same site to serve the irrigation project. Pursuant to this claim, proof of completion of works was filed five years later in the amount of 1,726 cfs for irrigation only, and was accepted by the Idaho State Engineer.[83]

The project itself was authorized in April 1904 and Congress set aside $2.6 million to pay for its construction. In 1905-06 the government surveyed and laid out several town sites into lots with streets, diagonal avenues, public squares, and reservations for municipal and other purposes. Land office shanties and tents pitched by prospective purchasers who rushed in from all parts of the West and

---

82 United States Reclamation Service, A History of the Minidoka Project, Idaho to 1912 inclusive, Complied by P. M. Fogg, Engineer, Volume 1, August 1915 (History of the Minidoka Project, Vol. 1), 2-6. Box 294 Minidoka, Project Histories, Feature Histories and Reports, 1902-1932, Entry 10 (PHFHR, 1902-1932), Records of the Bureau of Reclamation, Record Group 115 (RG 115), National Archives at Denver (NARA-Denver); William D. Gertsch, "The Upper Snake River Project: A Historical Study of Reclamation and Regional Development, 1890-1930" (PhD diss., University of Washington, 1974), 111-112, 119-121, and 133-135; and Stacy, *Legacy*, 22-24.

83 History of the Minidoka Project, Vol. 1, 153-155; Box 294; Minidoka; PHFHR, 1902-1932; RG 115; NARA-Denver. No Idaho State Water Right was issued for the power application because no power plant was developed at that time. The government later built a power plant at Minidoka Dam and on January 9, 1909, an application was filed with the State Engineer by the government for a power right for 2,500 cfs at Minidoka Dam. Interior Secretary R. A. Ballinger to Attorney General G. W. Wickersham, February 4, 1910, 7. Folder: Reclamation Service, Minidoka Project – Idaho, Idaho State Water Rights – Snake River, Dec. 30, 1908 to March 27, 1930 (Folder: Reclamation Service), Box 1619 8-3 Minidoka, Central Classified Files, 1907-1936 (CCF, 1907-1936), Department of the Interior, Office of the Secretary, Record Group 48 (RG 48), NARA II.

Southwest, some bringing building materials with them, occupied the sites in advance of land sales. Within months colonists intending to settle on the Minidoka Project had erected scores of houses, stores, and dwellings in and around the booming town sites. Development was rapid and the district surrounding the new towns was soon intersected by hundreds of miles of canals running through irrigated farmland. The Reclamation Service completed Minidoka Dam and most of the canals and laterals by the end of 1906 and soon thousands of acres of formerly uninhabited sagebrush desert was being irrigated in the vicinity of the new communities of Rupert, Acequia, Paul, Declo, Burley, and Heyburn. Each of the towns was connected to the Oregon Short Line to facilitate shipment of supplies and agricultural products.

The original project was divided into two divisions known as the Gravity Division, subsequently operated by Minidoka Irrigation District, and the South Side Pumping Division, operated by Burley Irrigation District. The Minidoka diversion structure had a maximum height above the river bed of 86 feet; it raised the water level about 46 feet in the river and acted as a combined storage, diversion, and power dam. The reservoir, known as Lake Walcott, stored 107,200 acre-feet of water.[84]

When the Bureau built its Minidoka Project the hydroelectric power industry was entering a formidable era of expansion, having only recently developed the technical ability to transmit electrical energy with alternating current over long distances with minimal losses. As the Reclamation Service embarked on its construction of regulating and storage dams for irrigation purposes on western rivers, it became aware of the possibilities of combining irrigation storage with use of falling water at dam sites to generate hydroelectric power for its projects.

With passage of the Town Sites and Power Development Act of 1906, Congress formally recognized the feasibility of generating hydroelectric power at federal Reclamation dam sites. The act authorized the sale of surplus power generated at these dams, giving

84 History of the Minidoka Project, Vol. 1, 2-6; Box 294; PHFHR, 1902-1932; RG 115; NARA-Denver; Department of the Interior, *Eighth Annual Report of the Reclamation Service, 1908-1909* (GPO 1910), 151-153; United States Department of the Interior, Bureau of Reclamation, Minidoka Project, Annual Project History, Report of Construction, Operation and Maintenance, vol. 36, Calendar Year, 1942, Burley, Idaho (Minidoka Project, Annual Project History, vol. 36), 17. Box 305 Minidoka, PHFHR, 1902-1932, RG 115, NARA-Denver; Hoyt, *Water Utilization*, 44; and Gertsch, "The Upper Snake River Project," 120-121; "Town to be Built on A Desert," *Central California Register*, November 15, 1905, 5.

preference to municipalities and rural electric cooperatives serving project areas. As important, it established the policy of using revenues from power sales to assist in financing its costly reclamation construction program.[85] In 1909, Minidoka Dam's power plant, the first federal power plant in the Pacific Northwest, was completed with five generating units.[86] The turbines were rated at 2,000 horsepower (hp) each and at highest efficiency each unit used a maximum of 425 cubic feet per second (cfs). The power station, located on the downstream side of the diversion dam, developed nearly 8,000 kilowatts (kW) of power, which was transmitted eleven miles to the three pumping plants used to raise water into the main south side irrigation delivery canals and to a re-lift plant at the west end of the gravity division. Construction on pumping stations proceeded simultaneously with the power plant. After being used to generate power at Minidoka Dam, the water was released downstream and diverted at Milner Dam to serve prior irrigation rights of the Carey Act canal companies located along the river below.[87]

In addition to Minidoka Dam, the Reclamation Service initiated construction on a dam and reservoir located some three hundred miles upstream at Jackson Lake to store water for the Minidoka Project. Located just below the Teton Mountains in the state of Wyoming, Jackson Dam was a temporary log-crib structure. Completed in 1907, it raised the level of Jackson Lake fifteen feet to store approximately 300,000 acre-feet of water for use by Minidoka Project irrigators. The agency believed that the temporary dam would supply sufficient river regulation for five years – at which time a permanent dam would raise the level of the lake so as to impound an additional 80,000 to 100,000 acre-feet of water.

When the Twin Falls North Side Land & Water Company failed in its effort to construct off-stream reservoirs on its project, the United States contracted with the company in December 1907 to deliver 140,000 acre-feet per year for temporary use on the company's north side lands. The government retained the right to terminate the rental agreement as soon as Minidoka Project demands required the stored

85 Act of April 16, 1906, 34 Stat. 116; 43 USC 561.

86 A sixth generating unit was added in 1927 and a seventh in 1942. See Minidoka Project, Annual Project History, Vol. 36, 17. Box 305, PHFHR, 1902-1932, RG 115, NARA-Denver.

87 History of the Minidoka Project, Vol. 1, August 1915, 61-63. Box 294, PHFHR, 1902-1932, RG 115, NARA-Denver; and Gretsch, "The Upper Snake River Project," 133-135.

water. After three irrigation seasons, the government issued a termination notice to the company in 1910 because the federal project was nearing completion and additional stored water was required to meet the needs of Minidoka Project lands. By this time, however, the North Side Project had become entirely dependent upon water storage space in the federal reservoir.[88]

Fortunately for the Twin Falls North Side Project, passage of the Warren Act by Congress in 1911 ushered in a new phase in western reclamation. This act provided that whenever storage or carrying capacity of federal works was in excess of that required by federal project lands the excess could be contracted to "individual corporations, associations, and irrigation districts" organized for the purposes of reclamation. The act further authorized the Secretary of the Interior to cooperate with water users' associations, corporations, and water districts for the construction and use of reservoirs and distributing systems so long as the water was not delivered "to any one landowner in excess of an amount sufficient to irrigate one hundred and sixty acres."[89] The Idaho Carey Act Projects, with their own acreage limitations, were compatible partners for shared irrigation storage facilities.

The federal government and the Twin Falls private irrigation companies thus became full-fledged partners in the business of reclamation on the Upper Snake River. When high water in the summer of 1911 washed out the temporary dam at Jackson Lake Reservoir, the Reclamation Service and the private Twin Falls companies rushed to erect another dam before the spring run-off of 1912. They completed the first stage of a permanent dam which was capable of storing 380,000 acre-feet. The Twin Falls irrigation companies contracted with the government for an enlargement of Jackson Lake Dam in return for storage space in the top portion of the federal reservoir. In 1916, the dam was raised to increase storage capacity by 400,000 acre-feet with the cost being borne by the Twin Falls companies in exchange for storage rights in the reservoir. In 1919 dredging of the outlet channel brought the total storage capacity to 847,000 acre-feet.[90] In addition to

88 History of the Minidoka Project, Vol. 1, 40-42; Box 294, PHFHR, 1902-1932, RG 115, NARA-Denver; *Eighth Annual Report of the Reclamation Service*, 81; and Gretsch, "The Upper Snake River Project," 126-127.

89 Act of February 21, 1911, 36 Stat. 925, 43 USC 523 et seq.

90 Gertsch, "The Upper Snake River Project," 124-132; and Hoyt, *Water Utilization*, 40.

the benefits accruing to the Twin Falls irrigators on the two large Carey Act projects, the USRS came to the aid of some thirteen Upper Snake River irrigation companies and districts by reserving 100,000 acre-feet of stored water for them in the expanded Jackson Lake Reservoir to supplement their natural flow rights.[91]

Initiated as separate and distinct irrigation projects, the federal Minidoka Project and state Carey Act projects in the Upper Snake River Valley were interwoven over the decades into a basin-wide cooperative irrigation venture. The projects initially shared space in the small storage reservoir at Jackson Lake and later at American Falls, and Palisades and reservoir spaceholders, whether federal, state, or private entities, contributed to the cost of the dams constructed at each of these three sites.

Over time, storage space in the various reservoirs was exchanged between the federal reclamation project, Carey Act projects and a host of individuals, canal companies, and irrigation and reservoir districts, who united to create a common plan for administering the operation of the Snake River, in order to more effectively regulate, conserve, control, and distribute water to the benefit of all water users claiming rights to stored water throughout the entire upper basin above Milner Dam.[92]

Water scarcity and court decrees

Development of these large private and federal irrigation projects that depended on stored water complicated water administration in the Upper Snake River Basin. With enactment of the March 11, 1903 water permitting statute by the Idaho state legislature, administrative procedures were instituted, allowing a claimant to obtain a water right license from the State Engineer's Office (Idaho Department of Reclamation after 1919). Rights initiated through beneficial use could be adjudicated in district court, in order to be finally fixed and their

91 Contract Form 7-523, July 25, 1917, between the U. S. Reclamation Service and State of Idaho for cost of measuring and distributing natural flow rights and stored water from Jackson Lake Reservoir for the year 1917, Contract with State of Idaho, 163-2A. Folder: Minidoka – Water Appropriations, Settlement of Idaho State Water Rights, Box 63 Minidoka 163-2-233, General Administrative & Project Records, 1902-19196, Entry 3 (GAPR, 1902-1919), RG 115, NARA-Denver.]

92 See for example, Exhibits A-F, in *Aberdeen-Springfield Canal Company, et al.* v. *Henry Eagle, Watermaster, District 36*, Supplemental Decree, March 12, 1969; in the District Court of the Seventh Judicial District, State of Idaho, County of Fremont.

relative priority determined.[93] In an adjudication, all of the water users on a stream or stream's tributary were united into one suit as either plaintiffs or defendants.[94] The state engineer, acting under the direction of the district court, made surveys and prepared plat maps, and then reported to the court showing the various ditches together with their capacities and the land served. Taking into consideration this evidence and any testimony submitted by claimants, the court issued a decree vesting rights in the various parties, as the evidence indicated. These district court decrees fixed the upper limit of a water user's right in terms of quantity of water, and established the purpose(s) of use and date of priority.[95]

One of the more important duties of the state engineer's office under the 1903 act was its supervision over distribution and use of water for irrigation and other purposes. While the state engineer had no judicial functions, his office was charged by law with the duty and responsibility of administering court decrees, and enforcing the orders of the court with reference to the use of public waters of the state. The 1903 water law divided the state into three water divisions with a water commissioner appointed for each; the commissioners served under the supervision of the state engineer.[96] The division commissioner had general control over the distribution of the public waters of the state within his division in accordance with the doctrine of prior appropriation. Each division was further divided into water districts so that each public stream and its tributaries, or independent source of water supply, constituted a water district. Within each water district, a watermaster, who was either appointed by the water commissioner or elected by local water users, was charged with handling and distribution of normal stream flow to the ditches and canals diverting from each stream in accordance with their priority dates.[97]

93 Prior to 1971, there were two ways of perfecting a water right under Idaho law: a person could apply for a permit or simply put water to beneficial use. The latter method required adjudication to establish the elements of a the right.

94 There are private adjudications that determine the rights among those water users joined in the adjudication and then there are general stream adjudications that are binding on the world. Many of the early adjudications were private adjudications.

95 Act of March 11, 1903. House Bill No. 146, 7th Session of the Legislature of the State of Idaho.

96 Act of March 11, 1903, 1903 Idaho Sess. Laws 223.

97 D. G. Martin, Idaho State Engineer, "Distribution of Water Among Appropriators," *8th Biennial Report of the State Engineer to the Governor of Idaho, 1909-1910,* 120-136; Samuel C. Wiel, *Idaho State Water Rights of the Western States*, Vol. II (San Francisco: Bancroft-Whiting Co., 1911), 1011; and Idaho State Water Codes, Pt. VI, Ch.49: 1188.

When the USRS entered the Snake River Basin to withdraw public lands for its irrigation project at Minidoka in 1902, individuals, private companies, and water users' associations owning and operating hundreds of ditches and canals already took water from the Snake River and its tributaries in Idaho above the Minidoka Project.

In the 1880s, Mormons from Utah established farms, and small individual and co-operative irrigation systems in the Upper Snake River basin. Over the next two decades, still more settlers, many of them non-Mormons, took up farming in the upper basin above American Falls. They also built temporary dams to divert the natural flow of the river into their private irrigation canal systems.[98] By the early 1900s, much of the land situated above American Falls with potential for irrigation by gravity flow from Snake River had been developed. Above American Falls, 275,000 acres were irrigated, resulting in the diversion of nearly all of the natural flow of the river during summer irrigation season. Below American Falls, long, large-capacity canals and expensive diversion dams built across the Snake River channel were required to deliver irrigation water to productive agricultural land. Nevertheless, in the early decades of the twentieth century, as a result of the Carey Act and the National Reclamation Act, new irrigation development came to center on areas below American Falls in the vicinity of present-day Minidoka and Twin Falls. To secure an assured surface water supply for irrigation, these lands depended upon large springs to replenish the Snake River below the mouth of Blackfoot River, the return flow to the river from upstream irrigation, and, ultimately, storage reservoir.[99]

Each of the existing diversions claimed natural flow rights in the waters of the Snake River prior to the USRS water filing, and among them were all of the pioneer diverters of water on the upper river and its tributaries. These water rights to the Snake River above Milner Dam were adjudicated in two state court decrees: *Rexburg Irrigation Company, et al.* vs. *Teton Irrigation Company, et al.,* District Court of

---

98 Arrington, *History of Idaho*, Vol. 1, 472-476.

99 Sufficient studies had been conducted by the US Geological Survey by 1903 to render reasonable assurance that even with the river dry above the mouth of Blackfoot River, the low stage of the river at Minidoka Dam would not average less than a flow of 2,000 to 2,100 cfs. If the Twin Falls projects successfully claimed a priority right to 3,400 cfs, it was clear that some storage reservoirs would become necessary to adequately supply Minidoka lands with water during the months of July and August. History of the Minidoka Project, Vol. 1, 1912, 5-6; Box 294, Entry 10; PHFHR, 1902-1932, Minidoka; RG 115; NARA-Denver.

the Sixth Judicial District of the State of Idaho, in Fremont County, filed in 1901; and in *Twin Falls Canal Company vs. Charles N. Foster*, filed in 1909 in the District Court of the Fourth Judicial District of the State of Idaho, in Twin Falls.

From the 1880s through the first decade of the twentieth century, irrigators in the Upper Snake River Basin increased in number and diverted increasing amounts of water. Not only were more acres brought into cultivation, but farmers switched from growing hay and grain to cultivating sugar beets, potatoes, and other row crops. Unlike grains, sugar beets and other vegetables required water not just in spring and early summer, but throughout the entire summer and into early autumn when the natural flow of the rivers diminished. In the absence of a clear and readily available record of water priorities and institutional mechanisms to enforce them, appropriators became competitors for water and in years of scarcity conflicts broke out. Outside of Mormon communities, water users on the Upper Snake River had no irrigation institutions or organizations whose duty it was to allocate water when it became scarce. In an attempt to clarify title to water rights on Teton River, a tributary of Henry's Fork of the Snake River, the Rexburg Irrigation Company in 1901 filed suit against the Teton Irrigation Canal Company. Four years later, the suit expanded to include irrigators and irrigation companies on the Snake River itself.[100]

The *Rexburg* suit, in essence, served as an adjudication of the appropriative water rights of participating parties who diverted water for irrigation uses from Henry's Fork and the main Snake River and its tributaries above Blackfoot. The suit resulted in issuance of the so-called *Rexburg Decree* setting out the water right priority date and amount of each plaintiff and defendant. Thereafter, the district watermaster, as provided for under the 1903 water law, oversaw the physical distribution of water among all the adjudicated claimants. It took nearly a decade from the time the case had been initiated for a final decree to be issued. The *Rexburg* suit involved upwards of 793,180 acres of irrigable land in Fremont and Bingham counties, a total of some 3,000 parties, and about 250 canals. These decreed rights

100 Arrington, *History of Idaho*, Vol. 1, 486-489; and *Rexburg Irrigation Company et al., Plaintiffs, v. Teton Irrigation Company et al., Defendants*, Decree, District Court of the Sixth Judicial District of the State of Idaho, In and For the County of Fremont, December 16, 1910.

included many of the oldest appropriative water rights on the Snake River.[101]

To a considerable extent the *Rexburg* suit – which involved the upper portion of the Snake River – was independent from the part of the river involved in the *Twin Falls* case; however, the quantity of water decreed to these various canal companies would materially affect the interests of the Minidoka and Twin Falls irrigation projects. In *Twin Falls Canal Company v. Foster*, the remaining persons, firms, corporations, and associations claiming the right to divert waters of the Snake River in Idaho between Milner Dam and American Falls in Cassia, Lincoln, and Twin Falls counties were brought in and made parties to an adjudication of the waters of Snake River diverted above Milner Dam. The case was brought by the owners of the South Side Twin Falls Canal against the United States and individual water users on the Minidoka Project, diverters on the North Side Twin Falls Project, and other appropriators diverting water below American Falls.[102] The suit involved about 2,500 parties joined as defendants, nearly one-half of whom were applicants for water under the Minidoka Project, and the water rights of more than one-half million acres of agricultural land. At the time the suit was filed, only relatively small percentage of the entire project lands at Minidoka and Twin Falls been watered yet both projects had plans for massive extension of their respective irrigated acreages.[103] Therefore, it was imperative that the rights of appropriators be established by a court decree to give these inchoate water rights legal standing in the state of Idaho.[104]

---

101 R. A. Ballinger, Secretary, Department of the Interior to The Honorable Attorney General, February 4, 1910, 1-3. Folder: Reclamation Service, Minidoka Project –Water Rights – Snake River, Dec. 30, 1908 to March 27, 1930, Box 1619 8-3 Minidoka, W CCF, 1907-1936, RG 48, NARA II. []

102 *Twin Falls Canal Company, A Corporation, Plaintiff, v. Charles N. Foster, Richard Ballinger, Secretary of the Interior of the United States, et al., Defendants*, Complaint, District Court of the Fourth Judicial District of the State of Idaho, In and For the County of Twin Falls, December 10, 1909. Folder: 163-2 Minidoka – Water Appropriations – Settlement of Water Rights to Dec. 31, 1910, Box 615 Minidoka 163-2-223, GAPR, 1902-1919, RG 115, NARA-Denver.

103 The South Side Twin Falls Project proposed a 600,000-acre expansion called the Bruneau Extension, while the Minidoka Project anticipated the 115,000-acre North Side Pumping Unit.

104 Prior to 1903 (House Bill No. 146, 7th Session of the Legislature of the State of Idaho), appropriations were made by posting notice at the point of diversion; recording the notice in county recorder's offices; and filing a copy of a notice in the state engineer's office. Hundreds of notices were posted with no or little work being done. To establish rights under these notices, beneficial use had to be made of the water and court decrees were always based on the amount of water beneficially used. The new law passed in 1903 specifically set the date of priority and the procedure to be followed before the state engineer in order to establish proof of completion of

The ultimate outcome of the *Twin Falls* law suit was delayed until June 1913 while the USRS and the Twin Falls companies tried to come to an agreement upon a permanent contract in connection with furnishing storage to the private canal companies from Lake Jackson Reservoir. In the interim, District Judge Edward A. Walters on June 30, 1910 called into his chambers an array of the best legal and engineering talent in the region, representing the major parties in the Snake River suit. He issued a temporary injunction for the irrigation season of 1910 to distribute water until the case came up for final determination in December. This decree stipulated that the North Side canal company and South Side canal company receive the first 3,400 cfs of natural flow after which the Minidoka Project had the right to 1,726 cfs; then the North Side Twin Falls Company was entitled to another 2,250 cfs; then the Minidoka Project came in again for enough additional water to supply the remaining needs of the federal reclamation project. With respect to the stored waters of Snake River at Jackson Lake and Lake Walcott, the court decreed that the stored water belonged to the Minidoka Project, but the North Side Canal Company under its contract with the United States was to receive two-fifths of the stored waters of Jackson Lake and another one-fifth was to be distributed to other parties as directed by the government agents. The Twin Falls Canal Company had no interest in the stored waters. D. G. Martin, Idaho state engineer, was appointed as commissioner of the court, authorized to give the injunction full force and effect, and to distribute to the respective canals the water specified. The parties to the suit, and their agents, were enjoined from diverting water from the Snake River at Milner and Minidoka dams in violation of the court order.[105]

In general terms, the final decree issued on June 20, 1913 adopted these temporary stipulations. The *Foster Decree* awarded the Department of the Interior use of the natural flow of Snake River at Minidoka Dam for power, light, manufacturing, and pumping purposes to the extent of 2,500 cfs with a priority date of June 15, 1909 and 200 cfs with a priority date of July 1, 1912, "said water to be returned to Snake River below the Minidoka Dam after being used for power

---

work and beneficial use to retain that priority.

105 *Twin Falls Canal Company v. Charles Foster, et al.*, Temporary Decree, District Court of the Fourth Judicial District of the State of Idaho, in and for Twin Falls County, June 30, 1910. Folder 163-2 Minidoka, Box 615, GAPR, 1902-1919, RG 115, NARA-Denver; and "Temporary Decree Issued," *The Twin Falls Times*, July 14, 1910.

purposes." These were the water rights associated with the original Minidoka power plant and its expansion of generating capacity during the early operation of the Minidoka Project.[106] An additional 2,500 cfs with a priority date of December 14, 1909 was awarded "for the purpose of filling Lake Walcott Reservoir above Minidoka Dam" so as "to be hereafter drawn out and used for irrigation and domestic purposes."[107] This right was associated with changes to the dam at Minidoka and its operation to make more effective use of the releases of stored water from Jackson Lake.

The decree also granted the Interior Department all of the stored water in both Lake Walcott and Jackson Lake reservoirs as they were "at present constructed." The water was to be distributed "to the Minidoka Project and canal system" under delivery contracts made with the Interior secretary. As to the additional stored water held in the proposed enlargement of Jackson Lake Reservoir, it was to "be distributed as provided in the contract between the United States, the Kuhn Irrigation and Canal Company, and the Twin Falls Canal Company." The court provided that the contract, when executed, "may be attached to this order."[108]

Finally, the *Foster Decree* made the Idaho state engineer – "until otherwise provided by Statute" – responsible for determining what constituted natural flow and what constituted stored water "flowing in Snake River at the Minidoka and Milner dams." Natural flow was defined as the amount of the river's water to which the various private Twin Falls projects were entitled "unaffected by the diversions or acts of the parties hereto…or by the releases of stored water."[109]

Defining natural flow in a court decree was considerably easier than divining it from stored water flowing in the same channel. Indeed, the

106 This ultimately was an issue in the SRBA.

107 *Twin Falls Canal Company v. Charles Foster, et al.,* Decree, District Court of the Fourth Judicial District of the State of Idaho, in and for Twin Falls County, June 20, 1913.

108 Ibid.

109 *Twin Falls Canal Company v. Charles Foster, et al.,* Decree, June 20, 1913; Samuel C. Wiel, *Idaho State Water Rights of the Western States*, Vol. II (San Francisco: Bancroft-Whiting Co., 1911), 1011; and Idaho State Water Codes, Pt. VI, Ch.49: 1188. District watermasters typically handled distribution of normal stream flow to ditches and canals. In 1909, new legislation was enacted to provide for appointment of a special deputy and assistants by the state engineer for the purpose of distributing stored water. Because natural flow and stored water were so inter-mingled on the upper Snake River, careful coordination and management of Snake River natural flow and storage rights was necessary, therefore, the duties of the two state officers were blended into a single individual who served as district watermaster and special deputy.

evolution of the irrigation and reservoir system of the upper basin severely complicated water administration. It introduced a fundamental problem that the *Foster Decree* highlighted – the problem of co-mingled natural flow and stored water – that irrigators, the state, and the federal government would have to resolve.

Co-mingling natural flow with stored water

During the early years of irrigation, when few streams were used to their limits even in the period of minimum flow, there was no great need for reliable information in regard to stream flow. There was plenty of land available and as long as there was surplus water in the streams, precise knowledge of the quantity was not required to insure success: There was water enough for everyone.

The large Carey Act projects and the Bureau of Reclamation's proposed project were planned to store spring flood flows for use during the summer irrigating season when normal flow might be too small to meet project water requirements. To ensure success, irrigation projects of this size required long-term stream-flow data, precipitation records, drainage area measurements, and weather stations to make annual water supply projections. Engineers, promoters, and financiers of the great irrigation projects along the Snake River would have wanted stream flow records at their precise points of diversion and proposed reservoir sites and to have the data carried over a number of years to determine reliable mean river flows, but events in the realm of irrigated agriculture moved too fast at the turn of the century to enable collection of this hydrologic data.

Run-off varied widely in the Snake River Basin from year-to-year and upstream irrigators above American Falls enjoyed the senior natural flow priorities. Downstream irrigators operating with water from the larger Carey Act and federal reclamation projects realized they could not rely solely upon their own, more junior, natural flow water rights – they needed holdover storage capacity as an insurance policy against subnormal run-off years. Consequently, construction of Jackson Lake reservoir complicated the water rights situation on the Snake River by creating two classes of water: stored water and natural flow.

With construction in 1907-08 of Jackson Lake reservoir and its subsequent reconstruction and enlargement in 1911-12, fifteen different canals in the upper valley and the irrigated areas below American Falls came to possess storage priorities as against normal flow irrigators in the basin above American Falls. The Snake River, in essence, became a "natural" three hundred mile-long canal that transported water released from storage in Jackson Lake on the headwaters in Wyoming during the summer season, past the head gates of more than 50 canals to Milner Dam where the two Twin Falls irrigation projects diverted the last of the stored water.[110] Stored water released into the Snake River channel, however, mingled with the natural flow of the river that had been awarded to other appropriators under various dates of priority as decreed by the courts. Their canals handled flow ranging from only one or two second-feet up to 3,600 second-feet. In a normal water year, total diversions amounted to about 6 million acre-feet, but it was difficult to measure individual diversions.

D. G. Martin, special deputy on stored water delivery for the Snake River above Milner, noted in 1912 that most of the headgates on upper valley canals were in poor condition. Too, there was "the entire absence of measuring devices of any kind, rendering it almost impossible to get any measure of the amount of water they might be drawing.[111] It was absolutely impossible to justly apportion irrigation water to the proper users without accurate stream and diversion measurements. The administrative difficulty of allocating the two intermingled types of water among eastern Snake River Basin diverters near Idaho Falls and Blackfoot with older natural flow appropriative state water rights, and the irrigators 300 miles downstream at Twin Falls and Minidoka with storage rights, made a situation ripe for conflict. All it would take is a "low water year" when a perceived failure to receive a just portion of the river's water flow resulted in crop failure.

By virtue of the water rights decreed in the *Rexburg* and *Foster* adjudications, after unstorable flood flows wasted past Milner in the winter and early spring, irrigators began diverting. As irrigation season progressed, these diversions depleted the flow of the river and practically dried up the channel of the Snake River below Milner Dam by diverting the entire natural flow of the river into irrigation canals.

110 *Nineteenth Annual Report of the Reclamation Service*, 11.

111 G. Clyde Baldwin, "Transmission and Delivery of Reservoir Water," in "Administrative Water Problems: A Symposium," *American Society of Civil Engineers Transactions* 94 (1930), 296-300; D. G. Martin to Hon. A. E. Robinson, State Engineer, October 8, 1912, *9th Biennial Report of the State Engineer to the Governor of Idaho, 1911-1912,* 145-155.

The manner by which the district watermaster distributed these decreed water rights was described by Thomas R. Newell of the U. S. Geological Survey:

> In Idaho, a decreed right specifies a quantity (if beneficially used) and, in all cases, a priority date. As river supplies decreased below the quantity necessary to fulfill all decrees in use, the river master refers to his chronological list of priority dates and declares the right with the latest date either partly or completely invalid and denies use thereunder. This process of eliminating one user after another is continued until the river stage reaches the summer low. Then as supplies increase, each right is restored in reverse order. [112]

The Carey Act and federal reclamation projects below American Falls had court-decreed natural flow priority rights dating to the early 1900s and commanded a significant quantity of natural flow rights on the river, but these rights were junior to many decreed rights in the upper basin. When their natural flow irrigation rights were exhausted, the Minidoka and Twin Falls projects called upon their stored water in Jackson Lake reservoir (and later American Falls) for the benefit of lands in the Minidoka and Twin Falls irrigation projects. The longer the period of denied use of natural flow rights, the longer the period of diversion from storage until that source was depleted or until the end of irrigation season lasting to the end of September. Thus, particularly prior to construction of American Falls Reservoir, Newell noted, summer minimum flow of Snake River above Milner "usually remains fairly constant for several weeks in the height of irrigation season." However, during this period the percentage of actual flow in the river represented by "natural flow rights" constantly decreased and late decrees had to be cut off in successive order. Conversely, the storage delivery rate from Jackson Lake became "quite large during the months of July and August," surpassing the volume of natural flow in the river.[113]

Idaho statutes provided for the administration of the water in a stream when the channel of the stream was used for conveying stored water from a reservoir for irrigation and other uses. The Idaho state

112 Thomas R. Newell's comments on Baldwin, "Transmission and Delivery of Reservoir Water," in "Administrative Water Problems: A Symposium," *American Society of Civil Engineers Transactions* 94 (1930), 322.

113 Ibid.

engineer, as the public official charged by state law with the duty and responsibility of administering the public waters of the state, was to determine what part of the water flowing in the Snake River at the Minidoka and Milner dams was stored water and what part was natural flow.

In 1909 new legislation was enacted that provided for a special deputy and assistants appointed by the state engineer, and paid for by the owner of the stored water, to distribute water released from storage reservoirs for irrigation purposes.

Because natural flow and stored water were so inter-mingled on the upper Snake River, careful coordination and management of Snake River natural flow and storage rights was necessary. Water District No. 36 (later renumbered District No. 1) was the most important territory in the state for the distribution of water and required special arrangements. The district embraced the waters of the Snake River and its tributaries from the headwaters of the stream to the irrigated lands serviced by the last irrigation diversions at Milner Dam. Responsibility for dividing and distributing natural flow rights to the various irrigation interests above Milner in strict accordance with decreed priorities as to time and amount fell to the watermaster of Water District No. 36. Elsewhere in Idaho, a special deputy state engineer was placed in charge of stored water distribution. However, in the Upper Snake River Basin, because of the extensive and early co-mingling of normal flow and storage water, the duties of the two state officers by 1919 had been blended into a single individual who served as both the district watermaster and special deputy.[114]

The job of distributing stored water seems simple conceptually. A certain amount of water was diverted from the reservoir and taken out below, undiminished in quantity except for losses due to natural causes during its transmission, while at the same time allowing those holding normal flow rights to draw their water entitlements. Equitable distribution, however, depended on a number of unknown factors and required judgments to be made by the engineer in charge. If the rate of flow was constant, the job may have been easier, but in the time of year that stored water was drawn on the Snake River, the river might be

---

114 "Distribution of Snake River Water During Greatest Drought," *Engineering News-Record*, November 11, 1920, 927-28; A. E. Robinson, Idaho State Engineer, "Snake River Distribution," *9th Biennial Report of the State Engineer to the Governor of Idaho, 1911-1912*, 131-155.

falling 200 to 500 second-feet per day. Uncertainty was compounded by return flow from irrigation, which augmented the normal flow of the river. Data on river flow at various reaches of the river was essential to equitable distribution of normal and stored water and the United States Geological Survey (USGS) and its hydrographers that were in charge of installing and maintaining the numerous gaging stations on the Snake River .

The job of District 36 watermaster/special deputy was as important as it was difficult, noted Idaho Commissioner of Reclamation George N. Carter. In presenting to the governor of Idaho his eighth annual report (1926), he commented on the position held by G. Clyde Baldwin, special deputy to the Idaho commissioner of reclamation and District No. 36 watermaster:

> It is doubtful if there is any other single item of interest of so great importance to as many people involved as a unit in Idaho as the equitable apportionment of the waters of the Snake River, and I assure you that the responsibility of carrying through an irrigation season, the faithful administration of the decree of courts and laws of the State, as they relate to the distribution of irrigation water in this district, is no light task.
>
> The diversity of the problems which are constantly presented in this work requires most careful consideration and correct solutions are necessary based on the soundest of judgment.[115]

Baldwin had held the position since 1919 when, following a terrible drought, the representatives of both stored and natural flow users appointed him to take charge of water distribution. At that time Baldwin had been district engineer for the USGS in charge of hydrometric work within the state of Idaho. Over time, it became the policy of the state engineer's office and department of reclamation, whenever possible, to appoint the acting watermaster on the Upper Snake River as the special deputy to deliver and distribute the stored water, along with his regular duties. In fact, by a cooperative arrangement effected between the state of Idaho and the USGS, it

115 George N. Carter, Idaho Commissioner of Reclamation, to Governor H. G. Baldridge, May 5, 1927, in G. Clyde Baldwin, "Water Distribution and Hydrometric Work in District No. 36" (typescript, 1926). Unsorted Water Distribution Reports; District 5: 1936, 1937, 1938, 1939; District 12A: 1915, 1919-1920, 1927; District 36: 1926, 1927, 1938, Accession no. 20071914 (Unsorted Water Distribution Reports, Acc. No. 20071914), AR 20 Department of Reclamation (AR 20) Idaho State Historical Society, Boise, ID (ISHS).

became possible to centralize, under the control of one person, both the distribution of normal flow and stored water and the stream measurement work on the Upper Snake River.[116]

The "Committee of Nine"

The year 1919, the driest year on record, left many farmers in the Upper Snake River Basin facing shortages of natural flow water rights in mid-June, not even mid-way through the irrigation season.

The canals were ordered to cut to their decreed amounts on June 4 and distribution on this basis lasted until June 10 when, in order to fill decreed rights, it was necessary to cease deliveries under state licenses and permits. On June 11, all diversions with decreed dates after 1902 were cut. On July 3 everyone with a priority after May 10, 1889 was ordered to cease diverting water. Meanwhile, users of stored water asking for early release of irrigation water from Jackson Lake had greatly depleted the reservoir. By early August all the outlet gates were raised clear of the water and the reservoir continued to drop "although at a continuously decreasing rate because of the limitation of the outlet capacity and lessening head." The flow of the river had become so small by late August that there was little benefit realized to the Minidoka Project from the release of storage at Jackson Lake. During the greater part of the remainder of irrigation season, the river was absolutely dry at a point about four miles below the town of Blackfoot, and at no time did the water passing Milner Dam exceed 100 cfs.[117]

Faced with increasingly complicated and contentious Snake River water distribution issues, a group of concerned water users met at Idaho Falls on April 23, 1919 for the purpose of devising more accurate and equitable methods of distributing normal flow and storage waters in Water District No. 36. Nine men, representing different geographic regions within the water district, was selected to act as advisors to the district watermaster. Three members (J. T. Fisher, Alfred Ricks and J. R. Thomson) belonged to the North Fork Protective Association, three

116 "Distribution of Snake River Water During Greatest Drought," *Engineering News-Record*, November 11, 1920, 927-28; W. G. Swendsen, Idaho Commissioner of Reclamation, "Water Distribution in District No. 36", *First Biennial Report of the Department of Reclamation State of Idaho, 1919-1920*(1921), 9-10.

117 "Distribution of Snake River Water During Greatest Drought," *Engineering News-Record*, November 11, 1920, 929.

(Christian Anderson, P. J. Davis and D. H. Blossom) to the Idaho Falls-based Farmer's Protective Irrigation Association representing irrigators above American Falls, and three (J. C. Wheelon, R. E. Shepherd and Frank A. Banks) from the Minidoka and Twin Falls projects representing the irrigators below American Falls.

At the first meeting of this advisory group, its members requested that Idaho's Commissioner of Reclamation Warren G. Swendsen appoint a new special deputy to oversee stored water distribution within the district. Swendsen himself had been appointed to the new position of Commissioner of Reclamation, head of the Department of Reclamation, by Governor David W. Davis in 1919. The Commissioner of Reclamation was the successor to both the State Land Board on Carey Act Projects and the State Engineer's Office. Swendsen would serve in the role until 1926.[118]

Following the recommendation of the advisory board, Swendsen appointed United States Geological Survey (USGS) district engineer Clyde T. Baldwin as the head of new joint federal-state office in Idaho Falls. The new office, opened on May 22, 1919. A three-member advisory committee to Water District No. 36 was subsequently appointed on March 1,1920. Its members were selected by their respective districts: the North Fork District, the South Fork District, and the Twin Falls District.

A new nine-member advisory panel for Water District No. 36 was organized a little less than four years later. At a meeting held at the Bonneville County courthouse, called by Swendsen, a "Committee of Nine" was organized on January 20, 1923. The members were selected by area: two from the upper or North Fork (Fall River – Teton area); four from the Central Section; and three from the Lower Valley (American Falls to Milner Dam). On February 12, 1923 the Committee of Nine was formally organized and made permanent with John W. Hart of Rigby selected as permanent chairman and John Lee of Grant as secretary. The leadership of Water District No. 36 remained incredibly stable over its long history. From 1923 to 1979 only five different individuals served as its chairman.[119]

---

118 W. G. Swendsen, *First Biennial Report of the Department of Reclamation State of Idaho, 1919-1920* (1921), 4, 8-9; and George N. Carter, *Fourth Biennial Report of the Department of Reclamation State of Idaho, 1925-1926* (1927), 4.

119 Joe L. Marker, "Committee of Nine: Watching Use of Snake River Water Since 1919," *The Post-Register* (Idaho Falls), March 19, 1976.

John Empey was the first watermaster of District No. 36; however, in every year subsequent to 1919 and until 1932, Baldwin served as USGS district engineer and was also elected district watermaster in charge of the distribution of normal flow decreed rights by the local water users. Thus, administration of normal flow and stored water in the Upper Basin was united under the control of Baldwin, who represented federal, state, and local interests and was also in-charge of stream measurements and all hydrographic work for thirteen years. When Baldwin retired in 1932, local water users and Idaho Commissioner of Reclamation R. W. Faris appointed Lynn Crandall, another USGS district engineer, to the post of special deputy/watermaster – an arrangement that continued to guarantee federal-state cooperation in the administration of the waters of Upper Snake River for twenty-six more years until his retirement in 1958.[120]

Each year before irrigation season opened, Baldwin (and later, Crandall), in consultation with the Committee of Nine, made adjustments in water delivery schedules to more efficiently and fairly distribute water as dictated by the various court decrees that governed distribution of normal run-off on the upper river and to release or transfer stored water to spaceholders in the various reservoirs. Precipitation, snow pack and forecasting run-off were also elements that were factored into decisions on how to manage storage reservoirs each season. As more knowledge was gained, further adjustments as to how division of the waters of the river were made included changing methods of division between natural and stored water at reservoir outlets, improving methods of measuring channel losses and gains, and altering the schedules of percentage transmission losses applied through successive river sections to transit storage. The watermaster and the Committee of Nine adopted compromise schedules annually in slightly revised form before each irrigation season, and as time passed, the refined schedules theoretically came more closely to represent true conditions on the river.[121]

120 Carlson, "The Snake River Story", 6-7.

121 Detailed accounts of water supply and how water was measured, stored, diverted, collected, and distributed according to the established Idaho State Water Right priorities on the river can be found in the district watermaster's annual reports, first prepared in 1919. See typescript annual reports by year, "Water Distribution and Hydrometric Work, District No. 36, Snake River, Idaho," AR 20, ISHS.

In praising the challenging work done by Baldwin as watermaster of District No. 36, Thomas R. Newell of the USGS in Boise commented in 1930 that the measure of his success over the previous decade in regulating and distributing the waters of the river was evidenced by the reduction in litigation over water rights, and uninterrupted service and negligible waste of water passing Milner Dam each year during "the regulation period."

> Transmission loss schedules and reservoir outlet divisions always must meet the tests of changing conditions and even changing ownerships. The small percentage of ultimate wastage (flow past Milner Dam during regulation period) is a real index of the excellence and efficiency of the control system of river operation as a whole. The fact that these major reservoirs [Jackson Lake and American Falls] have been operated ever since their construction in conjunction with natural water distribution from the same water-shed, through identical delivery channels and without the authority of a Court Commissioner, speaks well for the efforts of the administrative official and perhaps also of the tolerance of district water users.[122]

Despite severe challenges presented in the administration of water rights by disagreements over transmission losses, return flows, or the matter of equitable segregation of flowing commingled stored water and natural flow, water users in the Upper Snake River Valley through court decrees, creation of new irrigation institutions, compromises, and river operation agreements had discovered ways to manage and allocate the scarce water resources of Snake River while expanding irrigated agriculture in the early decades of the twentieth century and bringing radical changes to their region.

The need for more storage

Expansion of irrigated agriculture had transformed the landscape and brought astonishing changes to the eastern Snake River plain in the early decades of the twentieth century.

Between 1900 and 1920, Idaho's quantity of irrigated land had increased by *four hundred* percent, the largest increase of any western state. In the Snake River valley, the irrigation boom helped build a

---

122 See Newell's comments on G. Clyde Baldwin, "Transmission and Delivery of Reservoir Water," 321, 324.

string of more than a dozen prosperous towns along the river from Boise to Pocatello. Mostly due to the large-scale private and federal irrigation and reclamation projects, Idaho's population more than doubled between 1900 and 1910, rising to 325,000. By 1920, some 431,000 people resided in the state. During the 1920s growth continued, but at a slower rate. The six Snake River counties that benefited most from irrigation – Twin Falls, Cassia, Gooding, Lincoln, Jerome, and Minidoka – increased in population from only 5,700 in 1900 to nearly 70,000 in 1920. By that same year, about 2,164,000 acres in the Snake River Basin were irrigated, of which nearly 717,000 were irrigated by direct diversions from the main channel of the Snake River. In all of the United States, only the Missouri River Basin had more irrigated acres than the Snake River basin by 1930.[123] As the population of the Snake River valley rose, farm population increased, but so did inhabitants in the region's numerous small towns. For example, within the Minidoka Project by 1920 there were nearly 2,400 farms with a population of an estimated 9,500 persons. The number of towns stood at six, but the population of these towns at 9,000 inhabitants almost matched that of the farms.[124]

Carey Act boosters in southern Idaho were initially antagonistic to the USRS entry into Snake River Basin and viewed the Minidoka Project as a direct challenge to private irrigation enterprises. Within a decade, however, the stark distinction between government and private projects in the valley became somewhat clouded. Water shortages and drought had forced Minidoka and the Twin Falls irrigators to band together and cooperate to raise Jackson Lake dam and nearly triple the storage capacity of the only major storage reservoir on the Snake River. The growing alliance between private enterprise and the USRS served as a basis for subsequent efforts at consolidation and control of water resources along broad constructive lines. The great drought of 1919, the driest year on record in the Snake River valley, provided a powerful catalyst for united action by private, state, and federal irrigation

---

123 U. S. Bureau of the Census, Department of Commerce, Thirteenth Census of the United States, 1910: Agriculture General Report and Analysis (GPO, 1922), 845; Fourteenth Census of the United States, 1920: Irrigation and Drainage (GPO, 1922), 15 and 167; Fifteenth Census of the United States, 1930: Irrigation of Agricultural Lands (GPO, 1932), 15; and Hoyt, Water Utilization, 44.

124 *Nineteenth Annual Report of the Reclamation Service*, 158.

interests to organize new cooperative irrigation institutions and to build additional storage capacity for irrigation uses on the Snake River.[125]

In 1919, storage on the Snake River consisted of Jackson Lake reservoir and smaller reservoirs at Henry's Lake on the north fork headwaters and Lake Walcott, formed by the dam at Minidoka. The combined storage capacity of these reservoirs (1,032,180 acre-feet) were not enough to conserve the large winter and spring flows that ran down the river and were lost to irrigation uses each year. During the summer irrigation season, many of the upper basin's irrigators coveted these flood waters that had escaped past Milner Dam during the non-irrigation season. Irrigators reasoned that if this water could be conserved and saved by storage, they might become immune to the ravages of drought that periodically engulfed the region. Only by building additional storage reservoirs could a stable irrigated agricultural empire, envisioned by the Carey Act promoters, ever be fully realized. In addition, the United States contemplated opening settlement on its next phase of construction at the planned 110,000-acre North Side Pumping Division of the Minidoka Project. Bringing this acreage into production would require not only a large storage reservoir to secure a reliable water supply, but also additional hydroelectric power to operate electric pumps to lift this water up onto the irrigable bench lands located above the lands already served by diversion canals coming directly off the river.[126]

Between 1919 and 1927, the federal government negotiated with 40 or so irrigation companies and districts along a 300-mile stretch of the Snake River as well as the local private power company to acquire rights, finance, and build a new and massive storage reservoir at American Falls that would nearly triple the effective storage capacity of on-stream reservoirs on the Snake River above Milner Dam. Water stored behind the massive dam would provide irrigation water to an additional 110,000 acres of new land within the Minidoka Project and supplemental water to more than a million acres of land located both above and below the dam. The majority of the citizens of the region understood there was a crucial community of interests among the people of the basin that rested on the giant American Falls Dam.

---

125 Gertsch, "The Upper Snake River Project," 107-168, 176, 187-197; and *Nineteenth Annual Report of the Reclamation Service*, 15 and 157.

126 Gertsch, "The Upper Snake River Project," 187-197; and *Nineteenth Annual Report of the Reclamation Service*, 155-157, and 417-419.

The fate of the growing irrigated agricultural empire of the Upper Snake River Valley hung in the balance. Completion of the project would greatly enhance the economic stability of many private irrigation enterprises located between Heise and Twin Falls as well as on the federal Minidoka Reclamation Project. Failure to construct additional storage would mean uncertainty and a tenuous existence for farms and cities dependent upon the seasonal whims of mother nature.[127]

127 Gertsch, "The Upper Snake River Project," 187-197; USRS, Nineteenth Annual Report, 155-157, and 417-419; *Twentieth Annual Report of the Reclamation Service, 1920-1921* (GPO, 1921), 25, 157, 409, 411; *Twenty-first Annual Report of the Reclamation Service, 1921-1922* (GPO, 1922), 64-65, and 123; *Twenty-second Annual Report of the Reclamation Service, 1922-1923* (GPO, 1923), 6, 65-66, and 127; Department of the Interior, *Twenty-third Annual Report of the Bureau of Reclamation, 1923-1924* (GPO, 1924), 29, 37, 53-54; *Twenty-fourth Annual Report of the Bureau of Reclamation, 1924-1925* (GPO, 1925), 18-19, 39; *Twenty-fifth Annual Report of the Bureau of Reclamation, 1925-1926* (GPO, 1926), 6 and 15.

3

Cooperation and Conflicting Rights

Snake River water serviced not only irrigation, but also hydroelectric power generation – especially in the area below American Falls. Advancements in water power engineering, long-distance transmission of electrical energy, and the practical application and use of electric power during the first two decades of the twentieth century revolutionized urban and industrial development across the nation.

A particular type of electrical-power generating facility, using dams and falling water to turn turbines and generate electricity, developed much more rapidly in the West than in other regions of the United States.[128] That was notably true in Idaho, where falling water generated almost all of its electricity. What was unusual was the pace of expanding electric service to rural areas, which in the countryside of southern Idaho came decades earlier than in many rural parts of America. The US Reclamation Service and Idaho Power Company played key roles in electrifying irrigated farms and rural communities in the Snake River Basin. Use of hydroelectric power for heating, lighting, and manufacturing purposes in larger towns and cities of Idaho, such as Boise, created the initial interest in diverting water for power generation and even sparked some controversy at the State Constitutional Convention with respect to these power developments depriving water from irrigators.

Ironically, it was the power demands of electric irrigation pumps that largely propelled rural electrification in the lightly-populated southern Idaho counties within the Snake River watershed, which meant that irrigation and hydropower development proceeded in tandem.[129]

128 See Stacy, *Legacy*, 11-19; Williams, *Energy*, 170-177; and Hasson, "Electric Transmission of Power Long Distances," 49-72.

129 Stacy, *Legacy*, 1-99.

Hydroelectricity and irrigation

Factors such as the cost of electricity, heights of lift, and size and type of pumping plant became second only to availability of water in determining the success of the largest Snake River irrigation projects. In a 1913 article on waterpower resources in southern Idaho, the *Journal of Electricity, Power and Gas* noted that hydroelectric power companies delivering electrical power to pumping plants had been instrumental in populating the region with farms:

> Pumping plants operated with electric motors and made to irrigate tracts of land on a large scale are a comparatively new means of development. There are at least 150,000 acres of land in Idaho watered by this means alone, and there are at least 200,000 acres of land yet in sagebrush that can be irrigated by no other means. The power companies that have done most to bring about the tremendous development that has taken place are the Great Shoshone & Twin Falls Water Power Company of Twin Falls, Idaho, the Southern Idaho Water Power Company, the Government Reclamation Service of Boise, Idaho and the Idaho & Oregon Light & Power Company of Boise, Idaho.
>
> The largest pumping plant in the world is located near Minidoka, Idaho, and was completed three years ago by the Government Reclamation Service. Forty-eight thousand acres of land are watered under this system with lifts varying from 30 to 90 feet. The High Line pumping project near Twin Falls, Idaho, comprises 4000 acres of land, and is watered with lifts varying from 40 to 80 feet. The Twin Falls North Side pumping project, a Kuhn enterprise, is composed of 14,000 acres of land segregated under the Carey Act law. The lifts vary from 35 to 90 feet in height.[130]

Irrigation pumping requirements demanded more electrical power than any other use in southern Idaho in the early 1900s, so the successful conquest of the desert by irrigated agriculture was vital to the economic welfare of the region's electrical power industry. Electrical pumping plants lifted water up to the large irrigated tracts on the plains above the river during the summer irrigation season, while surplus power was used for general heating, lighting and other domestic purposes on farms and in rural towns the rest of the year. The market was served by a mushrooming number of small private electric companies in southern Idaho and by the federal government. The sale of surplus power on federal reclamation projects – typically through

---

130 E. A. Wilcox, "Water Power Resources of Southern Idaho," *Journal of Electricity, Power and Gas* 30:20 (May 17, 1913), 447-450.

local distribution companies that bought power wholesale from the Reclamation Service and retailed it to their members and customers – helped reduce the charges against the pumping load, making irrigation of more acreage economically feasible.[131]

Nature had provided a magnificent asset for hydroelectric power companies operating on the Snake River. The most remarkable opportunities for power development were located on the stream in the central portion of its semi-circular course across the Snake River Plain, precisely where immense irrigation tracts were under development by private enterprise and the Reclamation Service. As the river flowed from east to west, it tumbled over a succession of abrupt changes in the elevation of the riverbed – Idaho Falls, American Falls, Dry Creek Falls, Twin Falls, Shoshone Falls, Upper Salmon Falls, Lower Salmon Falls, Crane Falls, and Swan Falls. These settings afforded natural heads very favorable to power development and were located in the heart of irrigation country.[132]

Hydroelectric development on the Snake River consequently owed its success to irrigated agriculture. Yet, if hydropower and irrigated agriculture were to prosper together, power sites and possibilities on the river could not be considered independently of future storage and irrigation possibilities on the same stream. Some sites possessed excellent natural qualities for power development, but their commercial value could be compromised by construction of storage reservoirs meant for irrigation, which in order to fill, required storage of the winter flow of the river. Stored water became unavailable for run-of-the-river hydroelectric power plants such as those constructed in the early years of the twentieth century at various natural falls along the Snake River.[133]

131 This was certainly the case for the Minidoka Project. See *Nineteenth Annual Report of the Reclamation Service*, 155-157, and 417-419; *Twentieth Annual Report of the Reclamation Service*, 25, 157, 409, 411; *Twenty-first Annual Report of the Reclamation Service*, 64-65, and 123; *Twenty-second Annual Report of the Reclamation Service*, 6, 65-66, and 127; *Twenty-third Annual Report of the Bureau of Reclamation*, 29, 37, 53-54; *Twenty-fourth Annual Report of the Bureau of Reclamation*, 18-19, 39; and *Twenty-fifth Annual Report of the Bureau of Reclamation*, 6 and 15.

132 Hoyt, *Water Utilization*, 8, 35-36, 188-202.

133 At Milner, the river enters a canyon with a fall of 1,400 feet in 57 miles to Lower Salmon Falls. There were several concentrated vertical falls furnishing inexpensive power sites. If the river was regulated by storage for irrigation only, the value of these sites for power generation purposes would be largely destroyed. On the other hand, if the regulation permitted a comparatively small flow on this section of the river, a large amount of power could be developed. The Idaho Power Company owned real property at many of the potential power sites downstream of the proposed American Falls dam site: American Falls, Twin Falls, Shoshone Falls, Upper Salmon Falls, Lower Salmon Falls, Crane Falls and Swan Falls. See Hoyt, "Report Relative to

With development of the Minidoka Project and Carey Act irrigation tracts at Twin Falls, the river immediately below Milner Dam was turned more or less dry in the summer months, and for this reason power sites below the last point of diversion at Milner could be without sufficient water when demand was greatest for electricity to pump irrigation water. In addition, if storage reservoirs were operated only for irrigation uses, these conditions would be exaggerated; reservoirs might store all but the flood run-off in excess of their storage capacity during the non-irrigation season. On the other hand, run-of-the-river hydroelectric power facilities with early appropriative water right priorities to the natural flow of the Snake River, if not subordinated to irrigation, might seriously curtail the expansion of irrigation dependent upon junior water storage rights.[134]

Electrical energy, moreover, differed from other industrial products. The amount of capital investment required to build central power plants, erect long-distance, high-voltage electric transmission lines, and maintain far-flung substations and distribution systems was enormous. Initially, it was beyond the capability of Idaho's small, scattered population of struggling farmers to create sufficient demand to justify such investments. Since the Snake River Basin did not develop any large-scale manufacturing or new mining centers through the 1930s, agriculture remained the state's main economic enterprise and fueled population, economic and energy growth. Although the population of the state doubled each decade between 1890 and 1910 and by 32.6 percent in the 1910s, most of the basin's population still lived in small communities of fewer than 1,000 people in 1920, or on farms and ranches, thinly scattered over a broad region covering an area measuring approximately 400 miles by 300 miles. These rural farm communities developed a large summer demand for electricity to operate irrigation pumps, but winter domestic uses were incidental to energy consumption for agriculture. Unlike manufacturers, power

---

Water Power Resources of Snake River," August 1922. Folder: D-100, 12-G-51, WPB, 1894-1961, RG 57, NARA II.

134 Board of Engineers to Honorable D. W. Davis, Governor of the State of Idaho, April 10, 1920, 5-6 and 23. No Folder, Box 728 Minidoka 301.1, GAPR, 1919-1945, General Files, 1919-1929, RG 115, NARA-Denver; and Morris Bien, "Memorandum for Mr. Cobb," May 28, 1912. Folder: Reclamation Service, Minidoka Project, Box 1619, CCF, 1907-1936, RG 48, NARA II. This conflict was played out in the 1920s in the context of several applications made to the Federal Power Commission for licenses to build hydroelectric plants along the Snake River (most notably the plant at Twin Falls), as well as in a 1928 revision to the state constitution (Article XV, Section 3) that regulated and limited the use of water in Idaho for power purposes.

companies could not stockpile their output in warehouses until the power market ripened. Once generated, electrical energy had to be transmitted and used quickly. The earliest power companies were thus limited in size and distributed their electrical output in small quantities to a confined market. Many provided power for a few towns, local mines, or a single industrial enterprise.

From the late 1880s to 1906 some 50 different electric operating companies came into existence within the area later served by Idaho Power Company. After the turn of the century, hydroelectric facilities associated with large reclamation pumping projects came to the Snake River. Both state and federal sponsored irrigation ventures had hydroelectric power components. The Carey Act and Reclamation Act projects stimulated rapid expansion of irrigated agriculture and growth of rural communities in southern Idaho. Eastern investors became interested in this emergent power market and capitalized on the hydroelectric potential of the Snake River and its tributaries.[135]

Development of alternating current and use of transformers – which allowed electricity to be generated at a low voltage, stepped up to a higher voltage for more efficient transmission, and then stepped back down to a lower voltage for use – made electricity safer to use and vastly increased the service area of an electric power plant. With the advent of more efficient electrical motors and pumps, electricity could be used not only for lighting, but also as a source of power for machinery, thus creating more demand for daytime power and putting the electric power industry on a much firmer load basis than it had been previously as a part-time industry supplying power for lighting in towns or for seasonal irrigation pumping in rural areas. As long-distance electrical transmission technology evolved, powerhouse location from a technological perspective became more independent of the locus of electrical power consumption. Central power stations rapidly superseded the small, isolated generating plants. Nevertheless, building an electric transmission network to connect central stations with local distribution systems remained costly.[136]

---

135 Stacy, *Legacy*, 1-33.

136 Thomas Hughes, *Networks of Power: Electrification in Western Society, 1880 – 1930* (Baltimore: The John Hopkins University Press, 1983), 262-84; George C. Young and Frederic J. Cochrane, *Hydro Era: The Story of Idaho Power Company* Boise: Idaho Power Company, 1978), 13-21, map facing 26.

Investment in construction of high voltage, high-tension transmission systems to tie together numerous power generating plants within the region was one method of balancing uneven seasonal loads so that power could be more effectively supplied to all of the communities served by the hydroelectric power industry in southern Idaho. Idaho Power Company responded to this need. To secure these results required a network of linkages among power plants and construction of high voltage transmission lines to carry power from one region into another.[137] As the generation and transmission network of IPC matured into the 1930s, the company extended its power grid two-thirds of the way across the State of Idaho, some 200 to 250 miles, and eventually extended its network of transmission lines to the neighboring states of Utah, Oregon, and Wyoming.[138] In contrast, the Minidoka power plant, operated by the Reclamation Service under the power guidelines of the Reclamation Act of 1902, continued to generate electrical power that was incidental to the primary purpose of the Minidoka Reclamation Project, which was to provide water for irrigation, including electricity necessary to operate pumping stations to bring water to higher benchland within the project lying many feet above the river.[139]

Minidoka Dam and powerhouse

Congress authorized the Minidoka Project in April 1904, and by the end of the year most of the public land reserved for the project and opened for sale had been claimed by prospective settlers. Construction had sufficiently advanced that by 1907 the Reclamation Service was delivering irrigation water to a portion of the project. Soon after, more than 110,000 acres of formerly uninhabited sagebrush desert on both sides of the Snake River was being irrigated in the vicinity of the several new communities laid out by the Reclamation Service along the Oregon Short Line Railroad under the provisions of the Reclamation Act. About 72,000 acres lying principally in Minidoka County were placed under a gravity ditch, and the remaining 48,000 acres on the south side of the river in Cassia County were serviced with irrigation water lifted by three pumping plants with an average lift of 68 feet.

137 Ibid., 84-99.

138 Ibid, 36.

139 Stacy, *Legacy*, 35-36, and 84-99; Hoyt, *Water Utilization*, 44.

Thus, the original project was divided into two divisions known as the Gravity Division, later operated by Minidoka Irrigation District (1917), and the South Side Pumping Division, later operated by Burley Irrigation District (1926). In October 1906, the Reclamation Service completed its Minidoka Dam, located upstream from the Twin Falls Land and Water Company's Milner diversion dam. The Service's diversion structure had a maximum height above the riverbed of 86 feet; it raised the water level about 46 feet in the river and acted as a combined storage, diversion and power dam. The reservoir, known as Lake Walcott, held 107,200 acre-feet of which some 40,000 acre-feet was active storage. The storage function of the dam was not used to any great extent, as the level of Lake Walcott was normally held to near its maximum capacity to secure a maximum head on the turbines of the power plant.[140]

The primary purpose of the Minidoka Reclamation Project was to provide irrigation water for the settlers who had acquired homesteads on the public lands of the United States withdrawn and reserved for the reclamation project. Water supply for the project was diverted from the Snake River at headgates located at the north and south ends of Minidoka Dam. Most of the water released into these canals was used on farm units for irrigation and domestic purposes, but several 160-acre withdrawals of public land were reserved on the Minidoka Project for "townsite purposes." All of the towns were supplied with project water for irrigation and domestic uses and surplus electrical current generated at Minidoka powerhouse furnished light, heat and power for the towns and farms of the project. The National Reclamation Act, as amended by the so-called "Townsite Act" in 1906, authorized the Secretary of the Interior, on projects where development of power was necessary for irrigation purposes, to lease any surplus power for non-irrigation uses for a term not exceeding ten years. Preference was given to municipalities and electrical co-operatives within the project with the money derived from the sale of power being credited to the project reclamation fund, but such uses were subordinate to irrigation demand.[141]

---

140 History of the Minidoka Project, Vol. 1, 1912, 2-6. Box 294; Minidoka Project, Annual Project History, Vol. 36, 1942, 17; Box 305, PHFHR, 1902-1932, RG 115, NARA-Denver; US Department of the Interior, *Eighth Annual Report of the Reclamation Service, 1908-1909* (GPO 1910), 151-153; Hoyt, *Water Utilization*, 44 and 124-125; and Gertsch, "The Upper Snake River Project," 120-121.

When the Reclamation Service designed Minidoka Dam, it made provision for ten penstock openings near the north end of the dam at the site of the anticipated powerhouse. The power station at Minidoka Dam was completed in 1909 with three of its initial five generating units installed. The two remaining 2,000 horsepower turbines and 1,400 kilovolt-ampere (kva) generators were installed in the spring of 1911, thus completing the equipment as planned to a capacity of nearly 8,000 kW of power. Under the *Foster Decree*, the United States was awarded 2,700 cfs for operation of these five units at Minidoka power plant with priority dates of June 15, 1909 and July 1, 1912.[142]

A powerhouse on the Minidoka Project was necessary to meet pumping requirements on those parts of the project that could not be irrigated by gravity flow. Because electrical current for pumping purposes is coincident with the duration of the irrigation season, it quickly became apparent that a large opportunity existed for commercial power development during the non-irrigating season. Power generation for domestic and municipal uses continued to be a secondary – albeit increasingly important– feature of the project. There was a steady increase in the demand for pumping and commercial power year by year and in 1926 the Reclamation Service added a sixth generating unit to the Minidoka power plant that raised its installed capacity to 10,000 kW. The United States was granted a water right license for an additional 825 cfs for the sixth generating unit with a claimed priority date of June 16, 1927.[143]

The original electrical transmission lines on the Minidoka Project were strung on telephone poles paralleling the railroad line from the power plant at Minidoka to the various project towns, with branch lines

---

141 History of the Minidoka Project, Vol. 1, 1912, 2-6. Box 294, PHFHR, 1902-1932, RG 115, NARA-Denver; and Reclamation Act, April 16, 1906, Ch. 1631, Sec. 5, 34 Stat. 116; February 24, 1911, Ch. 155, 36 Stat. 930. The act provided that "no lease shall be made of such surplus water or power privileges as will impair the efficiency of the irrigation project."

142 Ballinger to Attorney General, February 4, 1910, 7. Folder: Reclamation Service, Minidoka, Box 1619, CCF, 1907-1936, RG 48, NARA II; History of the Minidoka Project, Vol. 1, 1912, 14-15 and 53-56. Box 294, PHFHR, 1902-1932, RG 115, NARA-Denver; Water Right Nos. 01-0217 and 01-0218; IDWR Back Files: Idaho Department of Water Resources website, On-Line Data, Idaho State Water Right and Adjudication Search; http://www.idwr.idaho.gov/apps/ExtSearch/SearchWRAJ.asp; last accessed April 2, 2009.

143 Water Right No. 01-4024, IDWR Water Right Files. The water right was not claimed by the Bureau until March 27, 1980. A seventh generating unit was added in 1942 with a claimed water right for 1,700 cfs with a priority date of May 7, 1942 (Water Right No. 01-4025) and additional units were issued a permit in 1985 with a claimed right of 5,875 cfs and a priority date of August 7, 1980.

extending to pumping stations south of the river. This transmission line was supplemented in the spring of 1911 with an eleven-mile-long transmission line dedicated solely to delivering electrical power to the three south side pumping stations. The total rated discharge capacity of the three pumping stations was approximately 1,400 cfs and at the height of irrigation season, all of the pumping units operated at maximum capacity to meet irrigation demand. At the west end of the north side gravity unit, a pumping station, with two centrifugal pumps with a combined 50 cfs capacity, was added in 1912-13 to lift water 18.5 feet to service an additional 2,220 acres with irrigation water. Power was supplied to the station by adding generating capacity to the Minidoka power plant and extending the transmission line from the town of Heyburn, a distance of seven miles.[144]

The Reclamation Service viewed providing power to project towns for municipal and domestic uses as a desirable and profitable way to use any excess power capacity as long as it did not compromise its primary irrigation mission. In the winter months when there were no irrigation pumping requirements, a large amount of excess seasonal power was available if a market could be found. Sale of energy for heating purposes to farms and towns was an attractive seasonal market, but firm year-round power sold for commercial uses was more lucrative. With approval of the Secretary of the Interior, the Reclamation Service entered into contracts with town corporations, rural co-operatives, or other parties in each of the local project towns, selling them small amounts of power wholesale. These private entities, in turn, retailed the power to individual consumers. The contracts specified the maximum quantities that could be delivered during certain periods of the year to safeguard the primacy of power for irrigation uses. Electrical demand in the project towns rose more rapidly than anticipated and settlers on project farms also soon developed an interest in securing electrical current on their rural parcels. Electrical lines did not begin to extend into the countryside until about 1913. A Minidoka Project "Commercial Power Unit" was set up to provide bookkeeping services for electrical billings in 1915. As electrical demand and profits accumulated, extensions to the power transmission system required for

144 History of the Minidoka Project, Vol. 1, 1912, 57-60 and 71-72. Box 294, PHFHR, 1902-1932, RG 115, NARA-Denver.

such service were paid for solely from the surplus accruing from the sale of commercial power.[145]

Not long after the Reclamation Service completed the Minidoka dam and powerhouse, it began modifying and adjusting project operations to meet changing circumstances and unanticipated problems. Chief among these was the efficient delivery of storage water from Jackson Lake Reservoir through the Snake River channel. In 1906, the Reclamation Service constructed a temporary log crib dam across the natural outlet of Jackson Lake to provide about 300,000 acre-feet (af) of storage for use on the Minidoka Project. The Reclamation Service anticipated that this volume of stored water would be sufficient to meet project needs for four or five years, by which time accumulated precipitation and stream flow data would indicate the appropriate size for a permanent structure. The initial 300,000 af storage capacity was more than what was immediately required by the Minidoka Project.

On December 4, 1907, the Reclamation Service entered into a temporary contract to provide 140,000 af of stored water during low water season to Twin Falls North Side Land & Water Company. Subsequently, annual renewals of the contract extended water deliveries through 1912. Water from Jackson Lake flowed in the Snake River channel past dozens of headgates to the Minidoka and Twin Falls points of diversion. Junior appropriators required release of this storage water relatively early during the irrigation season. Transmitting this stored water within the Snake River channel during the height of irrigation season past upstream canal headgates with earlier natural flow rights without suffering severe diversion losses to other appropriators was difficult. If an adequate amount of water could be stored in Lake Walcott to supply the Minidoka Project until the natural flow of Snake River dropped to its low water stage, misappropriation of stored water released from Jackson Lake by upstream appropriators might be more easily controlled. The least expensive solution was to raise the spillway at Minidoka Dam by installing flashboards sufficiently high to permit raising the elevation of the lake five feet, thereby creating an additional 53,500 af of storage capacity. Doing so also enhanced power production. With the flashboards in place the normal operating head for

---

145 History of the Minidoka Project, Vol. 1, 1912, 61-64. Box 294, PHFHR, 1902-1932, RG 115, NARA-Denver; and "Report Shows Minidoka Dam Integral Part of Powerplant," *Minidoka County News*, October 24, 1929.

the power plant was raised from about 43 to 48 feet. This spillway modification was completed in 1909-1910.[146]

The Minidoka Project's two points of diversion that released irrigation water for project lands through the north and south side canals were located within the reservoir formed behind Minidoka Dam. The dam also controlled releases of normal flow and stored water through the reservoir to meet downstream irrigation requirements at Milner Dam. Coordinated with these irrigation releases, the surface level of the reservoir was fluctuated by the Reclamation Service to maintain a suitable head of water to produce more or less energy as necessary for project pumping requirements and other electrical demands within the project area.

Writing in December 1910, shortly after the spillway modification was completed, the Service's Chief Engineer, A. P. Davis, its supervising engineer, F. E. Weymouth, and the Minidoka Project engineer, Charles H. Paul, all believed that even during "the season of greatest demand" (i.e., the irrigation season) Lake Walcott could be held "four or five feet above normal" to generate sufficient power not only for project pumping requirements, but also to provide excess current for growing commercial sales within the project area. As the irrigation project power demands grew, however, operating Lake Walcott for power purposes became more challenging. To more properly control discharge over the spillway crest to satisfy contract obligations for water delivery to the Twin Falls Canals at Milner Dam, the Reclamation Service in 1912 built four sluice gates into the diversion structure.[147]

---

146 History of the Minidoka Project, Vol. 1, 1912, 40-42 and 65-67. Box 294, PHFHR, 1902-1932, RG 115, NARA-Denver. For the purpose of securing storage rights at Lake Walcott, in the reservoir created by raising the spillway crest of the diversion dam, F. E. Weymouth, USRS, filed an application on December 14, 1909 for 2,500 cfs of the Snake River Flow.

147 A. P. Davis, F. E. Weymouth, Charles H. Paul to Director, USRS, December 18, 1910. Folder: 308, Minidoka, Board of Engineers Reports, Box 312, and History of the Minidoka Project, Vol. 1, 1912, 65-67. Box 294, PHFHR, 1902-1932, Minidoka, RG 115, NARA-Denver.

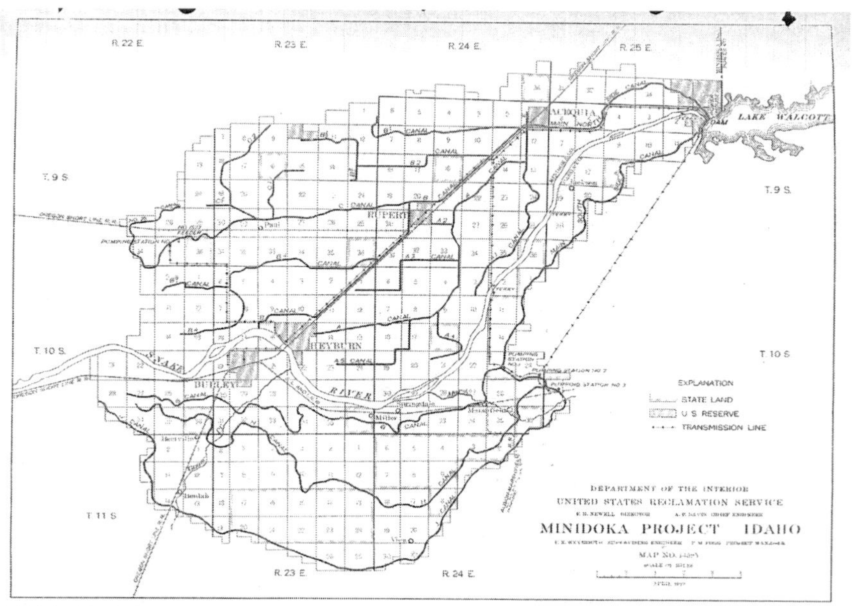

The Minidoka Project as of 1915

Source: "A History of the Minidoka Project, Idaho, to 1912 inclusive", United States Reclamation Service, Volume 2. Project Histories, Feature Histories, and Reports, 1902-1932, Minidoka, Box 294, Entry 10, RG 115, National Archives-Rocky Mountain Region (Denver)

In the years following execution of the temporary contract for storage water at Jackson Lake, the Twin Falls North Side Project became entirely dependent upon this storage space in the federal reservoir to meet its irrigation obligations.

Fortunately for the Twin Falls project, passage of the Warren Act by Congress in 1911 (36 Stat. 925) ushered in a new phase in western reclamation. The act provided that whenever storage or carrying capacity of federal works was in excess of federal project requirements the excess could be contracted to "individual corporations, associations, and irrigation districts" organized for the purposes of reclamation. The act further authorized the Secretary of the Interior to cooperate with water user associations, corporations, and water districts for construction and joint use of reservoirs and distribution systems. In the ensuing years, the federal government and the Twin Falls Carey Act irrigation companies became full-fledged partners in the business of reclamation. When high water in the summer of 1911 washed out the temporary dam at Jackson Lake Reservoir, the Reclamation Service and Twin Falls companies rushed to erect a joint-use dam before the spring run-off of 1912. They completed the first stage of a permanent dam capable of storing 380,000 acre-feet. The Twin Falls irrigation companies contracted with the government to enlarge Jackson Lake Dam in return for storage space in the top portion of the federal reservoir. In 1916, the dam was raised to increase storage capacity by 400,000 af with the cost being borne by the Twin Falls companies in exchange for storage rights in the reservoir. In 1919 dredging of the outlet channel brought the total storage capacity to 847,000 af. In addition to the benefits accruing to the Twin Falls irrigators, the Service came to the aid of some thirteen other upper Snake River irrigation companies and districts by reserving 100,000 af of stored water for them in the expanded Jackson Lake Reservoir to supplement their natural flow rights.[148]

As the Reclamation Service prepared to expand its storage capabilities at Jackson Lake, inquiries came into the Interior Department concerning the government's willingness to grant permits to private power companies for the use of this stored water to generate electrical power in the poorly served rural markets of the upper basin. While the Department believed it had the legal authority to do so under the Reclamation Act, as amended in 1906 and 1911, any contract was subject to limitations that "might prevent commercial development of power."[149] During irrigation season, for example, Morris Bien, the Service's supervising engineer, noted that the large flow of stored water might be used for power purposes without interfering with irrigation, if it was returned to the river above Minidoka Dam. During the non-

148 History of the Minidoka Project, Vol. 1, 1912, 40-42. Box 294. PHFHR, 1902-1932; Contract Form 7-523, July 25, 1917, Folder: Minidoka – Water Appropriations, Settlement of Idaho State Water Rights, Box 636, GAPR, 1902-1919,RG 115, NARA-Denver; *Eighth Annual Report of the Reclamation Service*, 81; Gretsch, "The Upper Snake River Project." 124-132; Act of February 21, 1911, 36 Stat. 925, 43 USC 523 et seq.; and Hoyt, *Water Utilization*, 40.

149 Act of February 24, 1911 (36 Stat., 930). This act amended Section 5 of the so called "townsites act" of 1906 providing for the withdrawal of public lands for townsite purposes in connection with irrigation projects under the Reclamation Act of 1902. The act gave the Secretary of Interior authority whenever the development of power was necessary for irrigation purposes under a reclamation project to lease surplus power for up to 10 years giving preferences to use for municipal purposes with the money derived from the sale of power being credited to the project in the reclamation fund, "provided, that no lease shall be made of such surplus water or power privileges as will impair the efficiency of the irrigation project."

irrigation season, however, the water held in storage had to be retained and saved for irrigation purposes; therefore, a smaller amount would be available in winter for power uses. In any contract, Bien noted, the Service "could not bind itself very closely as to the quantities to be made available because the flow of the river must be controlled principally for irrigation purposes and any power contract must be subordinate to the needs of water for irrigation." Power production was to remain incidental to the primary purpose of the reclamation project which was to provide water for irrigation.[150]

In addition to water right filings in support of irrigation enterprises on the upper river, appropriative applications for large amounts of water for power purposes had been made on the Snake River by the 1910s by predecessors-in-interest to Idaho Power Company (IPC) at American Falls above the Minidoka Project and at several existing and proposed power sites at natural falls not too far distant below Milner Dam.[151] New development or further expansion of existing commercial power facilities at these sites potentially endangered the water supply for anticipated upstream irrigation developments. Proposals by IPC to construct power plants at Twin Falls and American Falls in the early 1920s led to a prolonged debate over the subordination of the company's water rights at its proposed plants to Upper Basin irrigators.[152]

Power company consolidation and Idaho Power

Sweeping changes were about to take place in electric service to southern Idaho and eastern Oregon in mid-1916, at the moment when

150 Bien, "Memorandum for Mr. Cobb," May 28, 1912. Folder: Reclamation Service, Minidoka Project, Box 1619, CCF 1907-1936, RG 48, NARA II.

151 "The importance of the power filings," according to Interior Secretary's Ballinger's February 1910 letter to the US Attorney General – a letter in which Ballinger reviewed the status of irrigation on Snake River as well as recent court cases involving water rights on the upper river – "lies only in the fact that the depletion of the water supply in the river by irrigation projects above the power plants to a quantity below the amount necessary to furnish power required by their appropriation would be an infringement upon the rights acquired by such power appropriation if it be prior to the appropriation for irrigation purposes." Ballinger to Attorney General, February 4, 1910, 7. Folder: Reclamation Service, Minidoka Project, Box 1619, CCF 1907-1936, RG 48, NARA II.

152 Federal Power Commission, License on Government Lands, Project No. 18, Idaho, Idaho Power Company, June 11, 1934. File Folder 100-2, Idaho Power – Twin Falls, Project No. 18, Idaho Power Co., Formal, Part 2, January 1,1928 to December 31, 1948, Box 13, Accession No. 138-88-006, Federal Energy Regulatory Commission, Washington DC (FERC).

the Idaho Public Utilities Commission reported to Idaho Governor Moses Alexander on the negotiations between Idaho Power Company and the existing five struggling electric companies.

IPC proposed to consolidate the power market under a single electric utility, a development welcomed by the commission:

> The outcome of this merger is of considerable interest to the people of the state, giving, as it does, one large, well-organized, sufficiently capitalized company in lieu of the five over-capitalized and excessively bonded companies formerly serving southern Idaho. This merger should, and we believe will, result in a material reduction in expense of operation and in improved service and, under efficient control and supervision by the state, the consumers of electric current in southern Idaho will no doubt share in the benefits of the merger.[153]

Before Idaho Power was created, a host of independent companies had developed the hydroelectric power facilities located on the Snake River, as each local community sought to supply its own demand by developing independent local plants. Over the years, these local plants were superseded by more centralized stations constructed and operated by different companies. On the Upper Snake River, in southeastern Idaho near Idaho Falls, was the Idaho Power & Transportation Company which provided electrical service to that city and to Rigby, Rexburg, Sugar City and St Anthony. In southwestern Idaho, there were two principal companies, Idaho-Oregon Light & Power Company and Idaho Railway Light & Power Company, both headquartered in Boise. The former company operated the Horseshoe Bend plant on the Payette River, the Ox Bow plant on Snake River and leased power from the Barber plant on the Boise River. These plants distributed electrical power to Boise, Meridian, Emmett, Parma, and Weiser in Idaho and to Huntington and Ontario in Oregon. Idaho Railway Light & Power Company owned the Swan Falls power plant and distributing system at Caldwell and Nampa and served several other towns in the Boise Valley. It also provided electric power for mines near Silver City and for irrigation pumping within the Gem Irrigation District. The company's power plant on the Malad River, acquired from the Beaver River Power Company, distributed electricity to the towns of Gooding,

153 *Annual Report of the Idaho Public Utilities Commission* (June 30, 1916).

Glenns Ferry, Mountain Home, Boise, Caldwell, and other small towns.[154]

During the pioneer era of power development, independent hydropower companies likewise developed the area on the Snake River between Blackfoot and Glenns Ferry. From 1907 to 1915, some local distribution systems were connected by high voltage transmission lines to more centralized power generating stations. Several formerly independent plants were consolidated under the ownership of James S. and William S. Kuhn, who organized the Great Shoshone & Twin Falls Water Power Company and the Southern Idaho Power Company. Beginning in about 1907, the Kuhn interests built or took over the existing power plants at American Falls, Shoshone Falls, Twin Falls, Auger Falls, Upper and Lower Salmon Falls, and Thousand Springs. In 1910 the company declared that it had in operation, and was further developing on the Snake River "one of the greatest groups of water power plants in the United States." It was an era of exceptionally rapid economic development in southern Idaho and the power business grew with the economy. The company boasted to potential investors that its power facilities were located "in the center of the famous Twin Falls irrigation country" and its transmission network would extend to serve the entire region. Great Shoshone represented to its investors that its earnings in 1911 were estimated at three times its 1909 earnings, but this meant larger investments to keep pace with escalating demand.

Larger developments were undertaken than could be financed and in some cases sound engineering judgment was lacking. The Great Shoshone company fell as quickly as it rose in the volatile and competitive business. When it went into receivership in 1912, most of its assets would be acquired by the newly-formed Idaho Power Company in 1916. That same year, Southern Idaho Power Company was reorganized and also acquired by IPC.[155]

Incorporated in the state of Maine on May 6, 1915 as a subsidiary of a large holding company that consolidated dozens of large and small electric power companies in surrounding states, Idaho Power Company set out to absorb all of the separate, independent, non-governmental

---

154 Hiram T. French, History of Idaho: A Narrative Account of Its Historic Progress, Its People and Its Principal Interests (Chicago: Lewis Publishing Company, 1914), 480-485.

155 Hoyt, *Water Utilization*, 198; Wilcox, "Water Power Resources," 448; and "Great Shoshone and Twin Falls Water Power Company," *Newark Advocate*, July 21, 1910.

electric companies on the Snake River and fashion them into an integrated power system. On August 1, 1916, IPC took over ownership and operation of the electrical systems of the five major companies that served the populated portion of the Snake River Valley in southern Idaho and eastern Oregon. The region was largely dependent upon agriculture and related industries, but the progress of reclamation and agriculture was phenomenal during the early decades of the twentieth century. IPC executives understood this, and in the company's long-range planning co-operation with agriculture and reclamation interests was a high priority. By the mid-1920s, the power company supplied most of the state's privately-generated electric power and was on its way to becoming one of Idaho's largest and most powerful corporations. Consumption of electricity had increased dramatically since the 1890s and in a couple of decades had taken its place as an integral part of people's daily lives.[156]

Electric companies were a major factor behind much of the economic growth in the United States between 1890 and 1915. The way these companies grew was typical of a profound structural change in American business from family capitalism and partnerships to managerial capitalism. One of the first steps in the organization of corporations such as General Electric or Westinghouse, the dominant national corporations in the electrical field, was the horizontal combination of many smaller firms. IPC was organized to fulfill just such a role in southern Idaho. When it was created in 1915, the company was controlled by the Power Securities Corporation, which, in turn, was controlled by the Electric Power & Light Company, which through a supervision contract was controlled by the Electric Bond & Share Company (EBASCO), headquartered in New York City. EBASCO, a subsidiary of the General Electric Company and the world's largest electric utility "holding company," was created in 1905 to provide engineering advice on construction projects undertaken by small utility companies and to arrange for capital to finance them. It exercised tight control over its five subordinate holding companies with subsidiaries in 32 states and 13 foreign countries and assumed

156 *Idaho Power Company: A History of the Development of the Electric Industry in Southern Idaho and Eastern Oregon, 1887-1939 with Reference to Idaho Power Company and its Predecessors* (Boise: Idaho Power Company, ca. 1940), 43-54.

important roles in operating practices, engineering, construction, budgets, financial controls and financing.[157]

The "holding company" was a financial innovation developed by the major electrical equipment manufacturers – General Electric and Westinghouse. EBASCO accepted the relatively unattractive stock and bond issues of weak, local operating companies, held these securities as collateral, and offered the more preferred diversified and secure securities of the holding company to investors. In exchange the local operating utilities were required to buy their electrical generating and associated equipment from its parent company General Electric. Thus, through EBASCO, a direct union between General Electric Company and IPC, as an operating service company, was made. Having a financial stake in the operating companies, EBASCO itself offered management and engineering services that the local operating companies may not have been able to afford themselves. In doing so, the holding company hoped the operating companies would become valuable and their securities could ultimately be sold at a profit.[158]

Within a year of its organization, IPC was the largest operating company in the Snake River Basin. It owned power plants at five locations on the Snake River, divided into two independent groups of production and transmission facilities: an eastern group and a western group. The eastern group consisted of three power plants at American Falls serving the American Falls, Pocatello, Blackfoot and Aberdeen areas. The western group consisted of the Shoshone Falls, Lower Salmon Falls, Swan Falls, and Oxbow power plants serving southwestern Idaho. The Oxbow plant was located on the west bank of the Snake River in Oregon; all the others were in Idaho. In addition to these power plants, IPC owned other power-generating facilities on the Boise, Payette, and Malad rivers as well as at Thousand Springs – all of which brought electric lights and heating to towns, power to factories, electricity for irrigation pumps, and modern conveniences to farms in southern Idaho.[159]

---

157 Barrie A. Wigmore, *The Crash and its Aftermath: A History of the Securities Markets in the United States, 1929-1933* (Westport, Connecticut: Greenwood Press, 1985), 38-43; and David E. Nye, *Electrifying America: Social Meaning of a New Technology* (Cambridge: MIT Press, 1990), 170-171.

158 Nye, *Electrifying America,* 172; Business and Finance," *Time Magazine*, January 25, 1925; "Business: Bond and Share Defense," *Time Magazine*, November 23, 1936. .

159 Ibid., 44-45; and Hoyt, *Water Utilization*, 192, 198, 199-202.

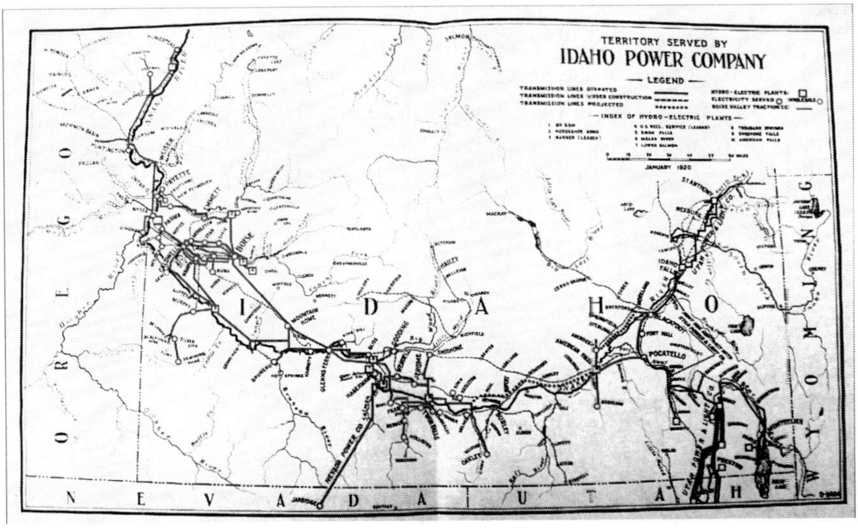

Territory Served by Idaho Power Company, January 1920. *Source:* Picture of map taken from the papers of Governor Charles Calvin Moore. See Folder 32: Public Utilities Commission of the State of Idaho, 1923-1926, Box 2, AR 2/13 Governors Papers: C.C. Moore, Idaho State Historical Society

Although not yet connected via transmission lines into a regional network, IPC power plants located on the Snake River were nevertheless connected by the river itself. All relied upon the surface flow of water in the river channel, supplemented to a greater or lesser extent by springs, seeps and side-flow, groundwater from the Snake River aquifer and return flow from irrigation. None of IPC's power plants on the Snake River was connected to a large storage reservoir: They were "run-of-the-river" plants, and they depended on the normal flow of the river, minus consumptive diversions by irrigators and other users upstream. IPC owned one small dam and reservoir with minimal storage capacity at American Falls, and held some flowage easements in anticipation of a future expansion of the American Falls reservoir site. As power demand increased, the company planned to expand the capacity of its East Side power plant at American Falls by adding several generating units to its existing powerhouse, and using the additional water stored at American Falls and released through this power plant to increase the firm capacity of its downstream plants at Shoshone and Lower Salmon Falls.[160]

As IPC noted in its 1943 corporate history, all the plants and plant sites it acquired "had been so designed and the natural characteristics of the sites were such as to permit the installation of additional capacity in relative[ly] small units." Thus, over the course of the twentieth century, the power company's system, as planned, would be equipped "to gradually increase its capacities at these plants and sites, paralleling the growth of load in the territory served."[161]

Building a power network

In the wake of its creation, IPC confronted a formidable list of challenges in rationalizing and integrating the operations of its predecessors into a coherent and efficient power network. The company faced duplication of transmission and distribution facilities and an uneven distribution of load among its major markets at Pocatello and Blackfoot, Twin Falls, and in the Boise Valley. In some locations there was insufficient power generated and in others a surplus. Overall the physical plant inherited from its bankrupt predecessors suffered from deferred maintenance. Finally, the company was saddled with a "below market" rate structure for all electric power purchased for irrigation purposes because these rates had been "frozen" by the newly organized Idaho Public Utilities Commission.

Nevertheless, EBASCO, the supervising company that controlled IPC, brought with it substantial financial resources, an experienced team of technical advisors, management expertise, and vast experience in the electrical industry. William T. Wallace, former general manager of an EBASCO-owned power company in Colorado, was appointed to straighten out the power situation for the company in southern Idaho. Susan M. Stacy, IPC historian, chronicled Wallace's major achievements in her 1991 history of the company:

> Wallace went to work. He ordered a comprehensive evaluation of the power plants, dams, substations, transmission lines, and myriad other equipment in order to prioritize repair, replacement, and expansion needs. The first and easiest task was to eliminate the

160 Hoyt, *Water Utilization*, 188-189, and 198.

161 Idaho Power Company, *History*, 62.

> duplicated equipment. The next priority was to do what Great Shoshone had envisioned when it plotted its invasion into the Boise area – connect the 27-mile void between Grandview and Swan Falls Dam. This work, done in 1916, connected the Twin Falls power plants with those of the western area by means of a 44,000-volt transmission line so that the whole system was available to serve the load wherever it was needed. Eventually, American Falls would be connected, but the need for that was not so urgent now. Altogether the new company's combined plants could generate about 21,240 kilowatts.
>
> Between 1915 and 1922 the company marched through equipment upgrades at each of its installations one by one.[162]

As of the early 1920s, IPC owned and operated ten hydroelectric power generating facilities on the Snake River and its tributaries between American Falls and Oxbow. Seven were on the main channel of the Snake River (three at American Falls, and one each at Shoshone Falls, Lower Salmon Falls, Swan Falls, and Oxbow), and an eighth was locate just off of the main river channel in the Thousand Springs area. These eight power plants leveraged only a small fraction of the power potential of the Snake River at each of the falls, as the sites were developed to meet current demand and were designed to accommodate expansion to take fuller advantage of the of the power generating potential at each location.

A general historical sketch of the construction and operation history of these eight power plants through the time of their conveyance to Idaho Power Company provides the backdrop for the emerging conflict between hydroelectric power interests and irrigated agriculture on the Snake River. The origins of that conflict extend back to the early years of the twentieth century and blossomed fully with the establishment of the Carey Act irrigation projects, the federal Minidoka Reclamation Project, and finally construction of the American Falls Dam in the mid-1920s.

Swan Falls

The Swan Falls power site was developed in 1900 by the Trade Dollar Consolidated Mining Company for generating power for its mines and mills at Silver City, located at a distance of approximately 26

---

162 Untitled Idaho Power Company document, undated, pages 221-23, IPC043297-043299. SWAN012 Disc, November 12, 2008, Idaho Attorney General's Office (Idaho AGO); and Stacy, *Legacy*, 28-29, and 45-46.

miles. Swan Falls was the first commercial hydroelectric power project completed on the Snake River and was quickly followed by other similar enterprises. The investment was made because of the relatively high cost of operating the company's mining machinery using steam generated by burning coal or wood. Once the power plant became operational, other mines in the region quickly contracted to be supplied with power from the electric plant at Swan Falls.[163]

Swan Falls, a natural fall of about four feet, is located where the river is crossed by a basaltic reef, about 30 miles due south of Boise and about 180 miles downstream from Milner Dam on the Snake River. The river in the Swan Falls reach had a considerable descent and the current was divided by an island that rose 60 feet above the reef. It was here that the Trade Dollar Consolidated Mining Company constructed its dam across the Snake River. A. J. Wiley, a highly regarded hydraulic engineer from Boise, served as the chief engineer on the construction project, but the project consulting electrical engineer, L. B. Stillwell, and hydraulic engineer, Thomas T. Johnstone, were from New York and Chicago. Johnstone's plan called for construction of a 400 foot-long concrete dam across the east channel, rising 12 feet above normal high water. The power generating plant, located on this channel, was designed to contain a total of four turbines, each generating a maximum of 750 horsepower under a 17 foot head. The west channel was to be dammed with a 420 foot-long rock-filled timber crib dam rising 12 feet with an apron extending 17 feet downstream. The crib portion of the dam was designed as the overflow section over which water would pour to a depth of 2.5 feet at low water stages.[164]

Photograph 4 (below) was taken of the Swan Falls dam and powerhouse in 1968 as part of a Historic American Engineering Report (HAER); the powerhouse is located to the right of the dam.

163 "Power plant on Snake," *Idaho Daily Statesman*, January 1, 1901; "One to be Placed on the Poorman at Silver City," *Idaho Daily Statesman*, August19, 1901; *Idaho Daily Statesman* August 20, 1901.

164 "Swan Falls Dam: Interesting Engineering Work in Progress on the Snake," *Idaho Daily Statesman*, May 19, 1900; "Power plant on Snake," *Idaho Daily Statesman*, January 1, 1901; "Water Runs Over Crest of Dam," *Idaho Daily Statesman*, April 4, 1901; "Not a Hitch in Swan Falls Power plant Since it was Started Up: Consulting Engineer Pleased with It," *Idaho Daily Statesman*, May 2, 1901.

View of Swan Falls Dam, Looking Downstream to South and West from the Canyon Road, September 1984. *Source:* Library of Congress, Prints and Photographs Division, Historic American Engineering Record, HAER ID.37-KU.V.1-4

James H. Hutchinson, general manager of the Trade Dollar Consolidated Mining Company, posted a notice of location on January 17, 1900 for 10,000 cfs of water at Swan Falls, the water to be diverted by means of a dam to be constructed across the Snake River and used for motor, manufacturing, mechanical, electric light, electric power and other useful and beneficial purposes and for irrigation near Silver City, Idaho and in the counties of Ada, Canyon, Owyhee, Elmore, Lincoln and Cassia and other places in Idaho. The same notice was filed with the state engineer the following month.[165] A few months later, Hutchinson conveyed the Swan Falls water right to Trade Dollar.

In 1901, the company installed its four wheels and three 300 kW generating units, applying 2,150 cfs to beneficial use. In October 1906,

165 Notice of Location at Swan Falls, January 17, 1900, Recorded February 2, 1900, Water Rights, Mill Sites and Tunnel Sites, Book 3, Owyhee County Recorder's Office; and Water Location No. 57 (1900), Idaho State Engineer.

Judge James H. Beatty of the U. S. Court of Appeals heard the case of *Trade Dollar Consolidated Mining Co., v. Fraser & Crussen* and on April 2, 1907, Beatty decreed 2,150 cfs to Trade Dollar, together with a valid legal right to develop up to 10,000 cfs at a large dam and diversion structure yet to be built at Swan Falls. The court gave Trade Dollar four years to complete its construction with due diligence and to put the remaining 7,850 cfs embraced in the notice of location to full beneficial use.[166]

In 1904, the Swan Falls Power Company had taken over construction efforts at the power site and constructed a new powerhouse that began generating power in 1906 using four additional wheels and two generators. Failure of equipment in the Swan Falls electrical power plant caused frequent shutdowns in the underground Trade Dollar mines at Florida Mountain, but between 1907 and 1910, the Swan Falls Power Company added 1,300 kW of capacity with two more wheels and two generators. On March 1, 1911, Swan Falls Power Company notified the Idaho state engineer that since 1906 beneficial use had been made of up to 3,300 cfs in addition to the 2,150 decreed by the U. S. Court of Appeals for a total of 5,450 cfs. The additional water use was recognized by the state engineer by adding another 1,840 cfs to the statement of beneficial use a few months later, and 1,460 cfs more on July 19, 1911; each of these additions (totaling 3,300 cfs) was assigned a priority date of July 17, 1900.[167]

In October 1911, Western Idaho Water & Power Company acquired Swan Falls Power Company and two months later sold it and its power site at Swan Falls to Idaho Railway Light & Power Company. In 1913, this company failed and went into receivership with its assets being acquired by Electric Investment Company at a court-ordered sale in 1915. In June 1916, IPC took over the property of the Electric Investment Company, including the Swan Falls power plant. Both the

---

166 In the Ninth Circuit Court of the United States, for the District Court of Idaho, Final Decree, *The Trade Dollar Consolidated Mining Company, a corporation, Plaintiff vs. Alfred A. Fraser and A. R. Crusen, Defendants*, April 2, 1907; Idaho State Water Right 02-0100 Back File; and *Idaho Daily Statesman*, June 26, 1900.

167 State of Idaho Water License, Idaho Railway Light & Power Company, Swan Falls Development; In the Circuit Court of the United States, For the District of Idaho, Final Decree, *The Trade Dollar Consolidated Mining Company, a corporation, Plaintiff vs. Alfred A. Fraser and A. R. Crusen, Defendants*, April 2, 1907, enclosed in Appendix of Franchises and Water Locations, Permits and Licenses of Idaho Railway, Light & Power Co., Idaho Power & Light Col, Idaho-Oregon Light & Power Co. and Great Shoshone and Twin Falls Water Power Co., pages 538-614. IPC043212-043216, SWAN012 Disc, November 12, 2008, Idaho AGO.

Swan Falls and the smaller Oxbow power plants serviced the electrical market in the Boise and Payette areas.[168]

Monthly records of the flow of Snake River from 1913 to 1920 measured at a gauging station a short distance below the plant recorded the flow as at least 6,900 cfs 90 percent of the time and at 12,500 cfs for 50 percent of the time. The absolute minimum flow was about 5,400 cfs, therefore there was more than sufficient flow in the Snake River to operate the powerhouse at full capacity in 1920. W. G. Hoyt, hydraulic engineer for the USGS, predicted in 1922 that once American Falls Reservoir was constructed the 90 percent flow would remain the same and the flow for 50 percent of the time would be reduced to about 7,500 cfs.[169] The Swan Falls powerhouse was located far enough downstream that the diversion of water at Milner Dam during the irrigation season would have little or no effect on the operation of the power plant at Swan Falls to the full capacity of the plant as it existed when IPC acquired it in 1916.

For its Swan Falls power development, IPC eventually secured four state water rights licenses. Three were pre-1919 rights, which as enumerated above totaled 5,450 cfs.[170] The fourth water right, Water License No. 14362 (02-2032), for 4,000 cfs with a priority date of July 29, 1919 was obtained by IPC on May 26, 1920.[171] In 1928, the Federal

---

168 Hoyt, *Water Utilization*, 201; Stacy, *Legacy*, 28-29 and 44; Idaho Power Company, *History*, 12-13, and 64.

169 Hoyt, "Report Relative to Water Power Resources of Snake River," August 1922, 69-70. Folder: D-100, 12-G-51, WPB, 1894-1961, RG 57, NARA II. Due to the inflow in the Snake River below Milner Dam and above this plant, the capability of the plant was not affected by the storage operations of the Bureau of Reclamation upstream at American Falls.

170 State of Idaho, Water License, Ada and Owyhee Counties, Snake River, Power, 1840 sec. ft. per second, Priority 1-17-1900, August 4, 1911. Idaho State Idaho State Water Right 02-4000 Back File; and State of Idaho, Water License, Ada and Owyhee Counties, Snake River, Power, 1460 sec. ft., Priority 4-17-1900, August 16, 1911. Idaho State Idaho State Water Right 02-4001 Back File; IDWR website, last accessed March 17, 2008. See also Copy of "Statement of Nature, Extent, and Ownership of Idaho State Water Rights," Application for License, Swan Falls Power Project (Idaho), Idaho Power Company, January 26, 1924. Idaho State Water Right 02-2032 Back File; and In the Circuit Court of the United States, for the District Court of Idaho, Final Decree, *The Trade Dollar Consolidated Mining Company, a corporation, Plaintiff vs. Alfred A. Fraser and A. R. Crusen, Defendants*, April 2, 1907; Idaho State Water Right 02-0100 Back File.

171 State of Idaho, License and Certificate of Idaho State Water Right, Water License No 14362, Water District No. 50, Amount 4000 sec. ft., Priority 7-29-19, May 26, 1920. Idaho State Water Right 02-2032 Back File. See also Copy of "Statement of Nature, Extent, and Ownership of Idaho State Water Rights," Application for License, Swan Falls Power Project (Idaho), Idaho Power Company, January 26, 1924. Idaho State Water Right 02-2032 Back File; and Application for Permit to Appropriate the Public Waters of the State of Idaho, Application No. 20149, Permit No. 14362, District No. 18, July 29, 1919. Idaho State Water Right 02-2032 Back File.

Power Commission licensed this facility as Project 503 in accordance with the Federal Power Act, and relicensed the facility in 1982.[172] By 1935, ten units would push the power capacity of the Swan Falls plant to 10,000 kW.[173]

Shoshone Falls

The natural attributes of Shoshone Falls were perfect for generating hydroelectric power. At the falls, the Snake River had cut its way through the horizontal lava flow, creating a spectacular waterfall plunging 212 feet over vertical rock ledges. Some 44 feet higher than the average height of Niagara Falls. The falls at Shoshone were known as the "Niagara of the West" by pioneers crossing the Snake River Plain. This spectacular work of nature, once slated for preservation at the turn of the twentieth century as a National Park, is the largest and most dramatic of the waterfalls on the Snake River.[174]

The Shoshone Falls power plant was located in Jerome County, about three miles northwest of the city of Twin Falls. The site was situated about 20 miles downstream from Milner where the Twin Falls irrigation companies would build their irrigation diversion structure across the Snake River. The plans of the Carey Act irrigation projects at Twin Falls called for diversion and use of the full flow of the river. When Milner Dam was closed during the irrigation season no natural stream flow would be allowed into the channel of the Snake River

---

172 The Federal Power Commission, License on Government Lands, Project No. 503 – Idaho, Idaho Power Company (Swan Falls Project), effective on January 1, 1928. Folder: Licenses for Idaho Power Projects by Federal Power Commission, 1928-1933, Box R 2/4, AR 20, ISHS; and Idaho Power Company, Project No. 503-000, Order Issuing New License (Major) (Issued December 22, 1982), FERC Reports 62,519. Idaho State Water Right 02-2032 Back File.

173 Hoyt, *Water Utilization*, 201; Stacy, *Legacy*, 28-29 and 44; Idaho Power Company, *History*, 12-13, and 64.

174 In August 1898 Idaho Senator George Shoup announced that the National Park Service had initiated withdrawal of some 30,000 acres for a National Park incorporating Twin Falls, Shoshone Falls, Blue Lakes and other natural features of interest in the Snake River Canyon. Withdrawal of these lands threatened use of the water resources in the river by the proposed Twin Falls irrigation projects as well as hydroelectric power projects. Local interests opposed the designation and the park proposal was withdrawn in 1900 in favor of land segregation for the Twin Falls Carey Act projects. The park proposal was reduced to include only Shoshone Falls and vicinity, but that proposed withdrawal was also rescinded in 1901 and the plan was abandoned. "Snake River Reserve," *Idaho Daily Statesman*, September 1, 1898; "Senator Shoup and Congressman Wilson Against It: Letters to the Governor," *Idaho Daily Statesman,* November 27, 1900; "Shoshone Falls Park," *Idaho Daily Statesman*, February 8, 1901; and "Shoshone Falls Best Twin Falls Attraction," *The Post-Register* (Idaho Falls), April 9, 1953.

below the point of diversion. Therefore, if the irrigation projects obtained superior appropriative water rights, during irrigation season the Shoshone plant would need to rely for its water supply upon what leaked through the dam, return flows from irrigation seepage, and underground flow from springs emerging along the along the canyon walls and from the riverbed. On the other hand, if power development at the falls preceded irrigation progress upstream, the plans for full development of the Carey Act projects at Twin Falls might need to be scaled back to accommodate the competing water demand of the downstream power plant.

The initiation of the water power development works at Shoshone Falls took place at a time when the entire countryside around Twin Falls was still in its original desert condition. Two competing notices of location for a power development at the falls were filed in 1901 under the Act of February 25, 1899, the Idaho statute governing water appropriations. One of the notices of appropriation was filed by Ira B. Perrine, and the other by J. C. Hollister. The following year, Areturas Conrad also posted a notice of his intent to appropriate water at the same point of diversion on the Snake River in Lincoln County, at Shoshone Falls. There were no conceivable outlets for hydroelectric power at the time, so construction work was only carried on intermittently for the next few years. Actual development work by Perrine was limited to sporadic work at a single tunnel site near the falls until James S. and William S. Kuhn became interested in the power project in 1905 and put the falling waters to actual beneficial use a few years later.

In March 1901, I. B. Perrine of Blue Lakes hired consulting engineer R. W. Faris to prepare a cost estimate for a proposed 325 horsepower project aimed at providing the village of Shoshone, 30 miles to the north, with electric lights and sufficient power to pump water for domestic purposes. Perrine filed a notice of appropriation for 1,000 cfs. The proposed development was aimed at a low-cost facility that would use only a portion of the available water supply in the Snake River at this location during the low flow period from August to April. Perrine was also interested in the Twin Falls irrigation project designed to irrigate more than 225,000 acres of land in the vicinity of Shoshone Falls. His limited concept of hydroelectric power development at Shoshone Falls offered some assurances that power and irrigation interests on the Upper Snake River would remain linked together and compatible.[175]

In contrast, the hydroelectric power project at Shoshone Falls proposed by H. L. Hollister, manager of the Croesus Mining Company, was backed by large eastern capitalists in Boston, New York and Pennsylvania, who formed the Idaho Consolidated Gold Mining Company to develop the gold mines around Hailey in Alturas County, the Thunder Mountain mining district on the Payette River, and in the copper mines in the vicinity of Contact, Elko County, Nevada. The main problem faced by these mining operations was lack of power to drive the machinery required for underground mining and processing ore. To overcome that difficulty, Hollister decided to organize a corporation to develop the falling waters at Shoshone Falls by construction of a hydroelectric plant to serve the surrounding mining districts. Backed by eastern investors and experienced electrical and hydrologic engineers from the East, he planned on organizing a separate development company to operate that branch of the business.[176]

The proposed layout of Hollister's power plant and its ancillary facilities required obtaining a lease of 50 acres of School Land under the jurisdiction of the State Board of Land Commissioners, the same state agency that managed land segregated for settlement under the Carey Act. The Board agreed in December 1901 to enter into a lease agreement for the purpose of aiding Hollister's mining enterprise in which the state agreed that upon the expenditure of $50,000 on the power plant at Shoshone Falls, the state would offer the site for sale to Hollister's company. The project prospectus called for spending some $325,000 on construction of three dams, a large tunnel, and initially one penstock connecting to a 4,000- to 5,000-horsepower plant, but expandable to four penstocks with an ultimate output of 20,000 horsepower. The electric power was to be used for stamp mills, smelters and other mining machinery as well as for domestic purposes in the mining camps. In addition, the prospectus called for production, transmission and distribution of electric power for development of a network of electric railways connecting to the Oregon Short Line Railroad and for municipal and industrial uses in the towns of Shoshone

175 "I. B. Perrine Here to Consult Over Blue Lakes Power Scheme," *Idaho Daily Statesman*, March 28, 1901; *Idaho Daily Statesman*, December 8, 1901.

176 "W. H. Dowe Talks of the Hollister Operation," *Idaho Daily Statesman*, February 26, 1902.

and Hailey. The proposed power development would require 1,400 cfs to operate the first two electrical generating units at full capacity. The ultimate development would presumably require at least double that flow.[177]

U. S. Geological Survey gauging records indicated that the average natural flow of the Snake River at Milner, some 20 miles above Shoshone Falls, was 2,090 cfs during the low water season in the month of August. The river's flow was supplemented by springs above Shoshone Falls which, according to measurements made by USGS hydrographers, contributed only an additional 100 cfs to the river between Milner and Shoshone Falls. Thus, establishment of a 2,800 cfs superior appropriative water right by Idaho Consolidated Gold Mining Company's power development at Shoshone Falls would threaten the viability of the large irrigation project at Twin Falls. The irrigation project was widely viewed as the most important economic development project in the state – an enterprise that would provide homes for tens of thousands of small farmers and make a great permanent increase in the wealth and tax base of the state.

Detailed surveys for the Twin Falls irrigation project had been completed and its promoters were actively engaged in financing the $1.5 million enterprise which was to be built under state supervision and control. The plans of the irrigation company required the diversion of 2,800 cfs at Milner Dam to reclaim all of the land under its proposed canal system. It was clear that if the irrigation company diverted all of the available water in the river during irrigation season there would not be sufficient water to operate the proposed Shoshone Falls hydroelectric plant located just 20 miles below the head of the canal system. The enormity of the conflict between the antagonistic water claims of the irrigation and power enterprises was summed up in an open letter submitted to the *Idaho Statesman* by hydraulic engineer Andrew J. Wiley, who had made the survey of the canals and diversion dam for the Twin Falls irrigation projects:

> If the time ever comes when both companies have carried out their plans there will unquestionably arise a conflict over the right to the

177 *Idaho Daily Statesman*, March 1, 1900; "Power at Shoshone Falls, *Idaho Daily Statesman*, November 28, 1901; "Boise-Payette Electric Power Company's Plans," *Idaho Daily Statesman*, December 8, 1901; "W. H. Dowe Talks of the Hollister Operation," *Idaho Daily Statesman*, February 26, 1902.

> water. It will take two years to effect the diversion of the water for irrigation purposes, on account of the construction of the great dam, while the power company can effect its diversion in a few months and at a comparatively small expense. There will then be a question if the prior right does not constitute a superior right. In which case the settlers upon the land would either be compelled to part with their water or to condemn and purchase the power plant.
>
> In any event the construction of the power plant at the Shoshone Falls will entail a loss of a large amount of money to one of the enterprises.
>
> The irrigation company was there first; it cannot move its land and the land cannot be watered from any other source. It claims that the state by virtue of its provisional contract is bound, morally if not legally, to protect the water right to what is virtually the state's own enterprise.[178]

The promoters of the Twin Falls irrigation project, as well as Wiley and state engineer D. W. Ross, suggested that the power company move its power development further down the river so as not to jeopardize the irrigation project. Just one year earlier, Wiley had completed construction of the hydroelectric power plant at Swan Falls, as chief engineer for the Trade Dollar Consolidated Mining Company. Five miles downstream at Thousand Springs, near Hagerman, he noted, enormous springs with a fall of 150 feet contributed enough water to the river to double the low water flow of the Snake River at Shoshone Falls and thereby created an ideal location for a hydroelectric power plant. The springs provided clear water, free of debris, an unchanging flow of water in all seasons, and was located closer to the point of ultimate use of the power to be generated. The power company could make full development of the entire discharge of the springs and the Snake River at this location, reasoned Wiley, without affecting irrigation above or below and without being affected by diversions for irrigation purposes.[179]

---

178 "Engineer Wiley on the Situation at Shoshone Falls," *Idaho Daily Statesman*, March 6, 1902. Wiley had a long and prestigious career as an Idaho hydraulic engineer. He served as assistant chief engineer for Idaho Mining & Irrigation Company on construction of a large irrigation system from 1888 to 1892; as chief engineer and manager of Owhyee Land & Irrigation Company constructing its large irrigation system, 1892-93; as chief engineer of the Swan Falls Power plant, 1900-02; as chief engineer of Boise-Payette River Electric Power Company, 1902-04; as consulting engineer on the various Twin Falls irrigation projects, 1900-04; and finally as a consulting engineer to the Reclamation Service in Idaho, Oregon, Washington and Montana. John W. Leonard, et al., *Who's Who in America*, Vol. 5 (Chicago: A. N. Marquis & Company, 1908), 2060.

179 "Engineer Wiley on the Situation at Shoshone Falls," *Idaho Daily Statesman*, March 6, 1902.

In the spring of 1902, Senator W. A. Clark and others claiming to own land on both sides of the river at Shoshone Falls secured an injunction to stop Hollister from proceeding with his big Shoshone Falls power project and moved to initiate condemnation proceedings. It was their intention to protect the falls and islands as "points of great scenic beauty" and they desired to keep the natural landscape at the site intact. They protested that Hollister had erected tents and other "unsightly buildings" on the premises and his men were blasting and destroying natural features on the islands. Judge Beatty, U. S. District Court, took the matter under advisement, but no judgment was forthcoming.[180]

By the summer of 1902, the *Idaho Daily Statesman* reported that a change of ownership of the Twin Falls Land & Water Company had infused additional capital into the company, with the result that the power project and irrigation enterprise had been "brought into harmony." Local investors Albert K. Steunenberger and James H. Lowell, who lacked the personal wealth to continue financing the giant irrigation project, favored dissolution of the Twin Falls Company and turning the irrigation project over to the state or federal government. Milner objected and offered to purchase their one-half interest in the irrigation company for $21,000 in cash and $80,000 to be paid over three years from the proceeds of the sale of land under the project. Having agreed to the sale of their capital stock to Milner, Steunenberger and Lowell withdrew from the corporation. The deal was brokered by William E. Borah, Boise attorney and future U. S. Senator from Idaho, and S. H. Hays, an Idaho attorney who represented both Mr. Milner and Mr. Hollister. The sale temporarily put Stanley Milner in control of the Twin Falls irrigation project. This led to a breakthrough in the stalemate between the competing proposed power developments by the eastern mining investors, and Perrine/Milner and made a plan to unite the power and irrigation enterprises possible.[181]

The two sides agreed to "work in harmony" on enterprises to promote the agricultural settlement of the Twin Falls region. The general plan included abandoning the mining-related power uses and refocusing on building a large and flourishing town at Twin Falls with a manufacturing base that supported the agricultural enterprises of the

---

180 "Not Enjoined," *Idaho Daily Statesman*, March 4, 1902; "Injunction Suit," *Idaho Daily Statesman*, April 11, 1902.

181 "A. K. Steunenberg and J. H. Lowell Sell Their Twin Falls Stock to S. B. Milner," *Idaho Daily Statesman*, July 10, 1902; *Shoshone Journal*, May 2, 1902 and July 11, 1902.

region and developing abundant, cheap hydroelectric power at Shoshone Falls and other sites primarily for municipal uses, for pumping irrigation water and other uses on Carey Project lands, and to build up a local transportation network using electric railways.[182] The new investors in Twin Falls Land & Water Company, F. H. Buhl and P. L. Kimberly, Pennsylvania financiers who were heavily invested in Utah iron mines and other western mining properties, were given majority control over the irrigation project through the purchase of Twin Falls Land & Water Company stock at $1.5 million. The capital invested was to be used to complete the dam and canal system and to reclaim the land in the Twin Falls area for agricultural purposes. As part of the deal, Milner and Perrine reportedly obtained control over the water power resources at Shoshone Falls.[183]

D. W. Ross, Idaho state engineer, was responsible for approving plans filed by irrigation companies for Carey Act projects. Ross had given his approval of the water supply for the initial Twin Falls irrigation project plan and the Interior Department had withdrawn lands relying upon his certification that the project was feasible. Subsequent to his approval in 1900, the available water supply had become depleted by competing claims for Snake River water for power purposes. Clear title to the water for irrigation use had been clouded by initiation of preliminary construction on a tunnel at Shoshone Falls and the lack of development work on the irrigation project. As late as the end of 1902, the Twin Falls Land & Water Company still had not entered into a contract with the state for the reclamation of the segregated Carey Act land and they possessed no right to initiate construction without a contract. Ross was no longer prepared to claim that there was sufficient water in the Snake River to irrigate all of the segregated lands.

With the development of a power plant at Shoshone Falls, Ross questioned "whether or not the settlers who would be compelled to purchase the canals would in the end have a right to divert the water for irrigation." He informed the State Board of Land Commissioners and the Interior Department in November 1902 of the withdrawal of his official approval of the plans and recommended that relations between the state and the irrigation company be severed. With the possibility of

182 "A. K. Steunenberg and J. H. Lowell Sell Their Twin Falls Stock to S. B. Milner," *Idaho Daily Statesman*, July 10, 1902.

183 "Chief Engineer R. W. Faris Brings Encouraging News from Salt Lake," *Idaho Daily Statesman*, December 1, 1902.

the proposed power development at Shoshone Falls establishing a superior appropriative water right, the state, Ross concluded, could no longer assure the company of a safe water right that could insure full use of available water supply for irrigation during the low water season.[184]

Angered by the state engineer's action, Governor Frank W. Hunt removed Ross from office, charging that he had undermined the whole program for reclamation of arid lands in southern Idaho that the state had taken such pains to promote since passage of the Carey Act. The controversy that led to his dismissal would be replayed again and again on the Snake River over the course of the twentieth century. Irrigators and power interests were frequently in conflict over rights to the limited water supply, yet the two sides were forced to co-operate for their mutual benefit.

It is ironic that the story of power development at Shoshone Falls begins with the site being developed by owners of the very irrigation companies whose operations would be challenged in the future by Idaho Power Company's operation of the Shoshone Falls power plant.

The Ross-Hunt conflict delayed but did not derail the power development at Shoshone Falls. The first power plant at the falls was ultimately developed by the Great Shoshone & Twin Falls Water Power Company, a company controlled by the Kuhn Syndicate which also was the major financial backer of several Carey Act irrigation projects in the vicinity of Jerome and Twin Falls. The facility was intended to generate electrical power for irrigation pumping and for domestic and commercial uses in towns and on irrigated farms in the greater Twin Falls area. The natural conditions of the riverbed and peculiar formations and faults set up difficult construction problems which took years to solve. By 1907, Great Shoshone had built several low concrete dams across the various channels at the head of the falls to control the waters and divert them into a 450-foot long tunnel and 176-foot long penstock dropping to a power plant at the base of the falls on the left bank of the river. The plant generated power using one 500 kW unit. IPC acquired the site in 1916 after Great Shoshone went into receivership. Within five years, IPC was able to boost the capacity of the Shoshone Falls facility by more than 11,000 kW through an

184 "After State Engineer," *Idaho Daily Statesman*, December 2, 1902.

expansion of the powerhouse, installation of a new penstock to replace the old, and refurbishment of the diversion dams.[185]

Photograph 5 is an early photograph of the Shoshone Falls power plant, circa 1907.

Shoshone Falls Power plant, ca. 1907

Great Shoshone constructed a web of transmission lines to distribute its electrical power. It built southerly to the towns of Kimberly, Twin Falls, Buhl, Filer and Hollister, and to the Salmon River Dam for its construction; northerly to the towns of Jerome and Shoshone; westerly to connect with the Lower Salmon Falls powerhouse; and easterly to Milner and Oakley, at the latter place providing power for construction work on Oakley Dam. Along with the Shoshone power plant, this transmission system and its sub-stations were acquired by Idaho Power Company in 1916.[186]

When it purchased the power system from Great Shoshone, in total IPC obtained two pre-1903 appropriative rights based upon claimants

185 Hoyt, *Water Utilization*, 200; Stacy, *Legacy*, 23-24; Idaho Power Company, *History*, 648.
186 "Idaho Editors Visit Twin Falls Power plant at Shoshone," *Twin Falls Daily Times*, November 12, 1922.

filing notices of appropriation and two state water permits for its Shoshone Falls facility issued in 1907 and 1920. The first two were acquired when the company purchased the power site: a right to 1,000 cfs with a priority date of November 11, 1901 based upon the notice of appropriation filed by I. B. Perrine in Lincoln County and an additional 1,000 cfs with a priority date of September 13, 1902, evidenced by the notice posted and filed for record by Conrad in the same county. It was questionable whether these notices were valid because of the failure to make beneficial use within the time prescribed by law.[187] In February 1907, the Kuhn Bros. completed a tunnel at Shoshone Falls and applied 130 cfs to beneficial use by August of that year. A new generating unit was added in 1909. These rights acquired by Great Shoshone were passed to the Electric Investment Company through a liquidation sale in 1916.

The state engineer issued a permit (No. 9264) for 2,000 cfs to D. C. MacWatters for additional tunnels and diversion works at Shoshone Falls filed upon by an application submitted to the Office of the State Engineer on April 30, 1913. The whole work contemplated under the application was to be completed within five years. The existing powerhouse was greatly enlarged and a 14,500 hp turbine connected to a 12,000 kva generator was installed. IPC succeeded to the rights under the 1913 application and 1920 permit under which the power company developed works of sufficient capacity to make beneficial use of 1,200 cfs.[188]

In the mid-1920s, in association with its construction of new facilities at Shoshone Falls, IPC secured another state water right. The company filed an application for a permit to appropriate 1,000 cfs from the Snake River for "power purposes" on August 9, 1920 (Water Right No. 02-2036). Under IPC's proposed plan of development, that water was to be directed over four dams varying between 15 and 20 feet in height and then through an intake tunnel and penstock. The company anticipated generating 14,000 hp (nearly 10,500 kW) under 205 feet of

---

187 Before the Federal Power Commission, Project No. 2258, Application for License, February 13, 1959, Exhibit C-D-E, Page 1-2 of 3. Formal File 100-2, Part 1, February 13, 1959 to December 23, 1973, FPC Project 2258, FERC; and Document No. 043276, IPC Disc, Swan 012, 11/12/08. No contemporary Idaho State Water Right License Numbers have been identified for these two rights. These water rights were held under the Act of February 25, 1899. Proof of work done may not have been adequate to prove up on these first two notices. IPC Disc, Swan 012, 11/12/08, IPC Document Nos. 043277

188 IPC Disc, Swan 012, 11/12/08, IPC Document Nos. 043277 thru 043279.

head. On August 24, 1920, Idaho's Commissioner of Reclamation, W. G. Swendsen, issued the company Permit No. 14881, which required the company to provide proof of completion of works by August 24, 1925 and proof of beneficial use of water by August 24, 1931. IPC made good on both requirements, and secured Water License No. 14881 (02-2036), for 1,000 cfs on December 10, 1925 with a priority date of August 9, 1920.[189]

Once the north and south side irrigation projects at Twin Falls were completed, nearly all the natural flow of the Snake River in a normal year was diverted during July, August and September for irrigation purposes at Milner Dam. During that season of the year, the water supply for the Shoshone plant was limited to what returned to the river between Milner Dam and the point of diversion at Shoshone Falls. As noted, Idaho Power Company added generating capacity to its Shoshone plant in the early 1920s and expansion of irrigation upstream re-ignited the controversy of whether power uses of water on the upper Snake River should be subordinated to the water requirements of the federal and state irrigation projects located upstream. When the upstream irrigators planned construction of a major federal storage reservoir at American Falls, the power company sought to protect its investment at Shoshone Falls through provisions in a June 15, 1923 contract with the Bureau of Reclamation. The provisions guaranteed certain minimum flows of water past the diversion dam at Milner by releasing stored water from the Bureau's proposed new storage reservoir. Conversely, the federal and state government sought to conserve and store as much water as necessary in artificial reservoirs constructed above Miner Dam and to meet all current and future irrigation requirements for the full development of irrigated agriculture in the Upper Snake River Basin.

Although the federal government began to regulate hydroelectric power plants in the broader public interest through the Federal Water and Power Act in 1920, those regulations did not initially extend to Shoshone Falls because it was not located on federal land and the Snake River in this reach was not considered a navigable stream. After the Federal Power Commission ruled in August 1955 that the entire stretch of the Snake River to American Falls was navigable, IPC was required

189 State of Idaho, License and Certificate of Water Right, Water License No. 14881, December 10, 1925; Idaho State Water Right 02-2036 Back File.

to obtain an FPC license for the facility in accordance with the Federal Power Act. However, not until 1979 was the Shoshone Falls plant finally licensed by the federal government for the first time as FPC Project No. 2778.[190]

American Falls

Between 1890 and 1916 at least a dozen power companies struggled to establish themselves at the small town of Pocatello, located twenty-five miles above American Falls. Once the industry solved its long-distance transmission problems, the power site at American Falls on the Snake River was recognized by pioneers in the Idaho hydroelectric power industry as an ideal site for a power plant.

In 1901 two rival companies, American Falls Power, Light & Water Company (AFPL&WC) and American Falls Water Power Company (AFWPC), requested a franchise to serve Pocatello with water power from American Falls. Pocatello's city council voted to grant a franchise to the company that could deliver power at the earlier date. AFPL&WC had purchased what became known as the Island site, while AFWPC was developing a site on the east side of the river. The former company overcame its logistical problems by lowering construction materials from the railroad bridge crossing at American Falls and floating them downstream to the island. Having completed its plant first, the Pocatello city council awarded AFPL&WC the franchise in July 1902.[191]

Photograph 5 (below) depicts American Falls prior to the construction of AFPL&WC's dam and Island Power plant.

190 Idaho Power Company, Order Issuing License (Major) (Issued June 13, 1979); File Series P-2778, Project No. 2778, Part 1, February 13, 1959 to November 6, 1987; Box 347, Accession No. 138-92-110; FERC.

191 "Right of Way for Pocatello Electric Line," *Idaho Daily Statesman*, January 28, 1902; "Wild West Receives Age of Electricity," *Idaho State Journal.*Centennial Edition, July 22, 1963.

American Falls Prior to the Construction of the AFPL&WC's Dam and Island Power plant
Source: Library of Congress, Prints and Photographs Division, Historic American Engineering Record, HAER ID, 39-AMFA, 1-1

James Henry Brady, who served the State of Idaho as governor (1909-1911) and later as U.S. Senator, was one of the incorporators of AFPL&WC. A former Kansas school teacher, newspaper editor and real estate agent, Brady came to Idaho in 1895 to enter the electric power and irrigation industries in the Snake River Valley, and resided at Pocatello. Brady's Island hydroelectric power plant was constructed by Boise civil engineer R. W. Faris, who later became general manager of the Twin Falls Land & Power Company (1906) and near the end of his career in 1931-38 served as Idaho Commissioner of Reclamation. The first unit of the Island plant was installed in 1901-02 and a second unit in the latter part of 1908; each unit had the capacity to use about 300 cfs. Although the diversion had been completed by Faris to a capacity of 1,000 cfs no more than about 600 cfs was ever used by the two installed generating units. The power plant was connected by a 33,000 volt transmission line to the town of Pocatello in 1901 and extended to Blackfoot in 1903.[192]

192 Hoyt, *Water Utilization*, 198; and Wilcox, "Water Power Resources," 448.

The figure below shows the layout of the AFPL&WC's Island Power plant as of 1902 and **Photograph 7** (below) depicts the plant as of 1976.

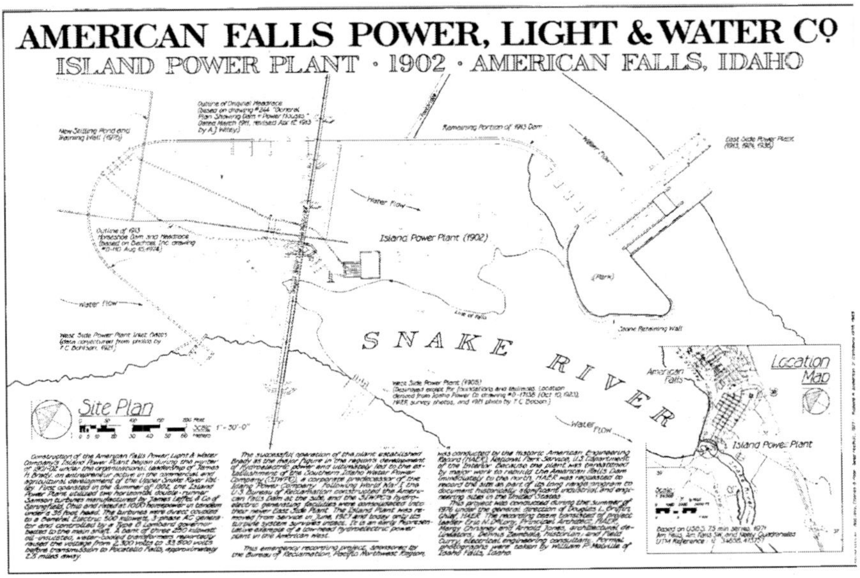

Site Plan for AFPL&WCo's Island Power plant

Source: Library of Congress, Prints and Photographs Division, Historic American Engineering Record, HAER ID-2

View of the Island Power plant Looking North (negative reversed), 1976. *Source*: Library of Congress, Prints and Photographs Division, Historic American Engineering Record, HAER ID, 39-AMFA, 1-8

Brady and AFPL&WC undertook construction of a second and potentially larger American Falls hydroelectric power plant on the west side of the river at the base of American Falls in 1901. Andrew J. Tarbet posted a notice of appropriation for 6,000 cfs to be used for power production at the plant on July 15, 1901. AFPL&WC built a second power plant at this location, called the West Side plant, between 1901 and 1905. In 1904 Idaho Consolidated Power Company (ICPC), organized by Brady and Alexander H. Tarbot, president of ICPC, acquired the West Side Powerhouse and completed its construction a year later. Although the diversion works were capable of carrying 2,200 cfs, only one 500 kW generating unit was installed with a capacity to handle only about 256 cfs. After 1907 when ICPC absorbed the Island and West Side plants, they continued to be operated by American Falls Power, Light & Water Company under an agreement with Idaho Consolidated Power Company.[193] For a number of years these two plants were operated only one shift per month to retain any water rights that they may have acquired. In 1911 these two American Falls power plants were acquired by Southern Idaho Water Power Company, a Kuhn enterprise.[194]

Photograph 8 (below) depicts the West Side Power plant as of 1921.

View of the West Side plant Looking Southwest, 1921
Source: Library of Congress, Prints and Photographs Division, Historic American Engineering Record, HAER ID, 39-AMFA, 1-13

In conjunction with the acquisition of the Island and West Side plants at American Falls and improvements to these facilities by its predecessors-in-interest, Idaho Power Company – organized in 1915 – would ultimately secure several state water right permits and licenses. The water rights were based upon the notices of appropriation posted by

193 *Idaho Power Company: A History* (1940), 17-20; Hoyt, *Water Utilization*, 198; and Wilcox, "Water Power Resources," 448; untitled Idaho Power Company document, undated, pages 209-210, IPC043285-043286, SWAN012 Disc, November 12, 2008, Idaho AGO

194 Untitled Idaho Power Company document, undated, pages 214-215, IPC043290-043291, SWAN012 Disc, November 12, 2008, Idaho AGO; Hoyt, *Water Utilization*, 198; and Wilcox, "Water Power Resources," 448.

the development companies. The first of these, in terms of priority, was Water Right No. A01-10382 Transfer #464, with a priority date of July 15, 1901 and Water Right No. A01-10383 Transfer #463 with a priority date of August 1, 1901 in the amounts of 253 cfs (West Side) and 611 cfs (Island), respectively.[195]

AFWPC had abandoned work on the East Side power plant when it became clear that there was no possibility of winning the franchise to serve the town of Pocatello in competition against AFPL&WC. A few years later, Brady purchased the East Side power plant from his former competitor and on June 7, 1905 he incorporated the American Falls Power Company (AFPC) to finance its development. AFWP had begun construction of the East Side power development at American Falls in 1901. Under a contract with Faris-Kesel Construction Company, Bradey's AFPC began work in the fall of 1905 on excavation of the powerhouse. High water forced temporary abandonment of the tailrace excavation. When the floods receded, work resumed, but in the fall of 1906 work was again postponed for financial reasons.[196] Work did not resume again until after the East Side plant also fell to the Southern Idaho Power Company in 1911.

Photograph 9 (below) depicts the East Side Power plant as of 1921.

195 "American Falls Power plant (East Side plant) owned by Idaho Power Co. and located adjacent to the town of American Falls," n. d ("American Falls Power plant"). Idaho State Idaho State Water Right 01-2046 Back File, Idaho Department of Water Resources website, On-Line Data, Idaho State Water Right and Adjudication Search, available at http://www.idwr.idaho.gov/apps/ExtSearch/ SearchWRAJ.asp, last accessed March 17, 2008. All water rights backfiles, cancelled and active, and all claim files were accessed via this website and on the aforementioned date; subsequent references will only include the specific backfile or claim file in question.

196 *Idaho Power Company: A History* (1940), 19; .; "Wild West Receives Age of Electricity," *Idaho State Journal: Centennial Edition*, July 22, 1963.

View of the East Side Power plant Looking North, 1921
Source: Library of Congress, Prints and Photographs Division, Historic American Engineering Record, HAER ID, 39-AMFA, 1-16

A number of water appropriations and permits for use of water at the East Side plant had been turned over to the new owner, but all had lapsed except Permit No. 4168 issued by the state engineer on November 28, 1908 to AFP for the full development of the site up to 6,000 cfs. In 1913 A. J. Wiley, superintendent of construction for the Southern Idaho Water Power Company, finally completed the first 6,500 horsepower (hp) unit of the East Side plant.[197]

By 1910, ICPC purchased all of the remaining electrical properties then operating in the greater Pocatello area, including Blackfoot Light & Water Company and AFPC's East Side Powerhouse at American Falls. In addition to his interest in the development of water power in southeastern Idaho, Brady was closely identified with the irrigation movement and Carey Act projects in the Upper Snake River Basin and was instrumental in organizing the private irrigation enterprises that successfully financed and completed construction of the Idaho Canal (Idaho Falls), Marysville Canal (Ashton) and the Fort Hall Indian Reservation Canal. He served as vice-president of the National

197 Untitled Idaho Power Company document, undated, pages 221-223, IPC043297-043299, SWAN012 Disc, November 12, 2008, Idaho AGO.

Irrigation Congress from 1896 to 1898 and was a member of its Executive Committee from 1900 to 1904, an era in which the national organization provided key legislative support for the passage of a national reclamation bill. Brady was also a dominant figure in Republican Party politics in Idaho during the first two decades of the twentieth century. In 1900 he was a delegate to the Republican national convention and served as chairman of the Idaho Republican State Central Committee, 1904-08; governor of the State of Idaho, 1909-11; and as U. S. Senator from the State of Idaho from 1913 until his death in 1918 in Washington, D.C.[198]

In 1911 the three American Falls power plants were acquired by Southern Idaho Water Power Company (SIWPC), a Kuhn Syndicate enterprise, which, as noted previously, finally completed the first 6,500 horsepower (hp) electrical generating unit in the East Side power plant in 1913.[199] The East Side plant was planned for an ultimate development of up to 40,000 hp. The electricity generated at American Falls was transmitted to Pocatello, Ross Fork and Blackfoot, and also supplied the town of American Falls. SIWPC made several improvements to its East Side power plant at American Falls to better serve its expanding power requirements of its southeastern Idaho market. The highest producing of the three American Falls facilities, the East Side plant, initially operated with only one generating unit, completed in 1913 under Idaho water right application (No. 5638) for 6,000 second-feet filed in 1908. Water License No. 4168 (01-2017), with a priority date of September 3, 1908, in the amount of 1400 cfs was issued for the East Side plant on February 5, 1920. The first unit of the East Side power plant had beneficially used only about 1,400 second-feet of water to generate power, and therefore a state water right

---

198 Idaho Department of Reclamation Records (AR 20), *The Carey Act in Idaho* (Boise: Idaho State Historical Society Library & Archives, 2004); "James H. Brady (Late a Senator from Idaho) Memorial Addresses Delivered in the Senate and House of Representatives of the United States, 65th Cong., 3rd Sess, Sen Doc. No. 449 (Washington, DC, 1919); and "Senator J. H. Brady Dead," *New York Times*, January 14, 1918. After he retired as governor, Brady again turned his attention to the agricultural development of southern Idaho and rehabilitated the irrigation works of the Mountain Home Irrigation Project.

199 The Kuhn Syndicate was heavily invested in southern Idaho power and irrigation developments in the early 20th century and controlled many of the important land and water companies in the region, including: Twin Falls North Side Land & Water Company; Twin Falls Salmon River Land & Water Company; Twin Falls Oakley Land & Water Company; the Great Shoshone and Twin Falls Water Power Company; and Twin Falls North Side Investment Company, a company that loaned large sums of money to settlers to aidd them in buying and stocking their farms.

for this amount was issued in License No. 4168 (01-2017), but the plant as designed had space for several additional generating units.[200]

Southern Idaho Water Power Company was reorganized in 1916 and was acquired by IPC, which added two units to the East Side plant, one in 1924, and a second unit in 1927. The three American Falls power plants provided IPC's entire power generation for the American Falls, Blackfoot, Pocatello, and Aberdeen areas. The power development at American Falls included a dam ranging in height from 6 to 23 feet and with a crest length of 1,138 feet.[201] The concrete dam conformed to the horseshoe-shaped crest of the falls, but raised the elevation of the water surface in the pool formed behind the dam only a few inches above the natural height of the falls.[202]

IPC made several improvements to its East Side power plant at American Falls to better serve the expanding power requirements of its southeastern Idaho market between 1919 and 1926. When acquired by IPC, the highest producing of the three American Falls power facilities, the East Side plant, had only one generating unit. Within a few years after acquisition of the American Falls facilities, Wallace and IPC expanded the East Side plant at American Falls by adding two more generating units designed to make use of an additional 4,600 second-feet of water. Work on the project was initiated in 1919 and the project works for the diversion and generation of power through the new units were completed and the water put to beneficial use by the summer of 1924.[203] Water License No. 14067 (01-2032), with a priority date of March 8, 1919, in the amount of 4600 cfs, was issued for the East Side plant on July 21, 1924.[204] Two years later, IPC secured an additional

---

200 State of Idaho, License and Certificate of Idaho State Water Right, Water License No. 4168, Water District No. 70, Amount 1400 sec. ft., Priority 9-13-1908, issued February 5, 1920. Idaho State Water Right 02-2017 Back File.

201 Untitled Idaho Power Company document, undated, pages 221-23, IPC043297-043299, SWAN012 Disc, November 12, 2008, Idaho, Attorney General's Office; Hoyt, *Water Utilization*, 198; and Wilcox, "Water Power Resources," 448.

202 Hoyt, *Water Utilization*, 198; and Wilcox, "Water Power Resources," 448; and untitled Idaho Power Company document, undated, pages 209-210, IPC043285-043286, SWAN012 Disc, November 12, 2008, Idaho AGO.

203 State of Idaho, No. 5638 Application for Permit to Appropriate the Public Waters of the State of Idaho. Certificate of Completion and Extension of Beneficial Use, February 28, 1917; State of Idaho, License and Certificate of Idaho State Water Right. Water License No. 4168, February 6, 1920; Affidavit of Publication, Notice of Proof of Completion of Works and Application of Water to Beneficial Use, Permit No. 14067, June 20, 1924. Idaho State Idaho State Water Right 02-2017 Back File.

water right permit from the state in the amount of 2000 cfs.[205] Water Right License No. 16635 (01-2046) was finally issued to IPC for the East Side plant on December 23, 1971, with a priority date of October 15, 1926, in the amount of 2000 cfs.[206]

The Reclamation Service purchased the Island and West Side power plants at American Falls from IPC in 1923 and built a new dam at the falls in 1927 to store water primarily for irrigation purposes.[207] The Bureau of Reclamation ceased to use and later abandoned the West Side and Island facilities. IPC retained ownership of the East Side plant and operated it as a commercial hydroelectric power plant without a federal power license until 1975, when it was finally licensed by the Federal Power Commission (FPC) as Project 2258.[208] That same year, a new IPC power plant – constructed as part of the rebuilding of American Falls Dam – replaced the old East Side plant (which was then decommissioned). It was licensed by the FPC as Project 2736.[209]

---

204 State of Idaho, License and Certificate of Idaho State Water Right, Water License No. 14067, Water District No. 36, Amount 4600 sec. ft., Priority 3-8-1919, issued July 21, 1924. Idaho State Water Right 02-2032 Back File.

205 State of Idaho, Department of Water Administration, License of Idaho State Water Right, License of Idaho State Water Right No. 16635/01-2046, Priority Oct. 15, 1926, Amount 2,000 cfs, issued December 23, 1971. Idaho State Idaho State Water Right 01-2046 Back File.

206 State of Idaho, Department of Water Administration, License of Idaho State Water Right, License of Idaho State Water Right No. 16635/01-2046, Priority Oct. 15, 1926, Amount 2,000 cfs, issued December 23, 1971. Idaho State Water Right 01-2046 Back File.

207 Hoyt, *Water Utilization*, 198. IPC also entered into a contract regarding the American Falls facility in 1923 that addressed the water rights – see the detailed discussion at pages below.

208 Federal Power Commission, Idaho Power Company, Project No. 2258 Order Issuing Major License (Constructed) (Issued March 1975), 1. Official Records, Part 2, December 24, 1973 to June 27, 1980, Federal Power Commission File Series Project 2258, FERC.

209 Federal Power Commission, Idaho Power Company, Project No. 2736, Order Issuing Major License (Unconstructed) and Permitting Withdrawal of Intervention (Issued March 31, 1975). Idaho State Idaho State Water Right 01-2046 Back File.

Lower Salmon Falls

Lower Salmon Falls was the first of the so-called "Middle Snake River" power projects. It is located in south-central Idaho near the towns of Hagerman and Bliss, about seventy river miles downstream from the last major point of diversion for irrigation at Milner Dam. The other Middle Snake plants are Upper Salmon Falls, located upstream about 10 miles from the Lower Salmon Falls project, and Bliss power house, situated about 20 river miles downstream of the Upper Salmon Falls project.

Many aborted efforts were made in the early years of the twentieth century to initiate a water power development at Lower Salmon Falls, but water permit holders failed to construct the works or make beneficial use of the water filed upon. Actual power development began at Lower Salmon Falls in about 1909 with a hydroelectric power plant designed and engineered by the Great Shoshone-Twin Falls Water Power Company.

For thousands of years Salmon Falls was an important Indian fishing and wintering area for the Shoshoni and Bannock Indians. Salmon could not ascend the Snake River above Salmon Falls in Hagerman Valley, so members of these tribes traveled a long distance to reach this important fishing site. Emigrants traveling along the Oregon Trail traded with the native people for salmon at the falls before leaving Hagerman Valley to ascend a long grade to the west in an approach to the Snake River crossing near the future site of Glenn's Ferry. Mining engineer A. D. Foote identified Upper and Lower Salmon Falls and Thousand Springs as promising power sites for the mines around Silver City in the 1890s, but actual development was postponed until the agricultural settlement of the Twin Falls area and advances in long-distance electric transmission technology made development of these hydroelectric power sites feasible.

The first hydroelectric power development at Lower Salmon Falls consisted of an overflow dam 375 feet long that created an operating head of 34 to 37 feet. The dam formed a small forebay at the east end of the falls leading directly to a reinforced concrete powerhouse. This power site was also built and operated by the Kuhn Syndicate's Great Shoshone & Twin Falls Water Power Company from about 1909 to 1915. The facility was intended to satisfy growing power needs of irrigation projects and rural communities at Twin Falls and in the King Hill and Glenns Ferry region.[210] Fentress Hill made an application for a permit to divert and use up to 6,000 cfs of the flow of Snake River at Lower Salmon Falls for power purposes on May 11, 1908 (Permit No. 36853).

The state engineer issued a permit that was later transferred to Great Shoshone-Twin Falls Power Company. The company began work on the plant in early 1909, and in 1910, initiated operation of a single 3,150 hp turbine and one 2,000 kva generator. In 1912, the company installed a second unit, identical to the first, to provide sufficient capacity to serve its rapidly-expanding transmission network. The company ultimately planned to install a total of 10 power generating units. Proof of completion of works to a capacity of 6,000 cfs was issued in 1913 with licensing dependent upon application of the water to beneficial use by May 20, 1917. In 1915 Shoshone & Twin Falls Water Power Company sold the plant to the Electric Investment Company under a decree of foreclosure at which time a maximum of 1,600 cfs had been used at high water and 1,330 cfs at low water.[211]

Photograph 10 below is a photograph of the Lower Salmon Falls and power plant, circa 1915.

---

210 Hoyt, *Water Utilization*, 200-201; and Stacy, *Legacy*, 23-24.

211 Untitled Idaho Power Company document, pages 101-103, IPC043265-043267, SWAN012 Disc, November 12, 2008, Idaho AGO.

Lower Salmon Falls and power plant, ca. 1915

On August 16, 1916, IPC assumed ownership and control over the existing Great Shoshone facilities at Lower Salmon Falls. A short time later, the company designed plans for installing a third turbine and generator, whereby the maximum possible use of water would increase to 6,500 cfs at low water and 7,200 cfs at high water. In 1919, IPC installed a third unit, comprised of one 6,200-hp turbine and one 3,600-kva generator.[212] Impact of this planned expansion upon upper basin irrigators was questioned, but IPC maintained that there was no material conflict with regard to the water rights of upper basin irrigators:

> There are no conflicting rights of any importance affecting this [Lower Salmon Falls] plant. We have assumed that substantially all the water of Snake River will be diverted at the Milner Dam for use for irrigation purposes under prior rights, and we consider as conflicting rights only such appropriations as divert or contemplate diverting the water between Milner dam and the plant at Lower Salmon Falls, and ... there are no adverse permits or appropriations of any importance

212 Application for License (Act of June 10, 1920, 41 Stat. 1063), Lower Salmon Falls Power Project (Idaho), Idaho Power Company, November 22, 1923, Exhibit Q. File Folder 100-2, Project No. 457, Formal, Part 1, December 19, 1923 to December 1946; Box 2, Accession No. 138-62A-0938, Idaho Power Company (Lower Salmon Hydro-Electric plant), Snake River, FERC.

> between these points that do not return the water to the stream above the point of diversion of the Lower Salmon Falls plant. [213]

Through its acquisition of the Lower Salmon Falls facility, IPC obtained two separate water rights under a single license issued on October 31, 1919, Water License No. 3685: the first (later designated 02-2001A) encompassed the use of water for the plant's first two units, conveying a right to 1,700 cfs with a priority date May 11, 1908, while the second right (later designated 02-2001B) concerned the use of water for the plant's third unit, conveying a right to 1,300 cfs with a priority date of July 2, 1908.[214] The senior upstream appropriative water right holders who diverted irrigation water at and above Milner Dam did not conflict with the water used to generate power at the Lower Salmon Falls plant.[215]

IPC later secured two additional licenses for Lower Salmon Falls as part of the plant's re-development in the 1930s and 1940s: Water License No. 18017 (02-2059), issued on August 20, 1937, which conveyed a right to 250 cfs with a priority date of March 30, 1935; and Water License No. 19924 (02-2060), issued September 29, 1950, which conveyed a right to 14,000 cfs with a priority date of August 8, 1946.[216] In association with the work contemplated in these two water right applications, the Federal Power Commission issued licenses in accordance with the Federal Power Act, first as Project 457 in 1928 and then as Project 2061 in 1946.[217]

---

213 Untitled Idaho Power Company document, page 103, IPC043267, SWAN012 Disc, November 12, 2008, Idaho AGO.

214 Idaho Department of Water Resources' Claim Files for Lower Salmon Falls, Idaho State Water Right 02-2001A Claim File and Idaho State Idaho State Water Right 02-20001B Claim File; Memorandum from J. Noyes to 02-2001B regarding Supporting data in State office files Idaho Power Co., January 29, 1985, and Memorandum from F. Luthy to 02-2001 Idaho Power Co. Lower Swan Falls [*sic*] regarding S. O. File Review, January 23, 1985. Idaho State Water Right 02-2001B Claim File; "Exhibit E," Application for License (Act of June 10, 1920, 41 Stat. 1063), Lower Salmon Falls Power Project (Idaho), Idaho Power Company, November 22, 1923. File Folder 100-2, Formal, Part 1, Box 2, FERC

215 See Untitled Idaho Power Company document, page 103, IPC043267, SWAN012 Disc,November 12, 2008, Idaho AGO.

216 State of Idaho, License and Certificate of Idaho State Water Right, Water License No. 18107, August 20, 1947; Idaho State Water Right 02-2059 Back File; and State of Idaho, License and Certificate of Idaho State Water Right, Water License No. 19924, September 29, 1950. Idaho State Water Right 02-2060 Back File.

217 The Federal Power Commission, License on Government Lands, Issued June 25, 1928, Effective Jan. 1, 1928, Project No. 457, Idaho, Idaho Power Company (Lower Salmon Falls Project). File Folder 100-2, Formal, Part 1, Box 2; and In the Matter of Idaho Power Company, Project No. 2061, Order Issuing License (Major) (Order Issuing License); Project No. 2061. Formal File 100-2, Part 1, October 1950 to January 31, 1968 (Formal File 100-2, Part 1), FERC.

Thousand Springs

The name "Thousand Springs" was coined by emigrants passing over the Oregon Trail viewing the south canyon rim of the Snake River in the Hagerman area where an array of springs gushed forth from a vertical bluff of lava rock. The source of the springs was a mystery as beyond the rim of canyon the plains were a dry sagebrush desert for sixty miles without a river running through it. Mining engineer A. D. Foote first identified Thousand Springs as a potential power generation site in 1894 when searching for a source of electricity for the Trade Dollar Mine in Silver City. The source was at too great a distance to transmit power efficiently at that time so instead Trade Dollar developed the Swan Falls site in 1901.[218]

Thousand Springs was, according to U. S Geological Survey hydraulic engineer W. G. Hoyt, "one of the most unique" power developments in the region, presenting special problems requiring great engineering skill to solve. The water at Thousand Springs originated in the mountains of central Idaho; the famous "Lost Rivers" of Idaho disappear through porous and permeable sediment once they reach the desert and flow laterally through layers of basalt lava, finally discharging in the Snake River Canyon where the river has eroded down through the basalt. Flow from Thousand Springs, Snow Springs and Sand Springs was collected in a great harness of cement and steel and used to generate power under a head of nearly 190 feet. Concrete retaining walls, flumes and pipes located along the face of the Snake River Canyon wall transported the water to penstocks which carried the water to turbines located on the canyon floor. The combined flow of the springs was about 600 to 700 cfs. The first two units at Thousand Springs, designed to harness some 2,700 hp each, were built from 1910 to 1912 by the Thousand Springs Power Company. The plant was connected to the transmission system of the Great Shoshone and Twin Falls Power Company. IPC acquired the Thousand Springs power development in March 1917 and added a third power station at Sand Springs in 1920-21 adding 8,000 hp to the system, making the entire plant capable of producing 13,300 hp. Much of the electrical power generated was used for irrigation purposes in the summer months for

---

218 Located about fifty-five river miles below Milner Dam, Thousand Springs relied upon the return flow from underground streams and seepage from the Snake River Plain. The volume of water at the springs was not directly connected to the amount of surface flow in the Snake River.

pumping water from the Snake River to fruit orchards, and alfalfa, wheat and grain fields.[219]

Idaho Power's Thousand Springs Facility, May 2007
Source: Photograph taken by author

In acquiring the Thousand Springs facility, IPC obtained Water License No. 9066 (36-2013) in the amount of 600 cfs with a priority date of November 20, 1912. The Thousand Springs Power Company had made application for a permit on November 20, 1912, but "transferred all its right, title and interest in and to said permit," to IPC on March 21, 1917. The State of Idaho issued a Certification of Completion of Works on May 10, 1918, and on February 11, 1922, IPC "made proof to the satisfaction of the Commissioner of Reclamation of

219 Hoyt, "Report Relative to Water Power Resources of Snake," August 1922, 57-63. Folder: D-100, 12-G-51, WPB, 1894-1961, RG 57, NARA II; E. A. Wilcox, "Electrical Development in Southern Idaho," *Twin Falls News*, January 9, 1913; and "Harnessing A Thousand Springs So the People May be Served," *Twin Falls Daily News*, May 24, 1920.

Idaho, of a right to the use of the waters of Thousand Springs and Snow Bank Springs for power purposes, under Use Permit No. 9066." That right was subsequently perfected, with Idaho's Commissioner of Reclamation, W. G. Swendsen issuing a License and Certificate of Water Right on July 5, 1923.[220]

Ox Bow

Located near Copperfield, Oregon, some 200 miles below Milner Dam, this plant was constructed between 1908 and 1914 by Idaho-Oregon Light & Power Company on a great ox bow bend in the Snake River. The company provided electric service via transmission lines connecting Horseshoe Bend, Boise and all the small towns in southwestern Idaho as far north as Weiser, as well as servicing Pendleton, La Grand and Baker City in Oregon.

Dominicus Winslow posted a notice of appropriation on November 3, 1906, at the site claiming 8,000 cfs for power purposes at a dam to be erected across the Snake River with two pressure tunnels connected to a power plant.[221] These water rights were subsequently assigned on February 1, 1907, to Idaho-Oregon Light & Power, a corporation chartered in Maine and led by William and Sinclair Mainland of Oshkosh, Wisconsin. The company intended for its Ox Bow plant to be the centerpiece of an interconnected system serving eastern Oregon and southwestern Idaho, with nearly 30,000 hp (or little over 22,000 kW). At the site of the proposed development the Snake River, for a distance of about six miles, descended through a rock canyon shaped like the upper half of the letter **S**. The point where two reaches of the stream came closest together was about 1900 feet across – an ox bow. In this six-mile stretch, the river dropped 56 feet in elevation. At this location, the plan was to build a dam across the Snake River channel and drive tunnels through the granite rock formation so that the waters of the Snake River could be diverted through them providing a fall of 56 feet to generate power. It was estimated that the project would yield between 48,000 and 100,000 hp depending on the river stage. Uncontrolled demand, a cutthroat market, financing difficulties, and

220 State of Idaho, License and Certificate of Water Right, Water License No. 9066, Water District No. 63, Amount 600 sec. ft., Priority 11-20-1912. Idaho State Water Right 36-2013 Back File.

221 "Notice of Appropriation," November 3, 1906. *Water Rights*, Book F, page 466, Baker County Recorder, Oregon.

disagreements with the original construction contractor, however, all delayed the construction project. Ultimately, Idaho-Oregon was forced to develop a considerably smaller plant than originally intended: the resulting powerhouse had a single generator rather than the three initially envisioned and produced only 804 hp (600 kW).[222]

In 1916, with the consolidation of numerous power companies in Idaho, IPC acquired the Ox Bow plant, which along with the Swan Falls plant primarily serviced the Boise-Payette region.[223] The Oxbow facility would eventually be replaced by a larger and more sophisticated plant developed by IPC as part of its three-dam complex at Hells Canyon.[224]

In 1915 the five existing hydroelectric power corporations operating in southern Idaho came under the control of the Electric Investment Company (EIC) and merged to form Idaho Power Company. The Idaho Public Utilities Commission approved consolidation in the belief that substitution of one organization for five, interconnection of all distributing systems, and replacing small generating plants with larger ones would improve service and decrease costs to consumers. The installed capacity of the company's nine plants represented only 18 percent of the potential of the developed sites and only 12 percent of the total potential of both the developed and undeveloped sites. This meant that 88 percent of the capacity of the power sites owned by IPC was still untapped. This untapped potential gave the company a large reserve with which to supply its future market.[225]

In the same year that IPC took possession of the Ox Bow, Swan Falls, Malad, Boise-Payette, Lower Salmon, Thousand Springs and Shoshone hydroelectric power plants on the Snake River and its tributaries, it came into ownership of additional undeveloped power sites at Twin Falls and Upper Salmon Falls. Beginning in the early 1920s, amidst considerable controversy, IPC would seek to develop both the Twin Falls and Upper Salmon Falls sites.[226]

---

222 Hoyt, *Water Utilization*, 202; Idaho Power Company, *History*, 24-26; George C. Young and Frederic J. Cochrane, *Hydro Era: The Story of Idaho Power Company* (Caldwell, ID: Caxton Printers for Idaho Power Company, 1978), 35-36; and "Charles J. Perkins on Idaho," *Twin Falls Times*, October 8, 1908.

223 Hoyt, Water Utilization, 202.

224 Young and Cochrane, *Hydro Era*, 35-36.

225 "Idaho Power Company," *Electric West* (August 1962), 3-6.

226 Hoyt, *Water Utilization,* 192-193 and 198-202; Idaho Power Company, *A History*, 43-55, and 61; George C. Young and Frederic J. Cochrane, *Hydro Era*, 39-40; Stacy, *Legacy*, 30-33; and "To

The electrical transmission infrastructure

Despite having acquired the major independent power plants and their transmission and distribution systems in 1915-16 on the Snake River from American Falls to Ox Bow, Idaho Power Company's hydroelectric power system remained divided for a number of years into two separate and distinct power markets in its eastern and western service areas of southern Idaho.

As of 1922, the various power plants within each group were interconnected with transmission lines, but no high-voltage transmission facilities linked the company's plants below Milner – Shoshone Falls, Lower Salmon Falls, Swan Falls, and Oxbow – with its plants located upriver at American Falls.[227] In addition, located between the two IPC power markets was the Reclamation Service's irrigation project at Minidoka. The Reclamation Service operated a power plant at Minidoka that was used primarily for pumping water for irrigation of lands on the south side of the Minidoka Project. Surplus electrical power not needed for pumping was used for incidental public utility purposes within the Minidoka Project area, including electrical service to the towns of Burley, Paul, Rupert, Heyburn and Acequia. IPC viewed the Reclamation Service, and its successor, the Bureau of Reclamation, as a potential competitor in the southern Idaho power market and throughout the middle decades of the twentieth century zealously opposed federal entry by Bonneville Power Administration into the commercial power industry in southern Idaho.

Nevertheless, the power company's own transmission interties and substations linked to federal power plants; for example, power was transmitted to and from the company's Milner substation to government plants at Black Canyon and Minidoka. Lease and purchase agreements for government power developed on the Boise Project assisted the company in meeting the demands of its eastern and western loads Meanwhile, IPC put its transmission network together piece by piece. Over a period of fifteen years, it extended the backbone high voltage line until it ran from one end of its service area to the other. In 1922, Idaho Power initiated construction of a 132 kilovolt (kv) transmission line connecting Thousand Springs to Caldwell, bringing energy

---

Enter Combine: Great Shoshone and Southern Idaho Power Company in Big Merger," *Twin Falls Times*, October 22, 1915.

227 Idaho Power Company, *A History*, 44-45; and Hoyt, *Water Utilization*, 192, 198, 199-202.

developed in the Twin Falls area to the Boise Valley. In 1927, a comparable line was completed linking Shoshone Falls and American Falls; ten years later that line was extended to Thousand Springs. The final section of the company's high capacity transmission line between Shoshone Falls and Thousand Springs was not completed until the spring of 1937. The company also tied its system together with other private companies such as Utah Power & Light Company at Pocatello, American Falls and Blackfoot, and with Eastern Oregon Light and Power Company at Durkee to exchange power and supplement its own power supplies. [228]

Between 1916 and 1934, the federal government and IPC executed a series of contracts by which the company wholly or partially leased either the Boise Project's power plants or purchased government power produced at those plants as a means of supplementing its commercial hydropower production on the Snake River. Such arrangements proved beneficial to the Bureau, water users in southwestern Idaho, and IPC.

The first of these contracts, executed on July 1, 1916, between the Reclamation Service and the Electric Investment Company (the operational forerunner to IPC) leased the Boise River power plant to private power companies until 1921.[229] The Boise power plant was part of the Arrowrock Division of the Boise Project, originally called the Payette-Boise Project when first authorized in 1905 under the provisions of the Reclamation Act of 1902. When initially outlined, the Payette-Boise Project "comprised something over 300,000 acres of land in the Boise and Payette valleys and other smaller valleys"; upon its initial construction it contained a little more than 240,000 acres between the Boise and Snake rivers.[230] The project was subsequently divided

228 Idaho Power Company, *A History*, 67-69; and "Idaho Power Company," *Electric West* (August 1962), 6-8.

229 Stacy, *Legacy*, 33 and 26-27; and Department of the Interior, Bureau of Reclamation, "Annual Project History of Boise Project, Idaho for 1919," 168; "Annual Project History of Boise Project, Idaho for 1920," 174; "Annual Project History of Boise Project, Idaho for 1921," 193. Folder: Department of the Interior, Bureau of Reclamation, Boise Project Annual Project History, 1919-1921, Volume 5; "Annual Project History of Boise Project, Idaho for 1922," 174; "Annual Project History and Operation and Maintenance Report of the Boise Project (Arrowrock Division), Idaho for 1923, " 163; "Annual Project History and Operation and Maintenance Report of the Boise Project (Arrowrock Division), Idaho for 1925," 86; Folder: Department of the Interior, Bureau of Reclamation, Boise Project Annual Project History, 1919-1921, Volume 6; Box 257 8NN-115-90-039 FY 91, Engineering & Research Center, Project Histories, Boise (ERC, PH, Boise), 1919-1925, RG 115, NARA-Denver.

230 Gilbert H. Hogue, Assistant Engineer, "History of the Payette Boise Project Idaho (Boise Project)," March 1916, 2 and 15. Folder: Department of the Interior, Bureau of Reclamation, Boise

into two major divisions, shaped by the area's geography: the Arrowrock Division, centered on the Boise River, and the Payette Division, centered on the Payette River. Today, the Boise and Payette divisions combined boast six storage or diversion dams:[231]

Dam Name	Built	Location
Boise River Diversion Dam	1908	Boise River, 7 miles southeast of city of Boise
Arrowrock Dam	1915	Boise River, 22 miles upstream of city of Boise
Anderson Ranch Dam	1951	22 miles northeast of Mountain Home, Idaho, South Fork of Boise River
Black Canyon Dam	1924	Payette River, near Emmett, Idaho
Cascade Dam	1948	Near Cascade, Idaho, North Fork Payette River
Deadwood Dam	1931	25 miles southeast of Cascade, Deadwood River

In addition to these dams, three power plants – at the Boise River Diversion Dam, Black Canyon Dam, and Anderson Ranch Dam – produced more than 50,000 kW of power, and seven pumping plants, canals, laterals, and drains serviced nearly 400,000 acres.[232] **Map 6** (below) shows the locations of the various Boise Project Dams, as of 1944.

Project, Annual Project History, Through 1912 & Early History, Volume 1, Box 6 8NS-115-93-213 FY 93, ERC, PH, 1911-1919, RG 115, NARA-Denver.

231 These do not include Hubbard Dam constructed by the Bureau in 1902 as a re-regulation facility and Lucky Peak Dam constructed by the Army Corps of Engineers in the 1950s to provided for flood control. See Bureau of Reclamation, Dataweb, "Dams, Projects, and Powerplants," "Hubbard Dam," website: http://www.usbr.gov/ dataweb/dams/id00376.htm; and William Joe Simonds, "The Boise Project," Bureau of Reclamation History Program (Denver: Bureau of Reclamation, 1997), available online at: http://www.usbr.gov/dataweb/projects/idaho/ boise/history.html, last accessed March 17, 2008.

232 Bureau of Reclamation, Dataweb, "Dams, Projects, and Powerplants," "Boise Project." website: http://www.usbr.gov/dataweb/html/boise.html; and Simonds, "The Boise Project."

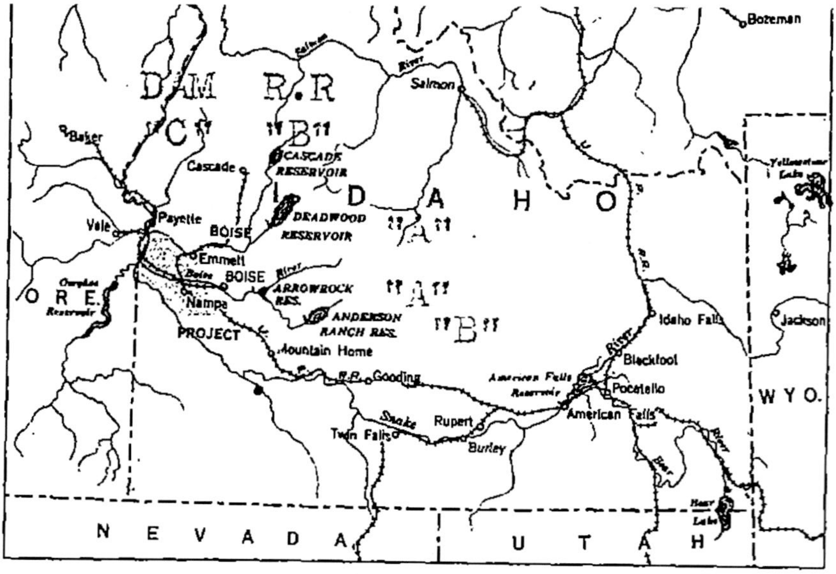

The Various Boise Project Dams as of 1944. Neither Cascade nor Anderson Ranch dams were completed but both were anticipated. *Source*: Boise Project, Idaho, Bureau of Reclamation, U.S. Department of the Interior, Boise Project History, 1944. Folder: Department of the Interior, Bureau of Reclamation, Boise Project, Annual Project History, 1943-1944, Box 254, 8NN-115-90-039Volume 12, FY 91, Engineering & Research Center, Project Histories, Boise, 1910-1911, RG 115, National Archives-Rocky Mountain Region (Denver)

The Boise power plant, and an accompanying two-line transmission system, was built in 1912 to provide electricity necessary to construct Arrowrock Dam.[233] With completion of the dam in October 1915, the Reclamation Service sought to dispose of the surplus power capacity at the Boise plant that was no longer needed for the massive Arrowrock Dam construction project.[234] Under the 1916 contract with EIC/IPC, the

233 Simonds, "The Boise Project"; and Department of the Interior, Bureau of Reclamation, "Annual Project History of Boise Project, Idaho for 1930," 12. Folder: Department of the Interior, Bureau of Reclamation, Boise Project, Annual Project History, 1930-1932, Volume 8, Box 256, ERC, PH, Boise, 1926-1932, RG 115, NARA-Denver.

234 Department of the Interior, Bureau of Reclamation, "Annual Project History of Boise Project, Idaho for 1919," 168; "Annual Project History of Boise Project, Idaho for 1920," 174; "Annual Project History of Boise Project, Idaho for 1921," 193-194; Folder: Boise Project Annual Project History, 1919-1921, Volume 5; "Annual Project History of Boise Project, Idaho for 1922," 174; "Annual Project History and Operation and Maintenance Report of the Boise Project (Arrowrock Division), Idaho for 1923, " 163; "Annual Project History and Operation and Maintenance Report of the Boise Project (Arrowrock Division), Idaho for 1925," 86. Folder: Boise Project Annual

power company operated the facility for the benefit of both its private customers and the federal government.[235] By means of high-voltage electrical transmission lines, the Boise plant was connected by IPC to the chain of larger private power plants that were supplying electric power to customers throughout its southwestern Idaho and Eastern Oregon power market." The government received electric power directly from IPC's transmission network for its Riverside Drainage, Arrowrock Dam and Boise River Diversion Dam at cost, or at the reduced rate of $0.0035 per kilowatt hour (kWh).[236]

By 1920, as IPC was emerging as the major supplier of hydroelectric power in southern Idaho, the Reclamation Service looked for the power company's assistance in meeting the summer power needs of the Minidoka Project, whose power demands for irrigation pumping exceeded the capacity of its only power plant at Minidoka Dam. On October 18, 1920, the Interior Department "approved a tentative form of contract" with the company that provided for IPC to supply power to the project; on April 1, 1921, First Assistant Secretary of the Interior E. C. Finney formally approved the new agreement, and this contract was subsequently consolidated with the earlier 1916 contract.[237]

Under the 1921 agreement, IPC returned operation of the Boise power facility to federal control but was permitted to make use of one-half of the plant's power output compensating the government for rent and operating costs. The other one-half of the output would be supplied

---

Project History, 1919-1921, Volume 6, Box 257, ERC, PH, Boise, 1919-1925, RG 115, NARA-Denver.

235 During the period May 1, 1915 through August 1, 1916, there was a proposed merger underway of the five major electrical companies that served southern Idaho and eastern Oregon, all of whom were in receivership because of their financial problems. Electric Investment Company, through a successful bid at a receiver's sale, had become the nominee to acquire title to the five properties and to operate them while the plan of reorganization and consolidation was completed and all of the properties could be turned over to IPC. On July 31, 1916, Electric Investment Company turned over legal title of the separate properties of the five predecessor companies and any operating agreements or contracts, including the Boise power plant, to IPC. Idaho Power Company, *History*, 43-53.

236 Department of the Interior, Bureau of Reclamation, "Annual Project History of Boise Project, Idaho for 1919," 168-169; Boise Project Annual Project History, 1919-1921, Volume 5, Box 257, ERC, PH, Boise, 1919-1925, RG 115, NARA-Denver. A separate contract in 1919 provided for the company to supply the government with power for construction work on the Boise River siphon of the Notus Canal. A different rate structure existed in this agreement: as Idaho Power supplied more electricity to the government, the rate per month declined from $0.035 to $0.02 for every 1000 KWH.

237 A. P. Davis, Director, to the Secretary of the Interior, March 11, 1921. Folder: 223.03, Boise, Correspondence re: Power Contracts with the Idaho Power Company, Thru 1929, Box 419 Boise 223.03-224.1, GAPR, 1919-1945, RG 115; NARA-Denver.

to the Boise and Minidoka projects to meet construction and irrigation pumping plant requirements using IPC's power network for a transmission charge. The power for Minidoka was to be delivered to the company's substation at Milner Dam, thereby requiring the Reclamation Service to spend $50,000 constructing a transmission line between the point of exchange at Milner and the government substation at Minidoka. The cost was justified as it would provide "permanent value in permitting the interchange of power between the systems of Idaho Power Company and the Reclamation Service." The contract was set to last until April 1, 1923, and, according to Reclamation Director A. P. Davis, would "secure a guaranteed additional supply of electricity amounting to 600 kilowatts during the peak irrigation months of July and August, 1921, and June, July and August, 1922."[238] This was to be the first of the three interconnections that IPC made with federal power plants prior to the 1940s that would help to expand the power company's service network.[239]

On April 1, 1923, with the expiration of the 1921 contract, IPC and the federal government executed another power exchange contract. This three-year agreement, as acting Reclamation Director F. E. Weymouth noted in his May 8, 1923, letter to the Secretary of the Interior, renewed the provisions of the prior agreement with one exception: the new contract "limited" the power demand of the government to:

> (a) one-half the capacity of the Boise power-house during periods when the Company's Horseshoe Bend plant must be operated, and (b) to 825 kilowatts on the Company's transmission lines at Milner, Idaho – the point of exchange with the Minidoka project.

---

238 Ibid. See also Department of the Interior, Bureau of Reclamation, "Annual Project History of Boise Project, Idaho for 1921," 194. Folder: Department of the Interior, Bureau of Reclamation, Boise Project Annual Project History, 1919-1921, Volume 5; Box 257; "Annual Project History and Operation and Maintenance Report of the Boise Project (Arrowrock Division), Idaho for 1923, " 163; "Annual Project History and Operation and Maintenance Report of the Boise Project (Arrowrock Division), Idaho for 1925," 86; Folder: Boise Project Annual Project History, 1919-1921, Volume 6, Box 257, ERC, PH, Boise, 1919-1925; and Davis to Secretary, March 11, 1921. Folder: 223.03, Box 419, GAPR, 1919-1945, RG 115, NARA-Denver.

239 Idaho Power Company, *History*, 68-69.

"These modifications," Weymouth believed, "will not interfere materially with the satisfactory meeting of the needs of the project."[240]

Irrigation demands in the Boise-Payette area continued to increase. On June 26, 1922, the Secretary of the Interior authorized construction of Black Canyon Dam. It was built primarily to supply irrigation water to the 22,000-acre Emmett Irrigation District and to the Payette Division of the Boise Project. The dam was essentially completed in late April 1924 and the design anticipated future power development. Gem Irrigation District, which encompassed 25,000 acres of irrigable land south of the Snake River and southwest of Caldwell, Idaho, appealed to the Bureau of Reclamation for relief from high power rates that IPC was charging them under an earlier agreement. The district, when first organized, had contracted with one of IPC's predecessor companies for electricity needed to power four pumping lifts on the Snake River which boosted the water some 126 feet to the irrigated plateau above the rim of the canyon. Poorly-financed and plagued by power costs, massive seepage losses in its canal system compounded Gem Irrigation District's difficulties, and the district soon found itself in arrears with IPC and on the verge of bankruptcy.

The Bureau of Reclamation (successor to the Reclamation Service) began construction on the proposed power facilities at Black Canyon in late 1924 following an agreement between the Bureau and Gem Irrigation District in which the district purchased power to operate the pumping units on its land. On November 26, 1924, the Bureau formally agreed to assist the district by supplying it with surplus power on a temporary (5 year) basis from its new Black Canyon power plant and to construct a storage reservoir on the Owyhee River to supply the district with gravity-delivered water. The service was faced with spending up to $100,000 of its Boise Project appropriation to construct a 55-mile transmission line from Black Canyon to the Gem Irrigation District, money that was needed elsewhere on the reclamation project. Once again the Bureau turned to IPC and entered into yet another power wheeling agreement with the district and the power company collectively.[241]

---

240 F. E. Weymouth, acting Director, to the Secretary of the Interior, May 9, 1921. Folder: 223.03, Box 419, GAPR, 1919-1945, RG 115, NARA-Denver.

241 Stacy, *Legacy*, 46-47 and 77-78; and "Power Supply for the Gem Irrigation District," June 20, 1926, 1-3. Folder: 320, Boise Project, Correspondence re Development of Power, Thru December 31, 1929, Box 436 Boise 320-400.04, GAPR, 1919-1945, RG 115, NARA-Denver.

Under the terms of this contract, executed on February 5, 1925, the Bureau agreed with IPC to connect a 66-kV transmission line from Black Canyon to the company's substation at Emmett, Idaho, a distance of five miles, and to have the company wheel power from Black Canyon to Gem Irrigation District. The power company charged a 4 percent transmission loss, while the United States "reserve[d] 1,000 K. W. of Black Canyon [power] for its uses at dam and for construction purposes." Moreover, surplus power not needed for irrigation pumping or other purposes by Gem Irrigation District or by the United States was offered to IPC for its use from April 15 to October 15 without cost and for the balance of the year by payment of $600 per month. Additionally,

> In return for summer power, Power Company agrees to turn over to Gem District, when Black Canyon power insufficient, any unused available power at its plants; without cost to extent available from operating plants and at actual cost of operation from plants otherwise idle.[242]

To the Bureau's district counsel, B. E. Stoutemyer, the benefits were obvious. He estimated that the contract would save Gem Irrigation District about $12,000 per year on interest and depreciation on the transmission line and operation and maintenance costs. The federal government would likewise save about $75,000 on construction of a transmission line from Black Canyon to Gem Irrigation District.[243]

The federal government was also able to secure a "reservation" of 1000 kW of power for use by the United States for construction purposes. According to Stoutemyer, some of this power could be traded to IPC in exchange for its transmission service, or to supplement supplies during water-short periods such as the month of August. Stoutemyer went on to explain that since the contract provided "for exchange of surplus power between the Gem District and the Power Company during the summer season," IPC might "not have any surplus" in August when irrigation load caused a summer peak in

---

242 "Power Supply for the Gem Irrigation District," n. d., 3. Folder: 320, Box 43, GAPR, 1919-1945, RG 115, NARA-Denver.

243 B. E. Stoutemyer to Commissioner, Draft of proposed contract between United States, Gem Irrigation District, and Idaho Power Company covering construction of transmission line – Black Canyon power plant – Boise Project, December 31, 1924. Folder: 223.03, Box 419, GAPR, 1919-1945, RG 115, NARA-Denver.

irrigation demand. Then again, he noted, electrical interconnection among power plants presented additional opportunities to exchange power when unanticipated openings arose:

> … it is also quite likely that they may have some surplus and the contract contains a provision under which such surplus, if available from the plants of the Power Company could be secured by the District without additional expense in exchange for surplus power which may be available from Black Canyon power plant at other times of the year.

There was then a surplus of winter power in the southern Idaho territory. For its part, the company secured through the contract "the right to the use of winter power which may be developed at the Black Canyon plant when needed." Its only cost for obtaining this right was "the extra costs of operating the plant during the winter." Although Stoutemyer believed that non-irrigation season electric power might one day become valuable to the Bureau, he observed that over the five-year life of the contract "there appears to be no other market available" but IPC.[244] Stoutemyer, therefore, recommended approval of the contract, which was granted on January 29, 1925.[245]

Unlike for the Bureau winter power was of some commercial value to IPC, noted U. S. Commissioner of Reclamation Elwood Mead, in a letter sent to the Secretary of the Interior recommending execution of the contract. In case of emergencies affecting one or more of its other plants, IPC could rely upon power from Black Canyon; it might also find it more economical to pay for the winter operation of the Black Canyon plant and shut down one or more of the older and less efficient plants.[246]

By early 1925, the power company was thoroughly interconnected with the Bureau's electrical generation and transmission system, purchasing and wheeling power from both Boise Diversion Dam (to Minidoka via Milner) and Black Canyon Dam facilities, and deriving

244 Ibid., and Elwood Mead, Commissioner to the Secretary of the Interior, January 26, 1925, and Mead to the Secretary, February 19, 1925. Folder: 223.03, Box 419. GAPR, 1919-1945, RG 115, NARA-Denver.

245 Stoutemyer to Commissioner, December 31, 1924, and Stoutemyer to the Secretary of the Interior, January 26, 1925, and Mead to the Secretary, February 19, 1925. Folder: 223.03, Box 419, GAPR, 1919-1945, RG 115, NARA-Denver.

246 Mead to the Secretary, February 19, 1925. Folder: 223.03, Box 419, GAPR, 1919-1945, RG 115, NARA-Denver.

the benefits that both offered. A little more than a year later, spurred by a low-water year, the Bureau revisited its existing contract with IPC for power from the Boise Diversion Dam power plant. In May 1926, B. E. Stoutemyer sent a memo to the Commissioner of Reclamation requesting advance approval for a one-year contract extension with the company to cover use of the government power plant at the diversion dam by IPC for standby purposes. In case of an extremely low water year and insufficient power production at the Black Canyon plant, the power company could wheel power developed at the diversion dam plant to the Gem Irrigation District.

Stoutemyer noted that because Boise River water was stored for irrigation use in either Deer Flat or Arrowrock reservoirs during the winter months, "very little power can be developed during the winter season." Although "the power plant is capable of developing a considerable amount of power through the irrigation season, which is available for pumping water for irrigation purposes," the intended recipient of this irrigation season power, the proposed Hillcrest Division of the Boise Project, had not yet been completed nor would it be completed for at least two years.[247] Additionally, since IPC would pay for the use of this short-term surplus power, the government stood to reap financial rewards from the arrangement.[248]

In early 1928, the expiration date for the three-party contract of February 5, 1925 loomed on the horizon. With the expected completion of Deadwood Reservoir in the spring of 1931, the Bureau anticipated availability of more power from Black Canyon and Boise Diversion Dam plants, making it available to settlers on the Owyhee reclamation project, near the Idaho-Oregon border. These factors caused the Bureau to contemplate a change in its transmission arrangement with IPC.[249]

247 Stoutemyer to Commissioner, Draft of proposed contract between United States and Idaho Power Company covering use of Government power plant at Diversion Dam – Boise Project, May 25, 1926.; Folder: 223.03; Box 419; Boise 223.03-224.1; GAPR, 1919-1945; RG 115; NARA-Denver.

248 P. W. Dent, Acting Commissioner to the Secretary of the Interior, June 30, 1926. Folder: 223.03, Box 419, GAPR, 1919-1945, RG 115, NARA-Denver.

249 See Chief Engineer to District Counsel, Transmission Line – Owyhee Project, February 29, 1928; District Counsel B. E. Stoutemyer to Commissioner, Transmission of surplus power from Black Canyon power plant to Oregon pumping districts, November 1, 1929; District Counsel B. E. Stoutemyer to Construction Engineer, Owyhee, Oregon, Transmission of surplus power from Black Canyon power plant to Oregon pumping districts – proposed contract with Idaho Power Company, December 24, 1929; and Acting Chief Engineer to District Counsel, Transmission of surplus power from Black Canyon power plant to Oregon pumping districts, December 30, 1929. Folder: 223.03, Box 419, GAPR, 1919-1945, RG 115, NARA-Denver.

As explained in his letter to the Commissioner of Reclamation, Stoutemyer believed it was "desirable to secure an early understanding with the power company…so that if we are not satisfied with the terms offered by the company, we will have time to construct our own transmission line prior to the time when it will be needed in the spring of 1931."[250]

On October 28, 1929, at a meeting between Stoutemyer, F. A. Banks, and IPC officials, Stoutemyer and Banks laid out the benefits of exchanging cash payments for transmission of Boise Project electricity to pumping plants in southwestern Idaho and eastern Oregon for winter power from the Bureau's project. According to Stoutemyer,

> we argued that after the completion of the Deadwood reservoir, the power company would be relieved of the responsibility which it now has for supplying a large part of the summer power used by the Gem district during August and September, particularly in low water years, and therefore that the winter power received by the company after the spring of 1931 would be a net benefit to the company without any offsetting responsibility for furnishing power to the Gem district; and that in view of this changed condition, the company could well afford to furnish all the transmission service desired by the government, in consideration of the winter power.

Given the benefits it stood to acquire, IPC was receptive to the idea.[251] Nearly two months later, after subsequent meetings, IPC provisionally consented to a contract that would have the company "transmit all the power generated" at the Boise and Black Canyon plants to "the pumping plants of the Gem district and to a point at the Ontario-Nyssa pumping plant" where the government's own power lines linked the plant, the Owyhee project, and the Ontario-Nyssa district in Oregon. In exchange, IPC agreed to accept as "full payment for the transmission service, the winter power available from the Black Canyon plant" – minus 1000 kW the government reserved "for construction purposes." This contract would run for five years, beginning in 1931. District Counsel B. E. Stoutemyer found this

---

250 Stoutemyer to Commissioner, Transmission of surplus power from Black Canyon power plant to Oregon pumping districts, November 1, 1929. Mead to the Secretary, February 19, 1925. Folder: 223.03, Box 419, GAPR, 1919-1945, RG 115, NARA-Denver.

251 Ibid.

arrangement "more favorable to the government" because it relieved the government of its current contractual obligations with the company that required the former to pay the latter for transmission service.[252]

Ultimately, the more comprehensive 1934 "water saving contract" between IPC and the United States – crafted during one of the most severe droughts in the Snake River Basin – would supersede this 1931 agreement. Nevertheless, over the nearly twenty years these power exchange and wheeling agreements were in effect, both the federal government and the power company profited from the arrangement. The Bureau, on the one hand, was able to dispose of excess electric power, avoid the costs of constructing a transmission system throughout southern Idaho, and manage to service not only its Minidoka Project but also irrigation-pumping districts such as the Gem Irrigation District.[253]

IPC, on the other hand, secured the firm power it needed to meet customer demand and the means to deliver that power to its divided service area. The importance of the latter can scarcely be overstated. Interconnections with the Bureau's plants afforded the company the means to balance the distribution of power throughout its entire southern Idaho service area even as the company worked on bridging the division between its markets. No less an observer than B. E. Stoutemyer, the man who had negotiated many of the Boise Project contracts with IPC, recognized the importance of these interconnections to the power company. In a May 3, 1924 memorandum to the Bureau's Chief Field Counsel in Denver – drafted in the midst of the controversy surrounding the power company's attempts to secure a federal license for a new power plant at Twin Falls – Stoutemyer pointed out "the deliveries of power at Boise from the Boise plants furnishes the power where it is most needed by the Company and lessens the amount of power necessary to be transmitted into Boise rather than increasing the power to be transmitted over the Company's lines."[254]

---

252 Stoutemyer to Commissioner, Transmission of surplus power from Black Canyon power plant to Oregon pumping districts, December 21, 1929. Mead to the Secretary, February 19, 1925. Folder: 223.03, Box 419, GAPR, 1919-1945, RG 115, NARA-Denver.

253 Idaho Power Company, *History*, 69.

254 Stoutemyer to Chief Field Counsel, Subject: Protest of Minidoka Project people against granting of application of Idaho Power Co. for Twin Falls Power Site – Minidoka Project, May 3, 1924, 5. Folder: 412.12 Idaho, Federal Power Applications (Application #18-Twin Falls Site on the Snake River), Idaho Power Company, 1930 thru (412.12 Idaho), Box 809 412.06_412.12 - Colorado – Idaho, General Correspondence File, 1930-1945 (GCF, 1930-1945), RG 115, NARA-Denver.

Over this same period of time, as IPC made use of its interconnections with federal power systems, the company was able to unify its service area. In 1922, the company built a 132-kV transmission line connecting Thousand Springs to Caldwell, bringing energy developed in the Twin Falls area to the Boise Valley. Five years later, IPC completed a line between Shoshone Falls and American Falls, and in 1937, finished construction of a line between Shoshone and Thousand Springs. By the end of 1939, the company could boast that it had "a total of 1,264 miles of transmission line as compared to approximately 688 total miles of transmission line acquired in 1916."[255]

In the years following its creation, IPC brought electricity to more and more of southern Idaho, growing alongside irrigated agriculture. The water resources of the Snake River, however, were finite, and the interests of irrigators, federal reclamation authorities, and the State of Idaho would clash with those of IPC on a number of occasions.

The first, and arguably most significant, came in the early 1920s. Following a devastating drought in 1919, irrigators, canal companies, private business interests, the State of Idaho, and the federal government sought to build another storage reservoir in the upper basin; they saw American Falls as the best possible site.

Construction of such a reservoir would destroy IPC's own plan for its future power plant expansion at American Falls. Only after protracted negotiations did the company agree to relinquish control of the site to federal reclamation authorities. Those negotiations resulted in a document of far-reaching consequence – the June 15, 1923 contract – a document that not only paved the way for a government-owned American Falls Reservoir operated in the interests of its irrigation spaceholders, but also confirmed the essential purpose of water administration in the Upper Snake River Basin for the remainder of the century.

255 Idaho Power Company, *History*, 67-68.

4

"Storage Reservoirs are a Necessity"

From the early 1900s, demand for irrigation pumping and commercial, municipal and domestic electrical needs had continued to rise in the Snake River Valley, as did demand for more storage reservoirs to conserve a greater portion of the winter flood flows in order to expand the water supply available for irrigation use. The power requirements to run irrigation pumps became the largest single factor in escalating the generating capacity of hydroelectric plants in southern Idaho.

This fact compelled the United States Reclamation Service, forerunner to the Bureau of Reclamation, to enter into electrical power production in 1906 on the Minidoka Project and to build the first federal power plant in the Pacific Northwest. Conquest of the desert by irrigated agriculture was also vital to the success of the private power industry in southern Idaho. Perhaps nowhere was both the conflict and cooperation of irrigators and private power interests more evident than in the competition to control the water storage possibilities and falling waters at American Falls on the Snake River. IPC's existing power rights and planned enlargement of its facilities at American Falls posed a potential barrier to full development of water resources for agricultural use in the Upper Snake River Basin. Negotiations between the Reclamation Service and IPC over the future development of the American Falls site were initiated in 1918 and ultimately resulted in the June 15, 1923 contract.

This agreement, which shaped water use in the Snake River Basin for generations, defined the rights of the federal government and the power company not only at American Falls, but also throughout the upper basin.

Drought

The immediate impetus for construction of the American Falls storage facility in 1926 was the devastating 1919 drought, although irrigation proponents had been aware for some time that more storage capacity was needed to meet existing irrigation water requirements.

Farmers along the Snake River had faced periodic water shortages since the 1890s. About once every five years, on average, water ran short before irrigators received all they needed. In 1905, the Snake River below Blackfoot went dry in a ten-mile stretch during the irrigation season and touched off legal battles leading to court decrees adjudicating water rights. Water was insufficient to mature crops again in 1910, and 1915, even on the lands served by the Twin Falls Canal Company, one of the most stable and self-sufficient private irrigation systems in the Snake River Basin.[256] In fact, in early 1919, before the drought disaster was imminent, the board of directors of the Twin Falls Canal Company had approached the Reclamation Service about co-operating in the development of a storage site that would yield the company an additional 100,000 acre-feet of storage as security against dry years.[257]

The drought of 1919 was – in the words of the U. S. Weather Service – "without precedent both in duration and intensity." In the spring of 1919 a cold spell prevented the snowpack in the mountains from melting, and rainfall in May and June was the lowest for those months on record. Apprehension gripped the valley as the river dried up and storage reservoirs were drawn down. On May 30, Jackson Lake contained only 547,000 acre-feet of storage, far below the normal average of 780,000 acre-feet for that date. No water was passing Milner Dam and the entire stream flow was being consumed for irrigation purposes. Because of the dry spring and summer, the natural flow of the river was insufficient to supply decreed rights on the river and it became necessary for the federal Minidoka project and the private Twin Falls project located below Milner Dam to draw on their

---

256 *Twin Falls Canal Company v. Charles N. Foster, et al.* (1913) involved a suit brought by the Twin Falls South Side Canal Company against the Twin Falls North Side Canal Company, the settlers on the Minidoka Project, and other irrigation interests on the Snake River below Blackfoot. Case files are found in the records of the Bureau of Reclamation. Folder 163-2 Water Appropriations - Settlement of Idaho State Water Rights, Jan. 1, 1911 to ___, Box 636 Minidoka 163.2 – 223, GAPR, 1902-1919, RG 115, NARA-Denver.

257 "More Storage is Favored by Canal Company," *Twin Falls Daily News*, February 25, 1919, 5.

storage at Jackson Lake as early as June 5 – six weeks earlier than normal.[258] Farmers struggled to cope with shortages, and soon altercations broke out along the river as individuals tampered with headgates in a desperate effort to get moisture for withering crops. In some rural districts vigilante committees had been organized and Idaho Governor David W. Davis was compelled to issue a proclamation asking farmers along the Snake River to refrain from violence. On July 1, the Twin Falls Canal Company announced it would furnish only 30 per cent of the normal water supply to its customers. By the middle of July, the water supply of several irrigation companies was exhausted.[259]

Drought-related losses gave rise to many measures, public and private, to relieve the suffering of individual farmers. Devastation was so widespread that Idaho congressman Addison T. Smith persuaded Congress and the General Land Office to adopt special regulations allowing homesteaders and settlers whose crops had been damaged by the unprecedented drought in 1919 to leave their homesteads for up to one year to obtain food and other necessities of life for their families. Upon making final proof with a corroborating affidavit, such an entry absence granted under this act was to be counted and construed as constructive residence by the homesteader. A person granted a leave under the act was to obtain credit to be applied towards continuous homestead residence and cultivation requirements during the time of his or her absence, as though actually living upon the entry.[260] In another generous act of private relief, stockholders of the Twin Falls Canal Company in the late summer of 1919 voted to relieve a drought situation by diverting sufficient water into their neighbors' canals on the North Side project in order to provide water for thirsty livestock and to fill farm cisterns with water for domestic uses.[261]

---

258 Weather Service quoted in US Department of Commerce, *Fourteenth Census of the United States Taken in the Year 1920, Volume VII: Irrigation and Drainage* (GPO, 1922), 167-165; W. G. Swendsen to The Water Users of Snake River, June 2, 1919 and W. G. Swenson to Acting Governor C. C. Moore, Departmental Report No. 2, from May 1 to June 1, 1919. Folder: Reclamation; Folder Divider: Department of Reclamation, 1919-1922, No Box, Governor Papers - D.W. Davis, 1919-1923, AR 2/12 (AR2/12), ISHS; Department of the Interior, *Nineteenth Annual Report of the Reclamation Service, 1919-1920* (GPO, 1920), 157; and B. E. Stoutemyer to Chief Council, Reclamation Service, June 26, 1919 and Stoutemyer to Chief Counsel, June 30, 1919; Folder: 163-2 Minidoka - Water Appropriations - Settlement of Water Rights, Jan. 1, 1911 to July 14, 1919, Box 636, GAPR, 1902-1919, RG 115, NARA-Denver.

259 "Call Governor in the Water Controversy," *The Idaho Register*, March 18, 1919,1; *Twin Falls Daily News*, July 1, 1919, 3.

260 "Announce Rules to Leave Homesteads," *Twin Falls Chronicle,* August 7, 1919.

261 "North Side is Grateful," *Twin Falls Daily News,* September 26, 1919.

Over the previous decade, irrigators and the companies that supplied them with water had looked to the Reclamation Service in times of need to relieve water shortages. The Service shared its existing facilities and enlarged them to conserve greater amounts of surplus water. However, the ability of the Service to provide extra water to irrigators was limited. Water stored in Jackson Lake and Lake Walcott supplemented natural flow rights in the river, but they could do little to stave off disaster in a critically dry year like 1919. By August 21, 1919, Jackson Lake Reservoir was empty. The drought continued until October. Canal corporations, associations, and individual farmers in the Idaho Falls area with water rights dating back to 1893 lost all of their crops because their water supply was cut off by the Snake River watermaster. Up and down the valley that year ranchers and farmers suffered $25,000,000 in crop losses alone, not taking into consideration the collateral impact on livestock, service industries and businesses.[262]

In the wake of the record drought, many irrigators, canal companies, and other business interests in the Snake River Basin, as well as Idaho state officials sought to prevent a repeat of the 1919 catastrophe by asking the federal government to provide additional water storage facilities on the river, specifically a large storage reservoir at American Falls. The idea of a 1.5 to 3 million acre-feet facility at the location to regulate the flow of the Snake River in the interests of irrigation was not a wholly new idea. As far back as the late nineteenth century, the United States Geological Survey and private irrigation interests had identified American Falls as a good potential storage reservoir site.[263]

The proposed dam project faced major obstacles of a type that made it difficult for any private enterprise to press forward with the project. Among the most challenging difficulties:

- Moving the existing town of American Falls.
- Realignment of the Oregon Short Line Railroad and its railroad bridge crossing the river just above the falls.

---

262 *Nineteenth Annual Report of the Reclamation Service*, 157; and Stoutemyer to Chief Council, June 26, 1919 and Stoutemyer to Chief Counsel, June 30, 1919. Folder: 163-2 Minidoka, Box 636, GAPR, 1902-1919, RG 115, NARA-Denver; and Mark Austin, et al. to Senator William E Borah, June 25, 1919, and Land Utilization File: Miscellaneous Agreement – US & Idaho Power Company, Box 833, Borah Papers, LOC.

263 Irvin E. Rockwell, *Saga of the American Falls Dam* (New York: The Hobson Book Press, 1947), 10-18; Gertsch, "The Upper Snake River Project," 173-176; and History of the Minidoka Project, Volume 1, August 1915, 3. Box 294, PHFHR, 1902-1932, RG 115, NARA-Denver.

■ Acquisition of approximately 70,000 acres of land needed for the reservoir flowage right, more than half of which belonged to the Shoshone-Bannock tribe on the Fort Hall Indian Reservation.

■ Negotiations with Idaho Power Company to extinguish its existing American Falls water rights, which proved to be the most difficult roadblock of all.

In 1916, the power company had acquired the American Falls site, its dam, three power plants (the Island, West Side, and East Side plants), and water rights, along with other assets of the Southern Idaho Power Company. Since construction of a federal reservoir would necessarily interfere with IPC's property and power rights at American Falls and at other points further down the Snake River, the Reclamation Service initiated negotiations with the company in late 1918, but for a variety of reasons an agreement was not signed until five years later in 1923.[264] Another three years passed before a 1.7 million acre-foot reservoir was completed at American Falls and began storing water.

On July 5, 1919, the Reclamation Service requested IPC to submit formal price quotes for various alternative proposals to transfer all, or a substantial portion, of the company's American Falls power facilities to the federal government.[265] Simultaneously, the Reclamation Service carried on negotiations with the irrigation districts and canal companies regarding funding mechanisms and cost sharing proposals for the project – which, in turn, would help determine the ultimate size of the storage facility.[266]

---

264 *Eighteenth Annual Report of the Reclamation Service*, 414; and *Nineteenth Annual Report of the Reclamation Service*, 417-418.

265 F. E. Weymouth to Idaho Power Company, July 5, 1919. Folder 301.1, Box 728 Minidoka 301.1, GAPR, 1919-1929, RG 115, NARA-Denver.

266 Hoyt, *Water Utilization*, 198; Wilcox, "Water Power Resources," 448; *Eighteenth Annual Report of the Reclamation Service, 1918-1919*, 414; and *Nineteenth Annual Report of the Reclamation Service*, 417-418. IPC sold the West Side and Island plants to the United States. The company had a 1908 priority power right of 1,400 cfs for generation of power through the first unit of the East Side plant; and, 4,600 cfs under an unperfected right initiated in 1919 for two additional units at the East Side plant eventually constructed in 1923-4 and 1926-7. "American Falls Power plant (East Side plant) owned by Idaho Power Co. and located adjacent to the town of American Falls," n. d. Idaho State Water Right 01-2046 Back File.

Planning the dam: Financing and storage

The Reclamation Service was primarily interested in the American Falls Project as a means to expand irrigated acreage on its Minidoka Project. From an economic perspective, the feasibility of the project depended upon building a large storage reservoir with a minimum storage capacity of 700,000 to 800,000 acre-feet.

Studies of the site revealed that the per acre-foot cost of the storage reservoir decreased significantly as the size of the reservoir increased; therefore, the Reclamation Service was interested in building the largest reservoir that could be justified to meet prospective demand. The 110,000 acres of irrigable land on the planned North Side Pumping Division of the Minidoka Project required a maximum additional water supply of about 520,000 acre-feet.

The Reclamation Service entered into negotiation with irrigation districts and canal companies who desired additional water to share in the cost of construction of a reservoir on a *pro rata* basis. The plan was originally presented to Congress on this basis, but by 1920 the projected size of the dam had grown from an estimate of 1.5 million acre-feet maximum capacity, to upwards of three million acre-feet, depending upon projections of where irrigation might expand in the upper or lower portions of the Upper Snake River Basin, and at what pace. In the meantime, the Reclamation Service entered into negotiations to contract with private canal companies and irrigation districts providing for the purchasers to obligate themselves to pay various sums of money to the federal government for construction, and in addition to make a good-faith payment of ten cents per acre-foot at the time of entering into a storage space contract.[267]

As for the funding plan, the Reclamation Service sought a direct congressional appropriation to fund the portion of the dam attributable to expansion of the Minidoka Project with the funds being reimbursable by the beneficiaries of the stored water. The problem of deferred payment, or non-payment, by settlers on earlier reclamation projects and declining receipts from sale of public lands left the Interior Department ill-positioned to tackle a major project such as American

267 *Nineteenth Annual Report of the Reclamation Service*, 15, 157, and 417-418; *Twentieth Annual Report of the Reclamation Service*, 25 and 411; *Twenty-first Annual Report of the Reclamation Service*, 64-66 and 123; and *Twenty-second Annual Report of the Bureau of Reclamation*, 65-66.

Falls reservoir out of the General Reclamation Fund. The primary cause for the failure of settlers on the Minidoka Project to meet repayment schedules was directly attributable to the 1919 drought and the unfavorable market for farm commodities. Following World War I, worldwide demand for food staples decreased. The U. S. market was flooded with an abundant harvest in 1920, and surplus crops in the Snake River Basin brought prices for some farm commodities below the cost of hauling the crops to market. With the farm recession in 1920-1921 following at the heels of the great drought the previous season, significant financial hardship confronted many settlers on the Snake River irrigation projects. The inability of farmers to meet their obligations to the General Reclamation Fund inclined Interior officials, including Interior Secretary Albert Fall, to question the wisdom of expanding irrigated acreage on a faltering Minidoka reclamation project by undertaking construction of an expensive new storage facility.[268]

Public sentiment in southern and eastern Idaho for a storage reservoir to meet current and future irrigation demand nevertheless remained strong. In the absence of any state program to construct storage reservoirs, the region's residents relied on the Reclamation Service to build one at American Falls, as they had earlier at Jackson Lake. The Reclamation Service replied that it would construct the storage works on the condition that water users who acquired storage space in the reservoir advanced the funds for their *pro-rata* share of the storage space. Congress would be asked to appropriate money under the Reclamation Act to provide supplemental storage water for federal project lands currently being served and for land within the authorized project area still under public ownership. Each independent canal company or irrigation district had to make its own arrangements to sell bonds to finance its proportional share of storage space in the reservoir. Other alternatives were also explored. Among them was the creation of a large reservoir (or "super-irrigation") district that embraced all of the lands of private owners outside the federal reclamation project that would use the reservoir for storage. The sole purposes of such a district would be to enable the sale of bonds on more favorable terms to fund construction of a larger dam and to enable "spaceholders" to enter into

268 *Nineteenth Annual Report of the Reclamation Service*, 43; and *Twentieth Annual Report of the Reclamation Service*, 11-12, and 160-161; A. P. Davis to Secretary of the Interior, November 21, 1921. Folder: Reclamation Bureau, Minidoka Project, American Falls Reservoir, Acquisition of Rights, Idaho Power Company, Box 1614, RG 48, NARA II.

more affordable water storage and delivery contracts with the United States. The matter was of such grave importance to Idaho that Governor David William Davis offered to call a special session of the state legislature to pass enabling legislation to authorize such a reservoir district.[269]

Meanwhile, the water shortage in 1919 and subsequent crop loss from drought piqued farmers' interest and created a strong immediate demand by farm organizations for additional storage on the Snake River. A meeting to discuss the subject at Idaho Falls on November 24, 1919 attracted 500 irrigators from the Snake River above the Twin Falls area. The Committee of Nine submitted a report at that hearing urging construction of a storage facility at either American Falls or Swan Valley (site of Palisades Reservoir). The assembled group of local irrigation districts, co-operatives and irrigation companies, and water rights attorneys, as well as state officials and representatives of the Reclamation Service, unanimously agreed that American Falls was the best location for an initial large storage reservoir. It was best situated to capture the flood flow going to "waste" down the river during the winter and spring months, as well as the vast spring flow and return flows from the Blackfoot area. American Falls was not only the optimal site for filling a large storage structure, but also a reservoir at this site controlled irrigation supply to both the Minidoka and Twin Falls areas, and therefore, would facilitate an exchange of stored water. Those in the upper basin above American Falls would supplement their water supply with "American Falls water" stored in Jackson Lake Reservoir, while the Minidoka irrigators and private water users downstream of the proposed dam would draw their "Jackson Lake" storage rights from American Falls. The federal reservoirs of the Snake River Basin would thus be operated together for the joint benefit of all the irrigators who participated in the American Falls Project – both those located above, as well as below, the actual reservoir site.

The plan called for the various irrigation interests, districts or companies to enter into individual binding agreements with the Reclamation Service using a standard form of contract, one form for

269 G. Clyde Baldwin, special deputy State Commissioner of Reclamation, to B. E. Stoutemyer, district counsel USRS, August 6, 1919; Stoutemyer to Baldwin, August 20, 1919; and Stoutemyer to Chief Counsel USRS, August 20, 1919. Folder: 301.1 MP, Construction & Engineering D—Dams, American Falls Dam & Reservoir, to April 30, 1920, Box 728 Minidoka 301.1, GAPR 1919-1945, RG 115, NARA-Denver; and Rockwell, *Saga*, 108-133.

those located above American Falls and another for irrigators diverting their water below the reservoir. These private entities and public water districts each were to advance the necessary funds to the federal government in proportion to the amount of stored water contracted to them in the reservoir. The Reclamation Service would then carry out the construction work and operate the reservoirs for the benefit of these "spaceholders," and in compliance with the normal flow rights and storage rights, as decreed by the courts and licensed under state water law, as well as in accordance with any contract rights, as specified in the agreements between the United States and spaceholders.[270]

With the terrible drought year barely behind them, water users in the Upper Snake River Basin were poised to take a major step forward to avoid future water shortages. Just before the drought struck in 1919, the Idaho Reclamation Association was organized at a meeting in Pocatello which attracted 300 delegates from 21 counties of the state. The short-term objective of the association was to organize support for Secretary of the Interior Franklin Lane's bill before Congress to appropriate $100 million in funding for reclamation of arid land and swamp land in the western states. On a broader scale, the formation of this statewide organization became a rallying point for the merger of local organizations along the river to unify farm interests to cooperate on issues related to reclamation and to actively lobby southern and eastern congressmen hostile to western reclamation. The leader of the Idaho Reclamation Association, Fred Reed of Boise, worked tirelessly over the next few years to get local and county reclamation organizations to adopt a broader perspective on agricultural issues pertinent to the Snake River Valley as a whole, and to become active in promoting reclamation legislation at the national level. The Idaho Reclamation Association became the leading advocate for the formation of a larger federation of thirteen western states into the Western States Reclamation Association. Idaho's governor Davis was elected chairman at its first meeting in Salt Lake City in November 1919 and immediately set to work lobbying Congress for a fund of $250 million to be set aside under the National Reclamation Act for settlement of former service men on western reclamation projects.[271]

---

270 "Big Things for Valley Planned at Idaho Falls," *Burley Bulletin*, November 28, 1919. Folder 301.1, Box 728, , GAPR, 1919-1929, RG 115, NARA-Denver.

271 Gertsch, "The Upper Snake River Project," 177-187; "Reclamation Booster Urges Hard Struggle for Irrigation," *Twin Falls Daily News*, January 19, 1920; "Western Men Hold Out for

The drought of 1919 was a pivotal event in the history of the Upper Snake River Basin and also stimulated a broader interest in basin-wide solutions to regional water shortage problems. Fred Reed and his Idaho Reclamation Association called for water conservation and for federally built storage reservoirs as a solution.[272] As early as July 1919, the *Idaho Register* called for a meeting of all upper basin water interests to make plans for the formation of an organization dedicated to "building of storage reservoirs to insure against future crop losses from water shortage."[273] At mass meetings attended by hundreds of stockholders of the Twin Falls Canal Company in December 1919 and January 1920, directors of the company were instructed to enter into contracts and co-operate with the Reclamation Service and other Snake River irrigation projects to build the American Falls Reservoir. The canal company anticipated receiving some 100,000 acre-feet of stored water from the reservoir at a cost to stockholders of about $5.00 per acre-foot, payable in installments.[274]

On January 16, 1920, Interior Secretary Franklin K. Lane authorized the Reclamation Service to enter into negotiations with any of the forty or more irrigation districts, canal companies, and individual irrigators diverting water from the Snake River that desired to cooperate with the United States under Section 2 of the Warren Act of 1911 to secure additional storage at American Falls.[275] Surveys conducted by the Committee of Nine in late 1919 had projected total storage demand by private users, irrigation districts and canal companies at about 744,000 acre-feet. Secretary Lane and the Reclamation Service set their own target for reserved space in the reservoir at about 740,000 acre-feet.

Each saw as critical the government's securing maximum cooperation of canal companies and irrigation districts, including those of the upper basin above American Falls, for the storage project to be economically feasible. For example, Reclamation Service engineers estimated that building a reservoir of 2.5 million acre-feet at American Falls was only one and one-half as great an expense as a reservoir of

---

Reclamation," *Twin Falls Daily News*, January 20, 1920.

272 *Pocatello Tribune*, June 4, 1919.

273 *Idaho Register*, July 8, 1919.

274 "Canal Stockholders Vote for More Water," *Twin Falls Daily News*, January 19, 1920

275 Finney to Secretary of the Interior, American Falls Project, November 3, 1921. Folder: Reclamation Bureau, Box 1614, RG 48, NARA II.

600,000 acre-feet with less than one-quarter of the storage capacity. The economy of scale was significant.[276]

In March of 1920, Director of Reclamation A. P. Davis and Idaho Governor Davis designated a six-member joint federal-state Board of Engineers under the chairmanship of I. W. McConnell, a consulting engineer with the firm of Stone & Webster of Boston, Massachusetts, one of the country's foremost electrical engineering and construction companies. As a former supervising engineer for the Reclamation Service in Colorado and Wyoming (1905-1909) and vice-president and general manager of the Idaho Irrigation Company (1910-12), a company formed for the purpose of reclaiming 167,000 acres of land on the Big Wood River under the Carey Act, McConnell was familiar with water supply issues in Idaho, the American West, and on the Snake River specifically.[277] Other members of the Board of Engineers had equally distinguished careers in irrigation and reclamation in Idaho and other western states, with all having worked for the state, the Reclamation Service, or private irrigation interests in the Snake River Basin. Two of the engineers had worked for the Reclamation Service on the Payette-Boise Project. F. W. Hanna was the former project engineer of the Payette-Boise Project and had been responsible for construction of the original canal distribution system. After World War I, he worked with Secretary of the Interior Franklin Lane in the colonization program to place returning veterans on reclamation projects and in the late 1920s was appointed chief engineer of the Bureau of Reclamation.[278] James Munn was also a Reclamation Service engineer and had been superintendent of construction on Arrowrock Dam (1911-1915), a 348-foot tall concrete structure located twenty miles upriver from Boise. When completed in 1915, Arrowrock Dam was the tallest dam in the world.[279] Two other members of the board had experience with private canal companies in the Twin Falls area. E. B Darlington had been chief engineer for the Twin Falls North Side

276 A. P. Davis to Secretary Lane, January 16, 1920. Folder: Reclamation Bureau, Minidoka Project, American Falls Reservoir, Idaho State Water Rights – General, January 16, 1920 to July 9, 1925, Box 1616 8-3, A-B, CCF 1907-1936, RG 48, NARA II.

277 "Alumni Notes," *The Cornell Civil Engineer* (1913), 342.

278 Franklin K. Lane, "Work and Home for Soldiers," *The Rotarian*, 14:4 (April 1919), 152-55; Bill G. Reid, "Franklin K. Lane's Idea for Veterans' Colonization, 1918-1921," *Pacific Historical Review* 33:4 (November 1964), 447-461.

279 William J. Simonds and Brit Storey, *The Boise Project* (Denver: Bureau of Reclamation, December 2009), 12-21.

Land & Water Company before becoming superintendent of the Minidoka Project (1923-1933). John C. Wheelon had been associated for a number of years with the Carey Act projects at Twin Falls and in 1920 was manager of the Twin Falls Canal Company, water supplier to the Twin Falls South Side Project, one of the most successful Carey Act projects in the West.[280] The final member of the board was a representative of the State of Idaho, W. G. Swendsen, Idaho's first Commissioner of Reclamation and head of the Idaho Department of Reclamation.[281]

The board was organized to consider fundamental issues related to the proposed American Falls Reservoir:

- The number of acre-feet available in the Snake River water supply.
- The amount of storage space that may be available and contracted for by various irrigation agencies.
- The size of the American Falls Reservoir necessary to meet projected irrigation and other demands.
- Other power and irrigation interests in the Snake River Valley that might affect the American Falls Project.

At the first meeting of the Board of Engineers on April 5, 1920, Swendsen asserted that the State of Idaho was interested in "the formation of a definite plan for the development of the remaining resources of the Snake River water supply on a broad and comprehensive basis which would insure to the state the maximum utility of the possibilities of the stream." The board decided that wise "conservation" of the waters of Snake River was the principal solution to water supply problems in the Snake River Basin.[282]

---

280 Mark Fiege, "Creating a Hybrid Landscape: Irrigated Agriculture in Idaho," in Dale D. Goble and Paul W. Hirt, Northwest Lands, Northwest People: Readings in Environmental History (Seattle: University of Washington Press, 1999), 372; Andrew Jensen, comp., Latter-Day Saint biographical encyclopedia : a compilation of biographical sketches of prominent men and women in the Church of Jesus Christ of Latter-Day Saint, Volume IV (Anne Arbor, Michigan: University Microfilms, facsimile print, 1972), 173.

281 W. G. Swendsen, *First Biennial Report of the Department of Reclamation State of Idaho, 1919-1920* (1921), 4, 8-9; and George N. Carter, *Fourth Biennial Report of the Department of Reclamation State of Idaho, 1925-1926* (1927), 4.

282 Board of Engineers to Honorable D. W. Davis, Governor of the State of Idaho, April 10, 1920, 1-3. No Folder, Box 728 Minidoka 301.1, General Files, 1919-1929, General Administrative & Project Correspondence, 1919-1945 (GAPC, 1919-1945), RG 115, NARA-Denver.

They concluded that the "greatest use of the water of Snake River will be found in the dedication of the entire flow of the stream to irrigation in so far as the water can be economically appropriated." To this end, the board suggested building a reservoir with the requisite storage capacity to capture, control and make beneficial use of "a large percentage of all the waters originating in the watershed above Milner Dam" for irrigation purposes. The limiting factor in reaching the full agricultural potential of the basin was water supply, concluded the board, as there was an excess of irrigable land available to use whatever water was conserved. "The net effect of this condition," wrote the board, "will be to dry up the river below Milner Dam during the irrigation season, also to as great an extent as possible below American Falls Reservoir, during the non-irrigation season." Below Milner Dam the primary use of water was for generation of power, as diversions for irrigation from points lower in the canyon were not feasible. "The principle involved therefore," concluded the board, "is to secure as nearly as possible a total use of the waters for irrigation above Milner Dam, and to secure the greatest possible use of power below Milner Dam."

Although these two objectives appeared to be in direct conflict, the board noted that "large accretions to the stream below Milner Dam," in part due to return flow from irrigation upstream, to a large extent restored the water supply for power use at Upper Salmon Falls. Therefore, the board concluded if power was constrained on the Snake River to the reaches below Salmon Falls that injury to future development by power interests was "relatively not very great."[283]

The board compiled data to equate the available water supply of Snake River at American Falls with the needs of the land susceptible to irrigation from the Snake River.

The total quantity of water needed for all new and old irrigable areas far outweighed the quantity available even with a 3,000,000 acre-foot reservoir at American Falls. The irrigation drafts required to water the areas considered irrigable from the Snake River within proposed irrigation projects, according to statistics compiled by the Board, are shown in **Table 4.1**.

283 Ibid., 5-6.

IPC, unfortunately, had projected a power plant at Twin Falls and was in the process of preparing an application for a state water right permit and a Federal Power Commission license at that very time. Proposing construction of a hydroelectric power plant so close to the diversion dam at Milner invited conflict with the State of Idaho, local irrigation interests and the Interior Department, all of whom protested issuance of a license to the private power company.

Table 4.1. Irrigation Drafts on Available Water Supply at American Falls for All Old and New Areas[284]

Project	*Natural Flow (acre-feet)*	*Storage (acre-feet)*	*Total (acre-feet)*
Old			
North Side Twin Falls	613,000	600,000	1,213,000
South Side Twin Falls	762,000	200,000	962,000
Minidoka	325,000	350,000	675,000
Idaho Falls district	---------	450,000	450,000
North Fork district	---------	200,000	200,000
TOTAL	**1,700,000**	**1,800,000**	**3,500,000**
New			
Mindoka Pumping			520,000
Hansen Butte			110,000
Wood River			200,000
Bruneau			2,000,000
Mountain Home			2,000,000
Dubois			900,000
Idaho Falls Pumping			450,000
Shelley-American Falls			500,000
TOTAL			**6,160,000**
TOTAL OF OLD AND NEW PROJECTS COMBINED			**9,660,000**

The Board of Engineers based its study of water supply requirements on an assumed irrigation draft of 5,720,000 acre-feet on the available supply at American Falls, with a total reservoir capacity of 3,900,000 acre-feet distributed among the upper basin storage reservoirs as follows: 850,000 acre-feet at Jackson Lake; 80,000 acre-feet at Lake Walcott; 200,000 acre-feet at Bechler Swamp; and 2,800,000 acre-feet at the proposed American Falls Reservoir. For this draft and reservoir capacity, over a 24-year period of record (1896-1919) there would have been shortages for all users in 1902, 1903, 1905 and 1906 ranging from 12 to 20 percent. Increasing the amount of storage by 75 percent had

284 Ibid., 13.

little effect on water shortages. Only two of the water-short years would be eliminated by providing reservoir capacity to the extent of 6,800,000 acre-feet. To eliminate shortages altogether would require storage capacity to the extent of 9,800,000 acre-feet.[285] Under the assumed draft and storage for the board's study, there would be water wasted at American Falls in only seven of the twenty-four years, and no waste whatsoever from 1896 to 1907, inclusive, based on an empty reservoir at the beginning of 1896. The total "waste" over the dam at Milner, without any allowance for power, during the seven seasons with surplus water would have been 11,190,000 acre-feet.[286]

A number of the Carey Act irrigation projects, including McConnell's own Idaho Irrigation Company on Big Wood River, had grossly miscalculated the amount of water required to supply their customers. Nevertheless, irrigation companies ruthlessly competed for homesteaders and continued to sell Carey Project land while promising to deliver water from a supply that already had been exhausted. As a result, many of those who had contracted for water with irrigation companies were chronically short of water and under their contracts their supply was further diminished by having to share proportionately in any shortages.[287] In determining how to allocate the available supply for irrigation purposes, the board preferred to give preference to supplementing the present inadequate water supply of existing irrigation projects. The total estimated draft for lands under irrigation, including supplemental supplies that they needed from storage, was 3,500,000 acre-feet. Deducting the estimated current and supplemental draft (3,500,000) from the total assumed maximum irrigation draft of 5,720,000, there remained 2,220,000 acre-feet for use in the new developments shown in **Table 4.1** wherever conditions dictated its most beneficial use.

To achieve the goal of maximizing use of the waters of the Snake River, the board was unanimous in concurring that a large dam at American Falls was necessary. How large a dam depended upon assumptions about the ultimate goal of the water storage project. The plan of the Reclamation Service was to build a dam that could

---

285 Ibid., 14.

286 Ibid., 13-15.

287 *Idaho Irrigation Company, et al. v. Gooding, et al.*, 265 US 518 (1924); Hugh T. Lovin, "Duty of Water in Idaho: A "New West" Irrigation Controversy, 1890-1920," *Arizona and the West,* 23:1 (Spring 1981), 5-28.

eventually be raised to achieve its "full capacity" of about 3,000,000 acre-feet. The Board of Engineers supported this approach. As of 1920, there were about 1,300,000 acres of land in Snake River Valley irrigated with water originating from the Snake River and its tributaries above American Falls. The board concluded that an ample water supply to supplement the existing needs for this acreage, plus an additional area of new land amounting to approximately 400,000 acres could be served from the proposed American Falls reservoir, operated in conjunction with existing water storage facilities, with a maximum shortage in any one year of 20 percent.[288]

The Board of Engineers noted that construction of a large dam at American Falls to be used primarily for storage of water for irrigation would eliminate the falls at the dam site as a source of firm hydroelectric power. Development of the reservoir for irrigation water supply would entail shutting off winter flood flow that formerly passed unimpeded through the American Falls powerhouses and down the river. The availability of a firm power supply was further complicated by the fact that diversions for irrigation at Minidoka or Milner dams were prior rights over the major power rights located lower downstream. Consequently, under natural flow conditions the power sites below Milner Dam reached the condition of lowest flow, and therefore their period of least firm power production, in the summer season, when practically all of the water was taken at Milner Dam. On the other hand, accretions occurred to the stream below Milner Dam from springs and irrigation return flow, to a slight extent at Twin Falls, and to a substantial extent between Shoshone Falls and Upper Salmon Falls. Therefore, the Board of Engineers concluded that storing winter stream flow at American Falls Reservoir to maximize irrigation uses would to a large extent eliminate flows useful for generation of power above Upper Salmon Falls, but at falls lower down the river American Falls storage would have less effect on the firm power supply. The development of Upper Salmon Falls and power facilities sited below this point on the river would leave the largest residue of water for

288 To calculate the available supply to meet irrigation requirements on these lands the Board looked at historic run-off records. These records suggested there was a minimum annual storable run-off available at Jackson Lake of 452,900 acre-feet and the mean run-off 958,800 acre-feet. Jackson Lake had a storage capacity of about 850,000 acre-feet. The minimum annual run-off at American Falls for the period of record, 1896 to 1919, was 4,147,000 acre-feet and the mean 6,656,000 acre-feet. Ibid., 7-8. See also *Nineteenth Annual Report of the Reclamation Service, 1919-1920*, 417-418.

irrigation in American Falls Reservoir, as these power plants, as conceived under current plans, required no supplementary flow from American Falls at any time of the year.[289]

Although the board favored restricting development of power sites on the Snake River to sites below Shoshone Falls, the long-term effect of irrigation return flows on potential water supplies for power production below Milner Dam was not predictable with any accuracy. Since it would take a long time for irrigators to make full use of the waters of Snake River for irrigation, the board cautioned that if additional power rights were granted in the interim then some surrender clause needed to be included in federal permits or grants of state water rights for power uses that subordinated these uses to future irrigation demands.

> Since the complete utilization of the Snake River water supply above Milner Dam for irrigation purposes will, under normal conditions require a considerable length of time, it is possible that power plants constructed in the near future for pumping purposes may, until such time as irrigation developments are complete, be used to supply commercial winter power, and by reason of such possible use winter power rights that initiate would mitigate later against a complete use of this water resource for irrigation purposes. In granting power rights in the future the Federal Government and the state should as far as possible provide restrictions requiring its eventual surrender when and as the waters are required for application to the land.[290]

According to the analysis completed by the joint federal-state Board of Engineers in April 1920, the water supply in the Upper Snake River Basin was not sufficient for the development of all the irrigable lands that could be served by diversions above Milner Dam. In addition to existing and authorized projects, two proposed irrigation projects – the Bruneau Project and the Mountain Home Project – contemplated the irrigation of about 200,000 more acres west of Twin Falls with Snake River water diverted above Milner. The board did not state an opinion on the economic or engineering feasibility of either project, but did note that sufficient water existed given its assumptions on irrigation drafts and reservoir capacity to provide an adequate water supply to a Minidoka pumping project of about 115,000 acres (North Side Pumping

---

289 Ibid., 5-6. This statement was made prior to the Idaho Power Company's FPC applications for facilities at Twin Falls and Upper Salmon Falls.

290 Ibid., 31.

Division) and furnish water for only one of the two big projects at Bruneau or Mountain Home.[291] The conclusions of the board with respect to reservoir size and irrigated acreage to be served by the American Falls Reservoir were largely embraced in the state appropriative water rights filings made by the Reclamation Service for the proposed new reservoir and associated irrigation developments in March 1921.[292]

During the remainder of 1920 and through 1921 there was a flurry of activity to make contractual arrangements for storage rights in the American Falls Reservoir between the Reclamation Service and individual canal companies, irrigation districts and other prospective "spaceholders" who wanted to store water for irrigation uses in the new reservoir. The Reclamation Service dispersed cost estimates on reservoirs of various sizes to individuals, canal companies and irrigation districts which might have desired to reserve storage space in the proposed reservoir and distributed sign-up forms to be used by applicants to reserve space.

By the end of 1920, as indicated in **Table 4.2** below, contracts had been signed with thirty-three individuals, ditch and canal companies and irrigation districts for a total of 359,020 acre-feet of storage space in the proposed American Falls Reservoir.

Table 4.2. List of Entities Contracting for Storage Space in American Falls Reservoir in 1920[293]

Name of Contractor	*Acre-feet*
Frank Paree	100
Blackfoot Irrigation Co	20,000
New Lavaside Ditch Co. (Ltd.)	4,000
Burgess Canal & Irrigating Co.	5,120
Lenroot Canal Co. (Ltd.)	9,000
Sunnydell irrigation district	8,000
Rudy Irrigation Canal Co (Ltd.)	2,000

291 Ibid.

292 State of Idaho, License and Certificate of Idaho State Water Right, Water License No. R-269; Idaho State Idaho State Water Right 01-2064 Back File; and State of Idaho, License and Certificate of Idaho State Water Right, Water License no. 15134; Idaho State Water Right 01-2040 Back File.

293 "General Statement concerning the American Falls Reservoir," prepared by Ottamar Hemele, November 18, 1924, 1-2; Folder 301.1 - MP Engineering & Construction, Dams, Box 729 Minidoka 301.1, GAPR, 1919-1945, RG 115, NARA-Denver. See also U.S. House, Subcommittee of House Committee on Appropriations, *Interior Department Appropriation Bill, 1925*, 68th Congress, 1st session (December 28, 1923).

People's Canal & Irrigation Co.	16,000
Antoine Poitevin	1,500
F. H. Churchill	500
Harrison Canal & Irrigation Co.	10,000
Enterprise Canal Co. (Ltd.)	6,000
Idaho irrigation district	25,000
M. Eugene Holt	150
Butte & Market Lake Canal Co.	9,000
Corbett Slough Ditch Co.	8,000
Minnie Schodde	900
E. H. Adams	500
Jane Carlsworth	500
W. H. Harris	10,000
W. J. Kramer	1,500
C. W. Mattson	2,000
C. R. Mattson	500
Lawton Mattson	500
C. R. Rand	1,000
Idaho Irrigation District	9,000
New Sweden irrigation district	25,000
Snake River Valley irrigation district	20,000
Murtaugh Canal Co.	20,000
Twin Falls Canal Co.	100,000
Martin Canal Co.	750
Woodville Canal Co.	5,000
Aberdeen-Springfield Canal Co.	37,500
TOTAL AMOUNT OF STORAGE CONTRACTED	**359,020**

In addition, the Minidoka Project North Side Pumping Division would require about 520,000 acre-feet, and additional supplemental water for irrigated tracts in the lower and upper sections of the Snake River Plain was estimated by the Reclamation Service at another 500,000 acre-feet. Thus, the minimum size of a reservoir capable of holding over in storage at least 1,380,000 acre-feet seemed to be justified by current demand alone. The Reclamation Service expected additional canal companies and irrigation districts to sign up for storage before a final decision on reservoir size was made. In the meantime, the Reclamation Service went forward with surveys, soil tests, acquisition of land for the dam and reservoir site, and selected and purchased a tract of land for relocation of the town of American Falls to replace the existing townsite which was to be submerged by the proposed reservoir.[294]

Farmers in the Snake River Valley continued to struggle in the wake of the terrible drought of 1919. In the latter part of 1920, a slumping national economy and low prices for farm products wiped out any profits farmers in southern Idaho hoped to make that year. During 1921 low farm prices once again compounded the continuing recession in rural Idaho without any recovery in sight. Irrigation and canal districts were unable to meet financial obligations due on their contract payments with the Reclamation Service for the new reservoir. Only a few of the more than thirty subscribers to the American Falls Dam made the advance payment due in February 1921.[295] The New Sweden Irrigation District and the Snake River Valley Irrigation District preserved their initial contracts, of 25,000 acre-feet and 20,000 acre-feet, respectively. The Murtaugh Canal Company also retained its original contract, which included the Milner Low Lift Irrigation District as an assignee, subscribing for 20,000 acre-feet.[296] Failure of others to pay subscriptions for storage space in the proposed reservoir continued into the following year. Financial conditions for farmers in the region did not show significant signs of improvement until 1923 when crop yields were good and prices for most farm products brought a fair profit.[297]

Elected in 1918, Idaho Governor Davis, a resident of American Falls, fought hard to revive Idaho's reclamation boom of the 1910s during his four years in office, and thereafter as the Commissioner of Reclamation (1923-24). New reservoir water was central to his program to strengthen the agricultural base of the state's economy. The year 1920 was a presidential election year and the work of the Western States Reclamation Association and its thirteen statewide affiliates had caused both major parties to include specific plans for a comprehensive policy on western reclamation. With the West in position to make

294 *Nineteenth Annual Report of the Reclamation Service, 1919-1920*, 157, 417-419; *Twentieth Annual Report of the Reclamation Service*, 1920-1921 (GPO, 1921), 157.

295 *Twentieth Annual Report of the Reclamation Service, 1920-1921*, 60-61.

296 "General Statement," 2-3. Folder 301.1, Box 729, GAPR, 1919-1945, RG 115, NARA-Denver. See also U.S. House of Representatives, Subcommittee of House Committee on Appropriations, *Interior Department Appropriation Bill, 1925*, 68th Congress, 1st session (December 28, 1923).

297 U. S. Bureau of Reclamation, *Twenty-third Annual Report of the Bureau of Reclamation* (GPO, 1923), 55-61.

certain political demands, important congressional committees concerned with public lands in the western states were induced to tour the agricultural districts of selected western states, including Idaho. The House Committee on Appropriations visited the federal irrigation projects between Boise and American Falls in July 1920. Governor Davis welcomed the committee members when they entered the state and traveled with them as they toured the reclamation projects. Local newspapers and the Idaho Reclamation Association believed the exposure of the committee members from mid-western, eastern and southern states to the advancements of irrigated agriculture over the previous fifteen years enhanced the state's efforts to secure funding for federal storage reservoirs in the Upper Snake River region. Senator Warren G. Harding's political support for western reclamation and his subsequent selection as the Republican Party's nominee for the presidency of the United States in the fall elections gave greater reassurance to the boosters of the American Falls reservoir.[298]

The image of Harding as a staunch supporter of western reclamation began to fade by the late summer of 1921 when he announced his opposition to further appropriations for reclamation development, including the reclamation bill for $250 million to fund a five to ten year development program sponsored by Oregon Senator Charles L. McNary and backed by the Western States Reclamation Association, which had been favorably reported out of the Senate Arid Lands Committee. Completion of American Falls Reservoir involved a huge financial commitment from the federal government, as well as the people of the Snake River Basin. Just as preliminary work was being undertaken to clear the reservoir site and acquire rights of way and flowage easements, the revolving fund that provided the financial assets for

---

298 "Idaho's Foremost Booster," *Franklin County Citizen* (Preston, ID), July 29, 1920; "Harding Stand on Reclamation Pleases Idaho," *Twin Falls Daily News*, September 2, 1920; "Stand By Program on Reclamation," *Franklin County Citizen* (Preston, ID), October 14, 1920. Davis served as governor of Idaho from January 6, 1919 to January 1, 1923. Governor Davis' status as a leader in the western reclamation movement led the Idaho state legislature and Senator Borah to endorse Davis for the Interior Secretary position in the Harding Cabinet. The endorsement was transmitted to the legislatures of all the western states and to Republican legislators in Congress. Although Davis lost his bid to become Interior Secretary, in March 1923 Secretary Albert B. Fall appointed Davis special assistant Secretary of the Interior to supervise the U. S. Reclamation Service. On June 18, 1923, the office of the Director of Reclamation was abolished and Davis became the first Commissioner of Reclamation on July 1, 1923. He left office on April 2, 1924. "Legislature Endorses Governor for Cabinet," *Twin Falls Daily News*, January 5, 1921; List of Reclamation Commissioners: Davis W. Davis, available at http://www.usbr.gov/history/CommissBios/ddavis.html, last accessed December 10, 2022.

continuing reclamation construction projects was breaking down and reclamation interests were forced to seek direct congressional appropriations through specific congressional bills to fund individual projects.

Non-payment by settlers on federal reclamation projects and declining receipts from public land sales contributed to the funding problem..[299] The former was particularly acute on the Minidoka Project because of the recent drought and collapse of crop prices. In 1921 Idaho Senator William E. Borah sought legislation to forgive the debt of settlers on the Minidoka Project. Borah's move incurred the wrath of Interior Secretary Albert B. Fall, who sought to have the federal outlay repaid by settlers, and imperiled the American Falls Reservoir project.

With the looming possibility of losing an opportunity to partner with the federal government to build American Falls Reservoir, Governor Davis invited representatives of the various contracting companies and districts to a conference to discuss funding for the reservoir. After extended discussion, a plan was formulated to save the project through the creation of a super district embracing the whole of the Upper Snake River from Rigby to Twin Falls. The sole purpose of the "American Falls Reservoir District," as it became known, was to finance the reservoir project; the district's large size made it easier to float securities on the market and obtain a lower interest rate than possible for smaller companies, districts or individuals.

Many of the canal companies and irrigation districts of the upper valley were reluctant to join an organization that would be dominated by larger lower valley interests. Efforts to convince them that the "reservoir district" was only a financing organization that would not in any manner interfere with operation and maintenance of any canal system within its boundaries failed and only a couple of districts in the Pocatello-Blackfoot area above the dam were convinced to join in the formation of the spaceholders organization known as the "American Falls Reservoir District." Before the end of 1921, a petition for the organization of American Falls Reservoir District had been signed by 1,388 persons, representing 142,000 acres of the more than 500,000 acres promoters had hoped to include in the proposed district. Most of

---

299 *Nineteenth Annual Report of the Reclamation Service, 1919-1920*, 43; and *Twentieth Annual Report of the Reclamation Service*, 1920-1921, 11.

the shareholders represented land in the lower part of the proposed service area around Twin Falls.[300]

On January 16, 1922, voters approved formation of American Falls Reservoir District, a district that included only five canal companies but represented about 400,000 acres, to secure 300,000 acre-feet of storage in the proposed American Falls Reservoir. The contract between this district and the United States was authorized on May 12, 1922 and bonds valued at $2.7 million were issued. This contract and similar ones with four other irrigation districts, amounting to 175,000 acre-feet, were executed by June 15, 1923.

Table 4.3 below lists the district names, sizes, and quantity of contracted water.

In addition to the water contractually allocated to the reservoir and irrigation districts, under the federal reclamation plan in place in 1923, the Reclamation Service was to obtain 520,000 acre-feet of water in the reservoir for use on its Minidoka Extension Project, serving approximately 120,000 acres.[301] Thus, only 995,000 acre-feet of storage space was needed to meet the needs of committed contractors and the Minidoka Project, far less than the proposed reservoir of 3,000,000 acre-feet. However, with the big reservoir district and a few other sizable irrigation districts having signed contracts in 1922, the Reclamation Service believed many smaller canal companies and irrigation districts in the Snake River Valley that required some storage water to supplement their present water rights would soon sign storage contracts. Ultimately, thirty-nine entities with diversion facilities on the Snake River between American Falls and Milner Dam participated in the American Falls Reservoir construction project.[302]

Table 4.3. Irrigation Districts Contracting with the United States for Construction of American Falls Reservoir, June 15, 1923[303]

300 "Representatives of 13 Western States Convene," *Twin Falls Daily* News, December 10, 1920; *Twenty-first Annual Report of the Reclamation Service, 1921-1922* (GPO, 1922), 64. See also *Twin Falls Daily News*, May 7, 1922 and August 27, 1922.

301 *Twenty-second Annual Report of the Bureau of Reclamation* (GPO, 1923), 65-66.

302 *Twenty-third Annual Report of the Bureau of Reclamation, 1923-1924* (GPO, 1924), 29. See also W. G. Swendsen, Idaho Commissioner of Reclamation, to Governor C. C. Moore, July 9, 1923. Folder 6: Reclamation, Department of: American Fall Dam, 1923, Box 3, AR 2/13, Governor Papers – C. C. Moore, ISHS.

303 W. G. Swendsen and D. C. Henny to Secretary of the Interior, November 10, 1923. Folder 6, Box 3, AR 2/13, ISHS.

Name of District	*Land Area in acres*	*Water Contracted for, in acre-feet*
American Falls Reservoir District	407,000	300,000
Empire Irrigation District	90,000	110,000
New Sweden Irrigation District	24,000	25,000
Snake River Valley Irrigation District	25,000	20,000
Milner Low Lift Irrigation District	9,400	20,000
TOTAL	555,400	475,000

Albert B. Fall, who was Secretary of the Interior under the Harding Administration from 1921 until his resignation in 1923, had deep misgivings about going forward with the American Falls Reservoir because of the inability of farmers on the Minidoka Project and Warren Act contractors on private projects to make payments on their obligations to the Reclamation Service. In addition, Fall was irritated at Idaho Senator William E. Borah, sponsor of the bill in Congress to forgive settlers on the Minidoka Project for the $600,000 that they were in arrears on their repayment obligation to the General Reclamation Fund. Secretary Fall was content to let the American Falls Reservoir project die by default. The chosen instrument was the contract with IPC which had been negotiated over the three previous years and wherein the government agreed to purchase the company's rights at American Falls within a stipulated period. The unsigned document was set to expire on January 13, 1922.[304]

The Idaho Power negotiations

In addition to obtaining private funds from spaceholders to share in the construction cost of American Falls Reservoir, the Reclamation Service was faced with pressure to reach contractual agreement with Idaho Power Company to obtain certain rights and property essential to prosecution of the reservoir project. The agreement required acceptance of the contract by the Secretary of the Interior and the State of Idaho, although the latter party was not a signatory to the contract.[305]

304 Rockwell, *Saga*, 5, and 23-27.

305 During the term of the negotiations with IPC four different people served as Interior Secretary: Franklin K. Lane (January 3, 1913 to March 1, 1920), John Barton Payne (March 15, 1920 to March 4, 1921), Albert B. Fall (March 5, 1921 to March 4, 1923) and Hubert Work (March5, 1923 to July 24, 1928).

The initial discussions between the Reclamation Service and IPC regarding negotiations to adjust the property rights and privileges of the latter's power plants and other rights at American Falls before the United States could build a storage reservoir at the site were initiated by the Reclamation Service in 1918 during the administration of Woodrow Wilson.

Since about 1910 the Reclamation Service had been studying the inter-relationship between water uses for irrigation and power on Snake River and had determined that IPC's power rights at American Falls were a key to future development of water resources on the river for irrigation purposes. In fiscal year 1919 the addition of a 5000 kW unit at Minidoka powerhouse was authorized. Although the proposed powerhouse expansion was later suspended, the Reclamation Service entered into negotiations with IPC to supply the service with power temporarily to meet its existing shortfall. In the process, the Reclamation Service learned that the power company had excess electrical capacity at American Falls. Reclamation was already considering the possibilities of building a storage reservoir and power plant at American Falls, and now contemplated immediate acquisition of IPC's two old power plants on the west side of the river which had a combined capacity of 1500 to 1800 kW.[306] On December 7, 1918 Interior Secretary Lane in conference with Reclamation Director Arthur P. Davis and Chief of Construction F. E. Weymouth, agreed to send a letter to Daniel F. McGee, vice-president of IPC, suggesting certain engineering studies for use in negotiations for the purchase of the American Falls power plants.[307]

The letter containing the first tentative proposition of the government was sent on December 10, 1918. Between the first proposal and the ultimate execution of the contract four and one-half years intervened during which time the two contracting parties and numerous other interested participants negotiated the complex agreement to share the site. Reflecting a decade later on the difficult decision to give up a portion of its property at American Falls, W. R.

---

306 Barry Dibble, Minidoka Project Manager, to Chief Engineer, December 19, 1921 and Chief Engineer to Director, December 20, 1921. U. S. Bureau of Reclamation, Region 1 Office, Boise, ID (USBR-Boise); and *Eighteenth Annual Report of the Reclamation Service, 1918-1919* (GPO, 1919), 414.

307 Hubert Work, Secretary of the Interior, to Senator William E. Borah, March 24, 1924. Folder: Reclamation Bureau, Box 1614, RG 48, NARA II.

Putnam, vice-president and general manager of IPC, wrote to Senator Borah:

> Much has been said relative to the consideration which the Power Company received for that portion of its property rights and privileges at American Falls sold to the United States. All the Company's rights and privileges at that point were extremely valuable, and the sale of any part represented a sacrifice for which money considerations would and did only but partly compensate the Company. It must be remembered that American Falls power site represented then, as it does now, a key situation so far as the electric servicing of the Southeastern part of Idaho is concerned.[308]

The power company wanted to hold on to its property, but considering the strategic location for a storage dam, it became evident that the United States had determined American Falls was necessary for its irrigation program, and the best the power company could hope for was "a give and take proposition." The government's interest in the site was clearly laid out in Paragraphs 2, 4, 5, 6, 7 and 9 of the contract. What the agreement sought to achieve from the perspective of the power company was expressed by Putnam in his letter to Borah:

> From the beginning a plan was sought to be evolved wherein and whereby that conflict could be avoided and the two vastly important interests, to-wit, irrigation and power, cooperate in effecting the larger use and the greater duty at the same strategic point. ... [F]or the government to successfully put through its proposition and accomplish the primary end which it had in view, it was necessary that these rights of the company at American Falls be adjusted for the mutual interests of the immediately contracting parties.

Adjusting the valuable rights and privileges of the contracting parties involved complex legal, engineering and economic issues. The two parties were only able to work together successfully, maintained Putnam, because each party kept in the forefront throughout the negotiations that the domains of the contracting parties were separate and distinct. The Bureau of Reclamation observed the distinction between the generation of power for irrigation purposes by the government for the Minidoka Project only on the one hand, and the

308 Putnam to Borah, August 8, 1928, 1. AFF001880, American Falls Power Situation, List of Various Reports & Letters, Idaho Power Company Records.

generation of power for commercial purposes by Idaho Power Company on the other hand. As for the power company, it sought to develop its commercial power business to the fullest extent possible "consistent with the primary use of water for irrigation."[309]

Into the early months of 1919 discussions with McGee, other IPC officials, and the New York engineering and legal staff of the Electric Bond & Share Company continued at the latter company's New York headquarters. Among the key participants for the power company were Putnam and McGee and several individuals who were largely responsible for writing the contract: F. B. Odlum, J. J. O'Connell, and A. C. Clogher.[310] The government proposed not only to build a reservoir at American Falls "for the full development of the waters of the Snake River for irrigation purposes" but also to "destroy," wholly or in part, "the commercial value" of IPC's power plants at American Falls. Ultimately, the government planned to replace the two West Side plants with a large 30,000-kilowatt, four-unit, government-owned and operated hydroelectric power plant primarily designed to provide electrical power during the irrigation season for pumping water to irrigated tracts on the Minidoka Project, specifically the North Side Pumping Unit, later to become A & B Irrigation Districts. The government could balance its electrical load by supplying electricity for domestic heating and lighting purposes on farms and in the reclamation project towns within the Minidoka Project.[311]

At the joint meetings in New York, the parties worked out some of the general terms for a contract which was submitted to a board of engineers at a subsequent meeting in Burley, Idaho. The board was comprised of key Bureau personnel who were familiar with the Minidoka Project: A. J. Wiley, a Bureau consulting engineer; F. A. Banks, Bureau construction engineer; Barry Dibble, Minidoka Project superintendent; James M. Gaylord, a Bureau electrical engineer from Denver; and B. E. Stoutemyer, the Bureau's district counsel. The proposed contract was to provide IPC a substitute power site at Shoshone Falls equivalent to American Falls with a transmission line

---

309 Ibid., 3-6.

310 Clay Boswell, Electric Bond & Share Company, to W. H. Putnam, November 3, 1927. AFF001839-41, American Falls Power Situation, List of Various Reports & Letters, Idaho Power Company Records.

311 F. E. Weymouth to Idaho Power Company, July 5, 1919. Folder 301.1, Box 728, GAPR, 1919-1929, RG 115, NARA-Denver.

from the Shoshone Falls plant to American Falls. In addition, the plan called for the United States to provide for 210,000 acre-feet of storage at American Falls and a 200,000 acre-foot regulating reservoir at Twin Falls in order to provide 8,000 cfs of flow at Shoshone Falls. Meeting this flow requirement would have required the release of about 1,000,000 acre-feet per year at Milner Dam, denying an irrigation water supply to about 200,000 acres of land in the upper basin. The United States would gain full control over the American Falls site, but would provide free power to IPC for a period of ten years. The United States would also pay $1,000,000 to extinguish all of IPC's power rights at the American Falls site.[312]

Perhaps the drought in southern Idaho caused some of the Bureau's representatives to question the wisdom of the proposed contract. By June 1919 at least two of the board members withdrew their previous support. A. J. Wiley wrote to the Bureau's Chief of Construction that Upper Salmon Falls, located more than thirty miles downstream of Shoshone Falls, would be a better replacement site for the IPC power development because it would not require any storage of water in American Falls Reservoir, nor would it require wasting any water past Milner Dam. Wiley also recalculated the amount of storage required at American Falls to make the output of the Shoshone plant comparable to IPC's American Falls plant; his independent calculation increased the amount of storage required from 200,000 acre-feet to 300,000 acre-feet in an average year and as much as 475,000 acre-feet in a dry year like 1906. District Counsel Stoutemyer also voiced similar opposition, adding that Idaho Governor Davis, who had not seen the proposed contract, was certain to object to the permanent loss of water for irrigation use to power development downstream. Furthermore, Stoutemyer noted when the proposal became public it would raise a violent protest all over the southern portion of Idaho.[313]

In the meantime, on June 10, 1920, the Federal Water Power Act became law and only seven days later IPC filed Application Nos. 18 and 19 with the Federal Power Commission for Upper Salmon Falls power site, as well as the Twin Falls site which was located only a short distance below Milner Dam. Both of these sites were on land

312 William Burpee, Interior Department Solicitor, "Informal Notes on the 1923 American Falls Contract with Idaho Power Company," (1962), 52-53. AF006222- AF00681, 05-14-2010 USBOR Response to IPC, AGO 06-17-2010-1, CD, Idaho AGO.

313 Ibid.

previously withdrawn by the United States for power purposes under the Reclamation Act in 1904. At about the same time, on July 1, 1920, Congress appropriated funding for the construction of the American Falls dam and power plant.

The United States continued its negotiations with the power company over the American Falls site. On July 5, Weymouth requested that McGee respond to three alternative written offers by the United States for the power company's property and rights at American Falls by making proposals of the amount of money it would accept for each proposition. Under the proposition favored by Weymouth, the United States would build a new reservoir and completely extinguish IPC's power right at American Falls by building a new government power plant or plants, and supply the load carried by IPC's current American Falls plants with any surplus being delivered to the company's transmission lines for its use free of charges for a period of 10 years. The second alternative provided the same proposal as the first, but offered the company power sites at Shoshone and Twin Falls with enough natural flow and storage to guarantee a perpetual flow of 1,000 cfs. The offer also included joint use of a 130,000-volt transmission line from American Falls to Minidoka. The final alternative offered all of alternative two, except the water would be provided not perpetually, but only until needed for irrigation.

Weymouth made clear in his correspondence with McGee that he found alternatives two and three undesirable. Alternative two would be most objectionable to irrigation interests because it would dedicate otherwise storable flows of the Snake River for exclusive power use below Milner Dam, and thereby condemn thousands of acres of irrigable desert land to perpetual aridity. Weymouth estimated that such a commitment of water for power generation could permanently reduce the potential population of the Snake River Valley by 16,000 to 18,000 residents. He anticipated that local communities would oppose such a solution and, he noted, the reduction in irrigated acreage and number of farms would rob the power company of a valuable future power market. He advised against this course of action. The third proposition was similar to the second, but was somewhat less objectionable because it provided for recapture of water in the future for irrigation purposes, but how this was to be accomplished was undecided.

As a counterproposal to all of the U. S. alternatives, McGee suggested that any American Falls power plant built by the government in connection with its dam be leased to IPC with power requirements of the United States being supplied by the company free of charge for 10 years; IPC was to be compensated by the surplus power which it could deliver to its own transmission lines and use for its own purposes. The United States objected because the reservoir as a Reclamation Service facility had to be operated in the interests of irrigation, not the generation of surplus power for the benefit of a private power company. [314]

Before entering into further negotiations with the power company, Reclamation Service engineers E. B. Debler and F. A. Banks studied the hydrometric records to determine in more detail the effect of the power company's water rights on the proposed American Falls development. The size of the reservoir chosen still rested upon the United States securing contracts from the irrigation districts and canal companies reserving storage space in the reservoir to finance their proportional share of construction costs. Generally, the larger the storage reservoir, the more rights needed to be acquired from the power company. The more second-feet of flow used by the power company at American Falls and its downstream power plants, the less water available for irrigation between American Falls and Milner Dam, and therefore the need for greater conservation of the remaining water supply. As a result of their studies, Debler and Banks reported to chief engineer Weymouth in September 1920, that irrigation demand by companies in the Snake River Valley justified initial construction of American Falls Reservoir to at least a capacity of 1,500,000 acre-feet. A dam of this size could be built, advised the engineers, "with but little cooperation on the part of the power company … If future storage demands made it seem advisable to make a considerable increase in the size of the reservoir, it would then be necessary to acquire further power rights at both Minidoka and American Falls."[315]

In October 1920, Weymouth and Director Davis informed IPC that the funds to proceed with the American Falls project were available and it was time to resume negotiations with the power company for

314 Ibid; and F. E. Weymouth to Idaho Power Company, July 5, 1919. Folder 301.1, Box 728 Minidoka 301.1, GAPR, 1919-1929, RG 115, NARA-Denver.

315 E. B. Debler and F. A. Banks to Chief Engineer Weymouth, September 4, 1920. Folder 301.1, Box 728, GAPR, 1919-1929, RG 115, NARA-Denver.

purchase of their American Falls power property and water rights. Later that month additional conferences were held in New York over a four-day period with McGee, Davis and Weymouth all in attendance. Following these meetings, in correspondence to IPC dated November 2, 1920, Davis outlined the terms of a possible contract between the United States and IPC for the American Falls development and purchase or condemnation of all, or a portion of, the company's power property and rights at American Falls. In the letter, Director Davis explained the overall implications of the storage reservoir and power developments at American Falls to the people of Idaho:

> It is a well-recognized fact that the welfare and growth of the state is dependent primarily upon the development of additional lands by irrigation and no part of the state offers better or larger possibilities in this respect than the Snake River Valley.
>
> The water resources of this river are, however, entirely inadequate to water all of the lands susceptible of irrigation and very large fertile areas of southern Idaho will have to remain forever in a desert state due to lack of water. The time has already passed when the natural flow of the river was sufficient to irrigate the lands under cultivation along its shores and storage reservoirs have been recognized as a necessity for many years past.
>
> Hydrometric studies indicate that the water supply of the Snake River when adequately controlled by reservoirs, will furnish a reliable supply to all lands now under cultivation and about 400,000 acres additional, but without additional storage no new areas at all can be made productive and habitable and even the present projects will suffer occasional serious losses. The power rights claimed by your company at American Falls will, if perfected and maintained, prevent absolutely the development of further storage reservoirs on Snake River, and consequently irrigation in southern Idaho.[316]

With storage provided by American Falls reservoir, the Reclamation Service planned to extend its irrigation system into the proposed Minidoka North Side Pumping Division, opening for settlement a compact block of 115,000 acres of fertile land located above (north and west) existing irrigated tracts within the Minidoka project. Since this development also required about 30,000 kW of electrical energy to lift water from the Snake River into the irrigation canals to serve the

316 The letter to IPC was also approved and endorsed by Secretary of the Interior John Barton Payne. See A. P. Davis, director USRS, to Idaho Power Company, November 2, 1920. Folder: Reclamation Bureau, Box 1614, RG 48, NARA II.

project, it called for not only additional storage upstream at American Falls Reservoir, but also development of more summer power capacity on the Minidoka project. Davis further noted that assuring the necessary storage of winter flows required for the Minidoka extension and supplemental water for other irrigated acreage below American Falls required condemnation of IPC's existing winter power rights at American Falls. Winter flows passing through the company's powerhouse at American Falls and disappearing down the Snake River into the canyon below Milner Dam to supply other hydroelectric power plants downstream were forever lost as far as irrigation was concerned. Every second-foot of water "wasted" past Milner Dam in winter, wrote Davis, meant about eighty acres of land was deprived of irrigation.

Thus, IPC's winter power rights at American Falls and below Milner restricted future growth of the state, and damaged the power company's ultimate market for its own electrical energy output. Furthermore, Davis noted that electrical power could be generated at many other places above American Falls, and by use of return flows to the Snake River below Thousand Springs, or at potential power sites on other streams in Idaho that would not compromise the use of water for irrigation purposes.[317]

In the meeting with power company officials, Davis had listened carefully to their concerns about maintaining a commercial power plant at American Falls, but reiterated a familiar refrain that the economic future and prosperity of the whole Snake River Basin was directly tied to expansion of irrigated agriculture. Davis' letter to IPC executives concluded:

> We are convinced that the ultimate development of irrigation in the Snake River Valley will require the elimination of the entire winter water right at American Falls and the storage of the entire winter flow and any plan that contemplates the purchase of less than the entire right is merely a temporary expedient.
>
> We hoped to be able to agree on a contract for the purchase of the entire right at this time, but understand that your objections to this plan are much more serious than they would be if we took at the present time only that part of the right over 2,500 second-feet.[318]

317 Ibid.

318 Ibid.

At this point in the negotiations, the power company had seemingly won an important concession allowing it to retain a private power site at American Falls, in addition to a water right that exceeded its current use. The United States' offer to IPC in November 1920 was to condemn or acquire, for the present, only that part of IPC's power right over 2,500 second-feet, or the rough equivalent of 5,000 acre-feet per day. The government anticipated building and operating its own more efficient, higher-head power plant connected directly to its large dam at American Falls to provide electrical power for Minidoka project pumping purposes. This early government proposal included a provision that after five years from the signing of the contract, the United States reserved the right during the non-irrigation season (October 1 to May 1) to deliver 192,000 kW a day of electrical energy to IPC from its American Falls power plant in lieu of the company's reserved winter water right of 5,000 acre-feet a day. Without this, or some other similar winter water-saving provision, the government was unwilling to enter into a contract leaving the power company with a firm unsubordinated power right to 2,500 cfs.

The government's opposition to a firm power right to 2,500 cfs arose from two concerns: 1) it would compromise storage for irrigation purposes, and 2) the company could construct additional power plants at places where it might later become necessary to condemn the plants in order to conserve this water for irrigation use. The government proposal offered to allow IPC to participate in the reservoir to the extent of 25,000 acre-feet of storage capacity, pay the company $250,000 in cash, and to connect the company's penstock on its East Side powerhouse to the new government dam so that it could benefit from the increased head to generate a large amount of additional summer power. [319]

IPC did not accept the offer and negotiations ceased for several months. The Reclamation Service considered proceeding with condemnation of the company's property, but feared the company might counter by influencing key members of Congress to oppose the appropriation bill for the American Falls project coming up for a vote in March of 1921. Not until February 16, did the two sides meet again in conference. After taking the matter under advisement for a period of

319 Ibid. Although the government planned to build a power plant on the west side of the river in conjunction with construction of American Falls Reservoir, the proposed plant was not built.

three months, the power company rejected the proposal made by Reclamation Service Director Davis back in November 1920. The company made a counteroffer (called Plan 2) which Davis later wrote was "unsatisfactory in detail." The company wanted winter releases from the reservoir to guarantee a flow of 9,000 cfs at Shoshone Falls; a summer right of 5,700 acre-feet at American Falls; a storage right of 50,000 acre-feet at American Falls and 12,000 acre-feet at Lake Walcott; ten years' free use of all storage capacity in American Falls in excess of 500,000 acre-feet; limiting winter diversions at Minidoka and Milner to 600 cfs; and the privilege of connecting its plant to the government dam.

Finally, IPC asked the government for a payment of $2.16 million. Price was a large factor in the quick rejection of IPC's proposal as the Reclamation Service previously had offered only $250,000 for the company's American Falls property. After several days of discussion and counter offers (Plans 3 and 4), the company and the Reclamation Service came up with a proposition (Plan 5) that "in its essential features coincides with the needs of the government," wrote Davis, but still required some modification. In terms of money, the company had requested $1.5 million under Plan 4, while the government was only willing to pay a maximum of $1 million for the concessions required. Davis noted that the company had a right, or inchoate right, to about 7,000 second-feet whenever it was flowing in the river at American Falls the year round. The proposition to limit this right to a maximum of 2,500 second-feet would substantially reduce and modify this right for several months of the year. The power company also owned land that the government needed for its dam, reservoir and power plant. Davis was given authority from Interior Secretary John B. Payne to make an offer of $500,000 to the company for its property. If necessary, noted Davis, the government might be willing to authorize a payment of $1 million, rather than proceed with a condemnation suit.[320]

The proposition which the two parties had tentatively agreed upon in February 1921, provided the basic structure for the agreement finally signed on June 15, 1923. The proposition was called "Plan 5," it being

320 A. P. Davis, director USRS, to Idaho Power Company, February 21, 1921 and A. P. Davis to Secretary of the Interior, February 21, 1921. Folder: Reclamation Bureau, Box 1614, RG 48, NARA II; and M. Gaylord and Barry Dibble to Chief Engineer U. S. Reclamation Service, March 28, 1921. Folder: 031 Water Appropriations thru 1929 (Folder 031). Box 672 Minidoka 023 – 031.4, GAPR, 1919-1945, RG 115, NARA-Denver.

the fifth proposal submitted by the government to the power company for purchase of its American Falls rights.

The major elements of the agreement were outlined in a letter from Director A. P. Davis to officials at IPC [EBASCO] headquarters in New York City, dated February 21, 1921. The proposition included basic concepts that became memorialized in the final agreement including: a "primary power right" for the company to divert up to a maximum of 2,500 second feet in the summer at American Falls for power generation; the government's "secondary power right" to a maximum of 6,000 second-feet in the summer for irrigation pumping purposes; and a "tertiary right" for the company to use for power purposes at American Falls all additional water released or wasted by the government during the summer months.

In addition, the power company was to be furnished with up to 300,000 acre-feet of storable water flowing into the reservoir after October 1 each year, of which only 40,000 acre-feet (later changed to 45,000) could be turned out of storage for use by the company in the irrigation season. The balance of 260,000 acre-feet was to be used during the non-irrigation season and before the reservoir filled each spring. The government was to have the right to store in American Falls Reservoir for its use, all of the balance of the water it was capable of storing in the reservoir, whose size was yet to be determined. In turn, the company was to convey to the United States all of its land and easements under the proposed dam and reservoir at American Falls, and the Island and West Side power plants, together with necessary rights of way for the government to erect a power plant, penstocks and tailrace on the west side of the river. Finally, the government was to receive "all the company's water rights and filings at American Falls and the right to limit the other rights of the company on Snake River in so far (if at all) as the rights allowed the United States to store and use water as herein provided might interfere with any rights of the company at any lower points on Snake River." The plan did not contain any concessions to the company regarding winter flows at Shoshone Falls [321]

321 Davis to Idaho Power, February 21, 1921. Folder: Reclamation Bureau, Box 1614, RG 48, NARA II. According to USBR records, files in the USBR Denver Office in 1953 contained microfilm copies of the drafts of several "Plans" that were considered prior to acceptance of the final contract in 1923. The final plan was apparently designated "Plan 6." The microfilm is no longer accessible through the archives in Denver. J. R. Riter, regional director USBR, to Chief,

Interior Secretary Payne approved Plan 5 and Director Davis and the federal negotiators returned to New York on February 24. What happened in New York is unknown, but on February 28, Davis drafted a new contract proposal designated Plan 6. Both the wording and sequence of articles were as they appeared in the final form of the contract. The major differences were the price of $500,000 offered by the government; 40,000 acre-feet of primary storage, rather than 45,000 acre-feet; and a requirement for the company to convey its property by warranty deed.[322]

During the negotiations in New York, the power company was aware that the Interior Department lacked any authority to grant the company storage or other water rights at American Falls as water rights were under the jurisdiction of the State of Idaho. According to the Reclamation Service's electrical engineer J. M. Gaylord and project manager Barry Dibble, in the course of the negotiations in New York IPC's proposal designated "Plan 2" had included certain specific representations with respect to the status of the company's water rights at American Falls:

> **Property Owned by the Company**
>
> The company has proven up on 1400 second-feet of water (equivalent to 1,013,000 acre-feet) for power purposes at American Falls. It is now using beneficially for power purposes, in addition to the above amount, 856 second-feet (equivalent to 620,000 acre-feet). It has prior filings for water-power rights to 4600 additional second-feet (equivalent to 3,330,000 acre-feet) which can be acquired by installing machinery so as to use this amount on or before June, 1924. The total of these amounts is 6,856 second-feet (equivalent to 4,963,000 acre-feet).[323]

These water rights correspond to the 1901 priority water rights held for generation of power at the West Side (253 second-feet) and Island power plants (611 second-feet). The 1908 priority power right of 1,400

Hydrology Branch, Project Planning Division, USBR, March 2, 1953. USBR-Boise.

322 William Burpee, Interior Department Solicitor, "Informal Notes on the 1923 American Falls Contract with Idaho Power Company," 91962), 61. 05-14-2010 USBOR Response to IPC AGO 06-17-2010-1 CO IAGO, AF006222- AF00681.

323 Gaylord and Dibble to Chief Engineer, March 28, 1921; Folder: 031; Box 672; Minidoka 023 – 031.4; GAPR, 1919-1945; RG 115; NARA-Denver. [D1-1-12] Neither Gaylord nor Dibble of the Reclamation Service believed that the power company had actually put as much water to beneficial use as was claimed in its water right applications.

second-feet for generation of power through the first unit of the East Side plant and 4,600 second-feet under a right initiated in 1919 for two additional units at the East Side plant located at the foot of American Falls.[324] All of these rights were natural flow rights, the largest of which were junior to the Minidoka project and the large Carey Act irrigation projects in the Twin Falls region. The power company did not possess any storage rights at American Falls.

Stoutemyer was aware of the defect in the proposed storage right that was contemplated under the contract. During the meetings in February 1921, representatives of IPC suggested that the company should make a water rights filing with the Idaho Commissioner of Reclamation for storage of the 300,000 acre-feet of water proposed to be retained by the company to "simplify the legal situation" by reserving water rights to be held under its own permit. In doing so, "it would not be necessary for the federal government to grant to the Company the rights proposed." If the government built its American Falls Dam, the power company was willing to agree that their water rights filing would be subject to the terms and conditions of the contract between the United States and the company. If not, the company could construct its own 300,000 acre-foot reservoir at the American Falls site. Stoutemyer proposed that the Reclamation Service prepare its own water rights filing for the American Falls storage reservoir together with a memorandum of agreement with IPC.[325] The memorandum of agreement was signed by F. F. Johnson, president of IPC on March 17, 1921. The agreement referenced the proposed contract and the desire of the United States to purchase certain water rights and property owned by the power company. The U.S. and IPC agreed to each prepare an application for state water rights to store water in American Falls Reservoir. The U. S. application was for a proposed American Falls reservoir of undetermined size up to 3,000,000 acre-feet and IPC for storage of approximately 300,000 acre-feet in the United States' proposed reservoir, or in a reservoir to be constructed by IPC. Under the terms of the memorandum, IPC and the United States agreed to the following:

---

324 "American Falls Power plant (East Side plant) owned by Idaho Power Co. and located adjacent to the town of American Falls," n. d. Idaho State Water Right 01-2046 Back File.

325 Stoutemyer to Chief Counsel, Washington, DC, March 18, 1921, Attachment 15 to redacted memorandum, "Information about Contract with the Idaho Power Company – American Falls," 1962, Boise Field Solicitor's Office, U. .S. Bureau of Reclamation.

> (a) The United States would delay filing its application until after Idaho Power Company filed its water rights application. The rights acquired by IPC under its application would be held "subject to the terms and provisions of the contract that may be entered into between the United States and the Idaho Power Company with reference to the construction by the United States of said American Falls reservoir."
>
> (b) If a contract for construction of the reservoir was not entered into within one year, the agreement was to have no effect. However, if the parties failed to agree upon the terms of a contract, and the government nevertheless chose to build the reservoir and commenced within two years, in any suit to condemn the property and rights of the Company under the power of eminent domain, the state water rights permit would be valued only at the amount of the filing fees paid by the company to the office of the Commissioner of Reclamation.[326]

In fact, IPC did not make one, but two, water right filings with the Idaho Commissioner of Reclamation on or about March 17, 1921: one for a diversion right for 1,200 second-feet for power purposes and the other a storage right for 300,000 acre-feet in American Falls Reservoir. The points of beneficial use were listed as existing and proposed powerhouses located at American Falls, Twin Falls, Shoshone Falls, Upper Salmon Falls, Lower Salmon Falls, and Swan Falls.[327] On being advised by W. R. Putnam that IPC had filed its application, Stoutemyer wrote to F. E. Weymouth, the Service's chief engineer in Denver:

---

326 B. E. Stoutemyer, district counsel, to Chief Engineer USRS, Denver, CO, March 19, 1921. Folder: 031, Box 672 Minidoka 023 – 031.4, GAPR, 1919-1945, RG 115, NARA-Denver. Stoutemyer mentions other correspondence on the subject including transfer of the signed agreement in other correspondence, specifically a March 12, 1921 letter from F. A. Banks to B. E. Stoutemyer and in correspondence dated March 18, 1921 between Stoutemyer and the Chief Engineer in Denver. See Stoutemyer to Chief Counsel, Washington, DC, March 18, 1921 with Attached Memorandum of Agreement, dated March 17, 1921 and signed by F. F. Johnson, president IPC, and Clarence T. Ward, Secretary IPC, Attachment 15 to redacted memorandum, "Information about Contract with the Idaho Power Company – American Falls, 1962, Boise Field Solicitor's Office, Bureau of Reclamation. A copy of this Memorandum of Agreement also appears in the Idaho Department of Water Resources' "back file" on the Bureau's Idaho State Water Rights License 01-2040. See Memorandum of Agreement. Idaho State Water Right 01-2040 Back File.

327 The first application for an Idaho State Water Right was covered by Permit No. 15201 (File 01-02041) and the second by Permit No. R-273 (File 01-02065) both with a priority of March 17, 1921. Apparently no license was ever issued for either of these water claims. As late as 1971 the disposition of these claims were still being discussed by Idaho Power Company and Idaho Department of Water Commission. It has been suggested that both were given up in lieu of storage and in flow rights in the name of the Bureau of Reclamation for American Falls Reservoir, see note by Ralph Mellin, October 8, 1996 in Application File for Permit No. 01-2046 (16635), attached to letter by C. Stephen Allred, deputy director Department of Water Administration, to James E. Bruce, Vice-President and Secretary, Idaho Power Company, February 23, 1971. Idaho State Water Right 01-2046 Back File.

> It is now in order for us to make our filing for the American Falls Reservoir. Mr. Banks has forwarded the necessary maps and descriptions of the land and data, etc. and has suggested a reservoir filing of 3,000,000 acre-feet and a diversion filing for the various projects expecting to secure water from American Falls Reservoir of 8,000 second-feet. The maps and lists of lands as furnished by Mr. Banks, include both the Bruneau and Mountain Home tracts, as well as practically all of the lands in the Snake River Valley under the old canals proposing to secure supplemental water from the American Falls Reservoir.[328]

On March 30, 1921, the Reclamation Service, through its district counsel, B. E. Stoutemyer, filed two applications for appropriative water right permits in the office of the Idaho Commissioner of Reclamation for 3 million acre-feet of storage and 8,000 second-feet for domestic, irrigation and power uses, as described in his letter to Weymouth. The first, Permit R-269 (01-2064) for American Falls Reservoir described the development as being comprised of a 100-foot high dam forming a reservoir with a storage capacity of 3,000,000 acre-feet of water, with the stored water being put to use for domestic, irrigation and power purposes. The amount of land to be irrigated was listed as "940,000 acres of old lands, 500,000 acres of new lands, said land being described in Use Permit mentioned in Paragraph 13 herein, which Use Permit is hereby referred to and made a part hereof." Paragraph 13 stated that the waters stored in American Falls Reservoir in accordance with R-269 were to be conducted to the place of use through the Riley Canal and other canal systems "under terms of Permit No. 15134" – the second of the two filings. Permit No. 15134 was also referred to and made a part of Permit No. R-269 by a second provision found in the reservoir permit application:

> Beneficial use of water stored in accordance herewith, to be made on or before April 23, 1926, as set out in 15134 for appropriation of the public waters of the State of Idaho, which is hereby referred to and made a part hereof.[329]

328 While his maps included both the Bruneau and Mountain Home tracts, Banks noted that irrigation of from 250,000 to 350,000 acres would be provided for on one or the other tract, but not both. Stoutemyer to Chief Engineer, March 19, 1921. Folder: 031, Box 672 Minidoka 023 – 031.4, GAPR, 1919-1945, RG 115, NARA-Denver.

329 Application for a Permit to Construct the American Falls Reservoir, and to Store the Unappropriate Waters of the State of Idaho, Application No. R 21513, Permit No. R 269, District No. 70, Approved May 18, 1921. Idaho State Water Right 01-2064 Back File.

Permit No. 15134 (01-2040) sought the diversion right of 8,000 cfs for the direct irrigation of 1,540,780 acres, and the indirect irrigation of 460,275 acres. The voluminous maps and lists of lands to be furnished water provided by F. A. Banks in the application included irrigation of some 250,000 to 350,000 acres in either (but not both) the Bruneau or Mountain Home tracts, as well as virtually all of the lands then irrigated from old canal systems in the Snake River Valley from the mouth of the canyon of the South Fork of Snake River, near Rigby, Idaho, downstream to Milner Dam. These lands were to receive supplemental water from the reservoir. Other new land to be irrigated included 115,000 acres on the Minidoka-North Side Pumping project, 35,000 acres on Michaud Flat, and 20,000 acres on Wood River.[330]

At the urging of Gaylord and Dibble, the electrical engineer and project manager of the Minidoka project, the USRS made a third water right filing with the Idaho Commissioner of Reclamation on April 21, 1921, for 9,000 second-feet of water for power purposes at American Falls dam. This filing was intended to secure a prior right for the government's proposed American Falls power plant against any future filing by IPC or speculative filings by other parties. The rationale for the 9,000 second-feet filing was explained by Gaylord and Dibble:

> The natural flow of the Snake River during part of the irrigation season at American Falls is about 2000 or 2500 second-feet. The projects located below American Falls which are at present in existence require 9500 second-feet, and the North Side Pumping Extension of this Minidoka Project will require an additional 2000 second-feet. There will therefore be released from the American Falls reservoir as at present contemplated, about 9000 second-feet of stored water. This at a head of 50 feet will produce approximately 30,000 kilowatts and therefore could all be used to advantage in generating power for use as indicated above.[331]

---

330 Permit File No. 15134, March 30, 1921. Idaho State Water Right 01-2041 Back File. See also Stoutemyer to Chief Engineer, March 19, 1921. Folder: 031, Box 672, GAPR, 1919-1945, RG 115, NARA-Denver.

331 Permit File No. 15237 (File 01-02041), April 21, 1921. See also Gaylord and Dibble to F. E. Weymouth, March 28, 1921; B. E. Stoutemyer, district counsel, to Chief Engineer, USRS, Denver, CO, March 30, 1921 and A. P. Davis, director, to Chief Engineer of US Reclamation Service, April 8, 1921. older: 031, Box 672 Minidoka 023 – 031.4, GAPR, 1919-1945, RG 115, NARA-Denver.

The Commissioner of Reclamation approved the application for the 9,000 second-foot power permit on July 29, 1921. Work on the power plant was to be completed within five years and the water put to full beneficial use on or before July 29, 1932.

The United States, however, never built this proposed government power plant at the base of American Falls Dam.

After filing for their storage and diversion rights at American Falls, officers of the government and officials of the company continued to meet in New York, Washington, D.C., and Idaho as the prolonged contract negotiations continued. Stoutemyer and a Mr. Burns prepared a draft contract for the United States dated April 28, 1921, to open final negotiations with IPC over its property and water rights at American Falls. Something in the proposal, whether technical, legal, operational or financial in nature, endangered further negotiations. IPC took a hard position during ensuing discussions making it clear to the government representatives that it was prepared to move forward with its own American Falls power project.

All of the details of the Stoutemyer/Burns draft contract are not known but a report on the proposed contract and subsequent negotiations with IPC written by Barry Dibble, superintendent of the Minidoka project, in May 1921, highlights several key provisions found in the final contract document over which there were still strong disagreements between the parties. Idaho Power Company maintained that its control over the American Falls power site was absolute and that the company was free to continue its own development, including its plan to raise the dam five feet, without interference from the United States or the Federal Power Commission. Dibble believed IPC was bluffing to enhance its negotiating position. Examination of title documents by the Reclamation Service suggested that a large amount of public land already was affected by IPC's existing development. Furthermore, in Dibble's estimation, there was no question that additional public land would be inundated by a five-foot rise in the surface elevation of the reservoir pool at American Falls, and therefore, the FPC would have jurisdiction over any new water storage project contemplated by IPC.[332]

---

332 Barry Dibble to Minidoka Project Manager and Chief Engineer, May 21, 1921, 1-3. B1_IPC 0011054. March 10, 2010, IPC First Production to State of Idaho, AGO – 06-10-2010-1 CO, Idaho AGO.

IPC's claims regarding the extent of its water rights at its three American Falls power plants were also challenged by the Reclamation Service. The company claimed it had proved up on 2,256 second-feet (1,633,000 acre-feet) and had prior filings for another 4,600 second-feet (3,330,000 acre-feet) for a total of 6,856 second-feet, equivalent to 4,963,000 acre-feet. Dibble disputed the amount of water being used and claimed by IPC. According to calculations made by W. H. Trenner, IPC's own chief engineer, the three American Falls plants had a maximum current capacity of 4,500 kilowatts. Dibble calculated that at 70 percent average efficiency the production of this amount of electricity would require approximately 1,675 second-feet, "equivalent to 300,000 acre-feet at 60 percent load factor during the five-month non-irrigation season." This amount was 681 second-feet less than IPC claimed it was currently putting to beneficial use and far less that the 2,500 second-feet maximum power right being discussed under the proposed contract. Furthermore, Dibble believed any addition to the present power capacity of IPC's power plants would require filing an application for a permit before the FPC. If his assumption was correct, Dibble concluded "the rights of the Idaho Power Company that we need for the reservoir are very materially lessened from what they claim and what we may have assumed in the past."[333]

Dibble also questioned the inclusion of two provisions in the proposed contract that intended to limit the amount of power that the Reclamation Service would be allowed to generate at its own powerhouse at American Falls. Dibble was particularly critical of the provision that power production be limited to irrigation pumping uses on the Minidoka project. He also urged removal of any provision granting tertiary power rights to IPC, especially the unlimited ones specified in the current version of the draft contract. Dibble noted that the proposed contract provided IPC with a primary right to 40,000 acre-feet of storage capacity and 260,000 acre-feet of secondary storage capacity for use during the non-irrigation season. He believed that the 40,000 acre-feet was "partly based upon the right which Idaho Power Company has claimed to raise its dam at American Falls to a height of five feet above the present crest." In the event that the company held no rights to the underlying land, their claim to this storage space would be greatly weakened. Nevertheless, Dibble concluded that conceding

333 Ibid., 4-5.

the right to store up to 300,000 acre-feet in the reservoir would "leave the rights of the Idaho Power Company in as good shape as they are at present, and granting them this water would obviate the necessity of giving them any other concessions."[334]

Idaho Governor Davis continued to oppose any dedication of water for power purposes to IPC which might otherwise be available for irrigation uses. Dibble noted that the 300,000 acre-feet made available to IPC translated into the subtraction of approximately 60,000 acres from the ultimate irrigable area in the Snake River Basin. This amount of land was large enough that it might result in the abandonment of at least one proposed irrigation project in the upper basin. Dibble believed that public opinion in Idaho would be critical of the Reclamation Service for sacrificing the ultimate irrigation development of the state to the demands of the power company. [335]

As for a provision in the proposed contract guaranteeing certain minimum flow at Shoshone Falls, Dibble noted that they were "in part at least based upon the claim by the Power Company that they were now developing that amount of water." Dibble accused the power company of having acted in bad faith because the power development at Shoshone Falls had been "initiated since these negotiations have been pending." Trenner or Putman of IPC had maintained to Dibble that the plant was designed to have its maximum efficiency at 75 percent load so that it would correspond to the minimum flow at Shoshone Falls.

The minimum flows specified in the contract exceeded this amount and would require the Reclamation Service to divert water at American Falls reservoir to regulate the flow at Shoshone Falls. F. A. Banks of the Reclamation Service believed that the right to run 45,000 acre-feet of primary storage and 255,000 acre-feet of winter right down the Snake River to Shoshone Falls was immaterial as long as the Twin Falls companies continued to divert stock or domestic water at Milner during the winter time. Enough waste spilled back into the river to meet the necessary winter flow at the head of Shoshone Falls without having to release any of the 255,000 acre-feet at American Falls. [336]

The contract negotiations moved forward over the following months with the Reclamation Service seeking further input on, or approval of,

334 Ibid., 5-7.

335 Ibid., 7-8.

336 Ibid., 8-9.

the draft agreement from local and state interests. At a meeting held on May 27, 1921, the Advisory Board, which represented the irrigation districts and canal companies cooperating with the United States in the construction of American Falls Reservoir, passed a resolution approving the form and language of a proposed contract between the United States and IPC. On June 8 and 9, 1921, a major conference was held in Washington D. C. regarding the proposed contract. The Reclamation Service was represented by Davis and Weymouth, as well as F. E. Banks, engineer; B. E. Stoutemyer, district counsel; and J. M. Gaylord, electrical engineer. The advisory board was represented by Edgar W. Moorman, president of the Murtaugh Canal Company, and B. B. Brayton, president of the New Sweden Irrigation District. The State of Idaho was represented by its state Commissioner of Reclamation, W. G. Swendsen. IPC was also well represented by three of its vice-presidents: McGee, C. E. Groesbech, and W. R. Putnam; attorney, F. B. Odlum; and consulting engineers, A. C. Clogher and J. J. O'Connell.[337] At the meeting in June the details of a final contract were worked out among the interested parties and work began shortly thereafter on preparing the final draft of the proposed American Falls power contract.[338]

The proposed American Falls contract was submitted to Idaho Reclamation Commissioner W. G. Swendsen, who had taken an active part in the overall negotiations, and was approved by him. The State of Idaho was not a formal party to the contract, but because of its pivotal importance to the state Interior Secretary Fall had consulted closely with the state during its development to assure the agreement was "in harmony with its desires," noted Swendsen. In fact, the secretary indicated to Swendsen and Idaho governor D. W. Davis that he required state approval before executing the agreement with IPC. Swendsen noted that while the power company was guaranteed certain water rights under the contract, the agreement protected Snake River irrigators and was consistent with the state constitutional provisions that declared

337 A. P. Davis to Secretary of the Interior, October 31, 1921; see also Finney to Mitchell, January 30, 1922. Folder: Reclamation Bureau, Box 1614, RG 48, NARA II.

338 The terms of the contract were settled upon at the June 9, 1921 conference. The earliest copy of the contract between IPC and the US in the JRP document set is dated September 17, 1921. The contract proposed at that time was compared to the language of the June 15, 1923 contract and they are in all material respects identical. The September 17, 1921 draft document is from Land Utilization File: Miscellaneous Agreement – US & Idaho Power Company, Box 833, Borah Papers, LOC. The only difference is in Section 36, wherein the 1921 draft identifies the Idaho Reclamation Commissioner by name, as well as by job title.

the rights of agriculturalists in water to be superior to other interests. In a letter to Governor Davis, he wrote:

> While it is true under the provisions of this contract the Power Company are retaining certain of their water rights for power uses, it is my belief that the present contract is so prepared as to safeguard the interests of the state in the best use of its water supply, and since under the provisions of our constitution the use of water for irrigation purposes is declared to be a superior use, it should be possible, if future demands for irrigation require, to have this water supply which under the provisions of this contract is retained by the Power Company for power uses, applied to irrigation use either by friendly purchase or condemnation. The facts are that under this contract the Power Company are giving up a considerable part of their rights in and to the use of water for power purposes so that the same may be used for irrigation, and the water rights retained by them for power uses will not be required to meet irrigation needs for some years to come. It would, therefore, be poor business judgment to purchase at this time all of the water power rights and thus burden the water users with an investment for which they would have no use for some time.[339]

Representatives of the spaceholding irrigation districts and companies who would help pay for the reservoir reviewed the proposed contract document and also requested that it be executed. On October 14, 1921, the State of Idaho Board of Land Commissioners, the state officials overseeing Carey Act irrigation projects, approved it. With all of the details tentatively agreed upon and the form of the contract prepared, IPC formally executed the contract on October 15, 1921, subject to the approval of the Secretary of the Interior, on or before January 11, 1922. On October 31, 1921, the contract was sent from Reclamation Director A. P. Davis to the Secretary of the Interior for his final signature and approval by the United States.[340] Assistant Secretary of the Interior Finney forwarded a copy of the contract to Secretary Fall, with the following comment:

> The attached contract proposes to acquire certain rights from the Idaho Power Company, now the owner of a dam, three power plants, and certain water rights at American Falls. The Government originally offered $250,000 cash and recognition of certain water rights, and the

---

339 W. G. Swendsen to D. W. Davis, October 14, 1921. Folder: Land Board, 1921, AR 2/12, ISHS.

340 Finney to Mitchell, January 30, 1922 and Hubert Work to Senator Burton K. Wheeler, March 10, 1926. Folder: Reclamation Bureau, Box 1614, RG 48, NARA II.

> company asked for $2,300,000. The attached contract proposes to recognize the water rights, as originally suggested, and to pay the company $1,000,000. It has, as you will note, the approval of the Board of Land Commissioners of Idaho.[341]

In late November 1921, A. P. Davis apprised Secretary Fall of certain "collateral facts" that his office should consider with respect to the pending contract between the United States and IPC. The economic feasibility of the dam project depended upon various irrigation districts and canal companies in the Snake River Valley, which desired additional water, to request subscriptions for water and to contract with the Reclamation Service for the cost of their *pro rata* share of storage rights in the proposed reservoir. Under the provisions of these contracts, the purchaser was obligated to pay various sums of money to the federal government at such time as they were needed for construction, in addition to a small payment of $.10 per acre foot at the time of entering into the contract. Twenty-four individuals and companies had signed contracts to date, wrote Davis, and the United States had spent about one-half million dollars on the project. The contracts had been entered into during an era of relative prosperity, which was succeeded by a period of depression leaving the farming communities unable to meet their obligations under the contracts. All but five of the contracts were now in default and most payments were more than 10 months overdue. The amount owed the government by the contractors in default was over $665,000. The General Reclamation Fund, which depended upon receipts from the sale of public lands and repayment of construction, operation, and maintenance charges by water users, suffered from a shortage of funds and did not have sufficient financial resources to undertake construction of the reservoir. Director Davis recommended that the reservoir be held in abeyance until farming communities and irrigation districts were able to meet their obligations and advance their respective shares under the existing contracts, or until the Reclamation Fund was reinforced by new legislation.[342]

As the expiration date of the contract neared, Idaho Representative Addison T. Smith's office and the Secretary of Interior's office were

341 Finney to Secretary, November 3, 1921; Folder: Reclamation Bureau, Box 1614, RG 48, NARA II.

342 A. P. Davis to Secretary of the Interior, November 21, 1921. Folder: Reclamation Bureau, Box 1614, RG 48, NARA II.

deluged with telegrams from individuals and organizations in Idaho urging approval of the contract with IPC.

However, Interior Secretary Fall did not sign the power contract by its expiration date, and therefore, it did not become effective. Fall was unwilling to contract with IPC and obligate the government and the Reclamation Service to spend $1 million over the next few years ($250,000 immediately) to acquire rights necessary for the Minidoka project because the settlers had fallen behind in their debt repayments to the Reclamation Service and there were no signs that this situation would change in the near future. Preoccupied with the Teapot Dome bribery scandal and investigation, without approving the contract, Fall escaped Washington, D.C. for a long vacation at a remote ranch in New Mexico, leaving E. C Finney as acting Secretary of the Interior. Working with the senators from Idaho, the state Commissioner of Reclamation, and other representative citizens interested in irrigation matters, Assistant Secretary Finney secured two extensions of time totaling 90 days to keep the contract alive until a plan could be formulated to fund the project.

Sidney Z. Mitchell, president of the Electric Bond and Share Company of New York, played a key role in seeing that the contract was executed. Mitchell was instrumental in organizing and financing the Idaho Power Company from its inception. A graduate of the Naval Academy at Annapolis, Mitchell served aboard ship in World War I and was discharged at Seattle, Washington. There he went to work for the Edison Company and built the first central power station in Seattle. Mitchell became the "sparkplug" for Edison Corporation's sales program in the northwest territories – Oregon, Idaho, Washington, British Columbia and Alaska. He joined General Electric Manufacturing Company which had sponsored the successful emergence of Electric Bond and Share Company as an organization that financed the expanding needs of electric service organizations nationwide. While working in his capacity as an officer of this corporation, Mitchell acquired the assets of seven utility companies lodged in federal bankruptcy court in Boise, Idaho and molded them into Idaho Power Company. [343] It was Mitchell who approved an initial extension of time for contract approval to February 1, 1922, and, when

343 Mitchell's biographical sketch is drawn from Rockwell, *Saga*, 184-196.

Secretary Fall again did not sign, secured an indefinite extension pending a special meeting of the board of directors.[344]

In the meantime, Idaho Commissioner of Reclamation W.G. Swendsen penned a memorandum on the proposed American Falls contract arguing for its approval by Secretary Fall. From the standpoint of the State of Idaho and its citizens, the importance of the large storage reservoir could not be over-estimated, opined Swendson. The reservoir would supply supplemental water to 500,000 acres of land already occupied by *bona fide* settlers who had suffered from past crop losses by virtue of drought. The protracted negotiations had continued over three years and resulted in a complicated agreement that won the approval of all participating parties, except the Interior secretary. The board of directors of the Trust Company of New York, however, had ratified the contract by "a very marginal vote" and was in danger of going into default if there were further delays. The irrigation future of the Snake River Basin was at stake, wrote Idaho's reclamation commissioner:

> It is essential to Idaho that this reservoir construction be undertaken without delay. The site occupies a very strategic position in the water-shed, and when the reservoir is completed it will afford a complete control for the entire Snake River discharge, thus making it possible to use to best advantage the entire flow of the stream.
>
> In the event a delay occurs in the building of this reservoir, the development of the Snake River resource will progress in a piecemeal fashion, by the construction of other small reservoir sites in the upper part of the drainage basin, thus tending to make the American Falls project unfeasible, with the result that ultimately the greatest good cannot be had from this water resource, but on the contrary a large portion thereof will of necessity be dissipated and wasted.[345]

On April 3, 1923, at a special meeting of the board of directors of IPC, an extension of time for execution of the American Falls contract on behalf of the United States was formally approved and extended to July 12, 1923. The contract finally was executed on behalf of the

344 E. C. Finney, acting Secretary of the Interior, to S. Z. Mitchell, EBASCO, January 11, 1922; S. Z. Mitchell to Finney, Jan 11, 1922; Finney to Hon. Addison T. Smith, January 14, 1922. Finney to Mitchell, January 30, 1922. Folder: Reclamation Bureau, Box 16[illegible]4, RG 48, NARA II; and W. G. Swendson to Gov. D. W. Davis, March 15, 1922. Folder: Land Board, 1921, AR 2/12, ISHS.

345 W.G. Swendsen, Memorandum Re: American Falls Reservoir Project, Contract between the U.S. Government and the Idaho Power Company, February 5, 1922, 2-3. USBR-Boise.

United States, not by Secretary Fall – who had been forced to resign his office over the Teapot Dome Scandal – but by his successor, Hubert Work.[346]

The U. S.-IPC contract regarding American Falls Reservoir was finally signed on June 15, 1923, by Secretary Work and F. F. Johnson, president of IPC.[347] The agreement conveyed the old West Side and Island power plants (but not their water rights), part of the IPC diversion dam, the bottomlands and flowage rights owned by IPC within the proposed American Falls reservoir site, and seventy town lots at American Falls to the United States, and exchanged certain natural flow water rights for storage rights in the reservoir. The United States planned to construct a 30,000-kW powerhouse on the west side of the river below the falls and to build a dam at American Falls of sufficient height to create a reservoir with a maximum capacity of up to 3,000,000 acre-feet of storage dedicated to conserving water primarily for irrigation purposes.

Ultimately, the reservoir at American Falls – operational in 1926 – was built to a much smaller capacity of 1.7 million acre-feet. The License and Certificate of Water Right, 01-2064, granted to the United States for the American Falls Reservoir by the Idaho Commissioner of Reclamation on July 9, 1942, assigned the United States the right to store 1.8 million acre-feet (including 100,000 acre-feet of bank storage) at American Falls with a priority date of March 30, 1921, for domestic and irrigation purposes.[348]

346 E. C. Finney, acting Secretary of the Interior, to S. Z. Mitchell, EBASCO, January 11, 1922; S. Z. Mitchell to Finney, Jan 11, 1922; Finney to Hon. Addison T. Smith, January 14, 1922; Finney to Mitchell, January 30, 1922. Folder: Reclamation Bureau, Box 1614, RG 48, NARA II; and Swendson to Davis, March 15, 1922. Folder: Land Board, 1921, AR 2/12, ISHS.

347 Herbert Work to Sen. Burton K. Wheeler, March 10, 1926. Folder: Reclamation Bureau, Box 1614, RG 48, NARA II.

348 The reservoir was constructed to only hold 1,700,000 acre-feet. The additional 100,000 acre-feet was considered part of the "bank storage" and accounted for the water exchange rate between the reservoir and surrounding aquifer. E. V. Berg, Commissioner of Reclamation to B. E. Stoutemyer, District Counsel USBR, July 13, 1942, enclosing License and Certificate of Water Right, R-269; B. E. Stoutemyer to Commissioner, July 15, 1942. Folder: Minidoka – General Correspondence re. Water Appropriations by the Service, 1930-1944 (Folder: Minidoka), Box 718 Minidoka Project, 031 – 032, Project Correspondence File, 1930-1945 (PCF, 1930-1945), RG 115, NARA-Denver.

The reservoir in practice

As noted on many occasions by Commissioner Swendsen and Governor Davis, construction of American Falls Reservoir and the June 15, 1923, contract were of vital importance to the State of Idaho and upper basin irrigators on the Snake River, not only for supplemental water to meet their current needs, but also for future expansion of irrigated agriculture in the Snake River Basin.

By 1922, the Reclamation Service under its Minidoka and Boise-Payette projects had reclaimed and irrigated some 105,580 acres of arid land in the Snake River Valley. As fast as this federal irrigated empire appeared, it was exceeded in size by several Carey Act projects that took shape in the Upper Snake Basin during the same period.

Sites for construction of additional large storage reservoirs to conserve the remaining flood waters in the upper basin for use on hundreds of thousands of unproductive but fertile acres of irrigable land had been identified for future irrigation development through federal reclamation funds and by private capital.[349]

In the 1920s the Bureau perceived IPC's plans for development of a storage reservoir at American Falls and additional power generating facilities at the foot of the dam, together with a proposed hydroelectric power plant on the Snake River a short distance below Milner Dam at Twin Falls and expansion of an existing power facility at Shoshone Falls, as a threat that would result in less water for these proposed irrigation developments. Therefore, under the terms of the June 15, 1923, contract the U. S. purchased the American Falls dam and reservoir site from IPC to constrain the development plans of the power company in the Upper Snake River Basin. The Bureau had originally wanted to condemn or purchase the entire power right of IPC at American Falls, but the power company successfully resisted this demand. According to Bureau construction engineer F. A. Banks, at the time the 1923 contract was approved, IPC had made proof of completion of works at American Falls on about 5,000 cfs, and had filings in good standing for an additional 3,000 cfs, or more. Collectively, these amounts of water, wrote Banks in a brief on the power company's relative rights in connection with American Falls,

---

349 *Twenty-second Annual Report of the United States Bureau of Reclamation*, 6, 64-66, 126-127.

constituted "all the flow past American Falls except for a month or so during the flood period."

Through the 1923 contract, however, Banks noted that the government was successful in reducing the company's power rights from a winter average of about 5,000 cfs to about 800 cfs (equivalent to the 255,000 acre-feet secondary regulatory right mentioned in the contract) and to curtail the summer power right to a maximum of 2,500 cfs during the height of irrigation season.[350]

Once the American Falls Reservoir was constructed, numerous Snake River irrigation districts and canal companies, and the American Falls Reservoir District became reservoir "spaceholders" based upon their contracts made with the Bureau for storage rights in the reservoir. IPC retained only falling water rights which were limited in quantity by the contract and subordinated to the rights of subsequent irrigators. In the future, the Bureau frequently urged the Federal Power Commission (and later the Federal Energy Regulatory Commission – FERC) to require subordination in all power licenses granted on the Snake River in order to protect future upstream irrigation expansion.

While it greatly helped in meeting the water supply needs of basin irrigators, construction of American Falls Reservoir did not mark the end of southern Idaho's water problems.

Indeed, the holdover storage it furnished proved insufficient to address irrigators' needs when drought swept through the region once more in the early 1930s. Still, American Falls Reservoir became the linchpin of a system of Upper Basin reservoirs that included Jackson Lake, Lake Walcott, and ultimately Henry's Fork and Palisades. In principle and practice, the federal government operated these reservoirs for the benefit of the irrigators along the Snake River. Hydropower generation did have a place at American Falls (and later, Palisades), and as specified in the June 15, 1923, that place was subordinate and incidental to the primary purpose of the reservoir, which was to store water for irrigation.

350 B. E. Stoutemyer to Office of District Counsel, Boise, ID, May 2, 1924.; F. A. Banks, "Power Company Contract American Falls Reservoir," attachment to letter from Banks to R. E. Shepherd, April 26, 1924. USBR-Boise.

5

June 15, 1923

June 15, 1923, is not a date typically cast in large type. It was a critical day in the history of Idaho, however, because on it was signed a contract between the United States and Idaho Power Company that not only paved the way for construction of American Falls Reservoir, but also established operational guidelines for the facility and shaped administration of water in the upper Snake River Basin for generations to come.

The need to conserve water for irrigation purposes by saving winter flood flows to storage in American Falls Reservoir restricted the supply of water available for power generation to those flows that were subsidiary to operation of the USBR reservoir system to meet existing and proposed irrigation needs.

With this in mind, the 1923 contract gave the United States most of IPC's existing water rights at American Falls. It also gave the United States more: The right to limit other water rights of the power company on the Snake River *downstream* of American Falls, to the extent that the use of the rights granted to the United States under the contract interfered at such downstream points, or as expressed in subdivision (e) of Article 9: "insofar as and no farther than the rights allowed and granted to the United States to store and use water as herein provided may interfere with any rights of the company at any lower points on Snake River."[351]

351 Department of the Interior, Bureau of Reclamation, Contract between the United States and the Idaho Power Company Relative to Power Rights at American Falls, Idaho, June 15, 1923 (Contract of June 15, 1923), Paragraph 9 (d) and (e). Throughout the contract, the water rights, filings or claims of the company to water rights are referred to, but nowhere are they specifically laid out by reference to notices of appropriation, permits, or deeds, or by dates or locations. As for the rights that would be interfered with downstream, the claim made by ICP was vague and coupled claimed existing rights at American Falls with all prospective or future claims downstream. The company claimed that the water it used at American Falls was or would be used "in its plants located and to be located at points below on the stream, so that the storing of water by the United States of winter flow at American Falls will interfere with certain power and other rights of the Company at American Falls and points below."

Construction of American Falls Reservoir had an appreciable effect on IPC's power operation and future plans for its power development at American Falls and to a lesser extent at other existing power sites downstream on the Snake River. The 1923 contract did not state precisely what water and power rights IPC possessed on the Snake River. However, in Section 4 of the preliminary recitals, the company made a remarkably elusive claim suggesting that it held broad water and power rights on the stream:

> … the water which [IPC] uses and claims for use at American Falls is or would be also used, to a large extent, in its plants located and to be located at points below on the stream, so that the storing of waters by the United States of winter flow at American Falls will interfere with certain power and other rights of the Company at American Falls and points below.[352]

Leaving aside the issue of the legal basis for claiming reuse of water in downstream power plants in 1919 and 1920, when the contract was being negotiated, the company did not have plants with the capacity to use the flows of water passing below American Falls. Before 1923, Idaho Power Company had in its East Side power plant at American Falls only one power-generating unit, a horizontal Francis-type turbine, installed in 1913. The company had a water right license for 1,400 cfs with a priority date of 1908, to operate this unit together with an unperfected supplemental permit for construction of two additional generating units using up to 4,600 cfs under a 1919, priority date. These two units were installed in 1924. At its next plant downstream – Shoshone Falls – before 1921, there were two generating units: one was a horizontal water wheel installed in 1907, and the other a similar unit built in 1909. Little if any water needed to be "wasted" over Milner Dam to provide the amount of water necessary to operate these two units at their maximum capacity. A third unit at Shoshone Falls was added in 1921, after IPC was fully aware of the government's plan for a storage reservoir at American Falls and after the initial congressional appropriations had been made for project construction. The only other power plant of IPC in existence in 1921 that could possibly be affected by the operation of American Falls Reservoir was the Swan Falls plant.

---

352 Department of the Interior, Bureau of Reclamation, "Contract between the United States and the Idaho Power Company Relative to Power Rights at American Falls, Idaho," June 15, 1923, Section 4.

It had ten generating units: two installed in 1910, two more in 1913, and the final six in 1918. Because of the large inflows to the Snake River below Milner and above Swan Falls, the power production at this plant was not affected by the government's storage operations at American Falls.

The contract between the U.S. and IPC was negotiated in an unsettled regulatory environment wherein the Federal Water Power Act was under discussion in Congress, but the contract was executed after the act took effect. Immediately upon passage of the Federal Water Power Act in June 1920, IPC filed applications with the Federal Power Commission for preliminary permits to construct additional hydroelectric power plants on the Snake River below American Falls. One was located only twenty-two miles below Milner Dam at Twin Falls and the other thirty-five miles further downstream at Upper Salmon Falls. Operation of the Twin Falls power plant, should it be approved by the Federal Power Commission and constructed by the power company, would clearly be affected by the proposed storage reservoir at American Falls. The Bureau of Reclamation and residents of the Minidoka project strongly opposed construction of the Twin Falls plant.

The recurrently vague or ambiguous language and seemingly contradictory provisions of the contract with respect to the rights and claims of the power company make it exceedingly difficult to interpret. To do so, it is necessary to understand the changing circumstances with respect to the stage of development of the river at the time of the negotiations, the desire of both the power company and irrigation institutions to integrate their facilities to make more effective use of available water resources, and the physical situation of the irrigation and power developments of the United States on the Snake River with respect to those of the power generating plants of Idaho Power Company. Finally, there was a sense of urgency about the matter generally. Even before the great drought of 1919, those who depended upon the natural flow of the river for their irrigation supply were exceedingly short of water. American Falls had been identified by both the irrigation and power interests as the key location for the next large storage reservoir to control the flow of the river. The June 15, 1923 contract became a primary means to achieve this goal, and the complicated arrangement of water rights between the two parties is laid out in the sections below.

Storage rights for the United States

The water rights acquired by the United States for the American Falls Reservoir were spelled out in the applications for water right permits filed with the Idaho Commissioner of Reclamation on March 30, 1921. The contract of June 15, 1923, attempted to define or limit the water rights and uses of water at American Falls for both the U. S. government and the power company.

It limited the flow available for winter power production purposes at American Falls by either the United States or IPC in favor of releases to reclaim arid lands through irrigation, and for incidental domestic and power purposes downstream.

Natural Flow Rights

Secondary Summer Power Right.[353] The North Side Pumping Division of the Minidoka project of the United States was proposed as a pumping project and required connection to a large power plant for the development of electric power for pumping water up onto high irrigable acreage on the plains north of the river. The federal government proposed to build and operate a 30,000 kW power plant on the west side of the river 200 feet downstream from American Falls Dam.[354] The limiting factor in the government summer power right under the provisions of the contract was not the quantity of water to be used, but rather average daily amount of power generated. However, in no case could the water used under this power right exceed certain daily

353 Ibid., Section 11.

354 Preliminary plans for the government's American Falls power plant provided for installation of four 7,500 kW units. During peak irrigation season demand, the proposed Minidoka North Side Pumping Project would require approximately 25,000 kW which would absorb practically all of the U.S. electrical capacity generated at American Falls. Before construction began on the power plant, Congress changed the Bureau's plan of development by first authorizing a gravity extension, known as the Gooding Division, located north of the river in the vicinity of Jerome, Idaho. Thus, the proposed Minidoka North Side pumping division was delayed, as was the immediate need for the American Falls power plant. See Elwood Mead to Secretary of the Interior, February 3, 1928; Solicitor E. O. Patterson to Secretary of the Interior, February 10, 1928; Hubert Work to W. R. Putnam, April 12, 1928; Elwood Mead, commissioner USBR, to Secretary of the Interior, April 23, 1929; Agreement of April 23, 1929. Folder: Reclamation Bureau, Box 1614, RG 48, NARA II.

averages. The government possessed a secondary right, inferior and subordinate to the summer power right reserved to IPC, to use the following maximum quantities of water at the government's power plant at American Falls to generate the following amounts of electricity:

- 1,500 cfs (6,700 kW) on April 1st, increasing uniformly to 5,000 cfs (25,000 kW) on May 1st, and to 6,000 cfs (30,000 kW) on June 1st;
- 6,000 cfs (30,000 kW) from June 1st and continuing to September 15th, inclusive;
- 6,000 cfs (25,000 kW) on September 15th, reducing uniformly to 1,500 cfs (5,500 kW) on October 31st.
- None, from November 1st to March 31st inclusive, except in connection with the production of a total of 10,000 kW in conjunction with the Minidoka power plant, as authorized.

Primary Winter Power Right.[355] This primary power right to develop not to exceed a half-hour peak of 10,000 kW of power allowed the United States to supply the winter season power demands and uses on the Minidoka project and its proposed extensions during the period November to March, inclusive. The water right, however, was intended to be supplemental and as much power as possible was to be generated at the Minidoka power plant without releasing water which otherwise might be held for irrigation. The balance needed, if any, would be generated at American Falls. In other words, the operations of the two were to be coordinated to produce 10,000 kW without wasting irrigation water over Milner Dam.

Storage Rights [356]

The United States acquired a superior or primary storage right, as against the power company, to construct, fill and use 3,000,000 acre-feet of storage capacity, including the capacity of the American Falls Reservoir anywhere on Snake River, in addition to the storage capacity

355 Contract between the United States and the Idaho Power Company Relative to Power Rights at American Falls, Idaho, June 15, 1923, Section 23.

356 Ibid., Section 10.

available in 1923 at Jackson Lake Reservoir and Lake Walcott.[357] As against the power company, the United States also claimed the right to store for irrigation and domestic purposes the water authorized to be used for summer and winter generation of power by the United States, whenever the Secretary of the Interior decided such water was necessary for irrigation purposes.

Water Rights for IPC

The contract stated that IPC was to convey all of its "water rights, appropriations, water filings and permits, at American Falls," either used or intended to be used in connection with its powerplants at American Falls. However, an exception was made. IPC expressly reserved certain specific "primary or tertiary" water rights out of those it conveyed to the United States under the 1923 contract.[358]

Natural Flow Rights

Primary Summer Power Right.[359] The company reserved the right to use, as against the United States and those claiming, or that may later claim under them, certain quantities of water for power production purposes at American Falls between April 1 and October 31 each year, which was the irrigating season on the Snake River. The right was limited to a maximum flow of 2,500 second-feet between May 1 and

357 At this time the ultimate size of the American Falls Reservoir had not been decided upon, but the alternative of constructing up to a maximum 3,000,000 acre-foot reservoir was still under discussion. Ultimately, it was constructed to only 1,700,000. Additional upstream storage facilities subsequently built in the ensuing decades supplemented the Bureau's overall upper Snake River storage capacity.

358 The terms "primary" and "tertiary" right related only to summer use of water, not storage rights or winter reserved water. IPC had no existing storage rights on the Snake River at American Falls at the time the contract was signed.

359 Contract between the United States and the Idaho Power Company Relative to Power Rights at American Falls, Idaho, June 15, 1923, Ibid., Section 10. During the summer months, the power company had the right to the use of the primary summer power right mentioned in paragraph 10 and also 45,000 acre-feet of stored water. The company never had storage rights at, or above, American Falls. Thus, it appears that Article 10 confines the company's summer power right to the natural flow of the river actually passing by the dam site. During low flow months of dry years a natural flow of under 2,500second-feet could limit the amount of water available to IPC to under the maximum flow of 2,500 second-feet. Recorded water measurements available from the U. S. Geological Survey at the time the contract was prepared indicate clearly that the flows at American Falls in July, August and September were less than the 2,500 cfs provided to the company as so-called "reserved" water rights under Section 10 of the contract. U. S. Geological Survey, *Water Supply Paper No. 1317* (Washington DC: Government Printing Office, 1951), Part 13, 134, 146.

September 15. Water use under this right began on April 1 with 100 second-feet being allowed and increasing uniformly to the 2,500 maximum on May 1. At the opposite end of the irrigating season, allowable use declined uniformly from 2,500 second-feet on September 15 to 100 second-feet on October 31.

This right had priority over the water rights of the United States at American Falls, but in no way affected the water supply for irrigation purposes as the passing of water through penstocks for development of electrical energy did not diminish the quality or quantity of water for irrigation purposes. Water used to generate power at American Falls during the summer was not lost to irrigation, but under the provisions of the contract as executed was available for diversion to the federal Minidoka project or the Twin Falls Carey Act projects for irrigation uses in satisfaction of rights that had already accrued.

The power company also reserved a primary right to use any water brought by it into American Falls from outside the Snake River watershed or water acquired from storage capacity constructed above American Falls, but agreed to subordinate its right to store any water originating within the Snake River watershed above American Falls to the primary storage right of the United States to build, fill and use 3,000,000 acre-feet.

Tertiary Summer Power Right.[360] The power company was granted a tertiary power right to the use of all additional water released, or wasted, from the reservoir by the United States from April 1 to October 31 (i.e., during the irrigation season), over and above the amounts required to furnish the primary rights reserved to IPC (up to 2,500 cfs) and the secondary rights granted to the United States (up to 6,000 cfs).

Reserved Water Right.[361] The power company reserved the right, for power development at American Falls and at its plants downstream of Milner Dam, the first 300,000 acre-feet flowing into the reservoir after October 1 of each year and actually storable within the reservoir.

360 Section 12 of the contract "reserved" to the company all of the power rights from water passing American Falls in excess of the rights designated to the United States under Section 11 and implies that the company already had the right to develop this power at American Falls. In fact, the company only had water rights to the water it had already developed under its state license and a future right applied for in accordance with applicable state water law. Therefore, this contract language "reserved" to the company greater rights than it actually possessed at American Falls. Ibid., Section 12.

361 Ibid., Section 19.

However, the government's winter power right took precedence over this storage right of the company. The 300,000 acre-feet under this reserved right was regulated by rules established in the contract governing the company's primary and secondary storage space (see A and B below, under 5.2.2 Storage Rights Granted).

Winter Waste Power Right.[362] The power company reserved the right, as against the United States, to use for power purposes at its American Falls plant water released, or wasted during the winter, between November 1 and March 31, in excess of the amount allowed to the United States under its winter power and storage rights granted by the contract. The amount of water available for wasting was determined by the Reclamation Service on a forecast basis using a set formula that took into consideration the minimum storable run-off of record and the necessity to provide the power company with 300,000 acre-feet while guaranteeing a reservoir full of water for irrigation purposes each year.

Storage Rights

The United States granted to IPC a total of 300,000 acre-feet of storage space in American Falls Reservoir beginning on October 1 each year to be used for generation of power at its American Falls plant and other plants downstream of Milner Dam. The right was divided into primary and secondary storage. Nearly two-thirds of the storage right had to be used during the non-irrigation season, between October 1 and April 1 and not more than 45,000 acre-feet could be carried over into the irrigating season beyond the time the reservoir was filled each year. The main purpose for the government investment in American Falls dam and reservoir was to store the high wintertime water flows and spring flood flows and to save these quantities of water for release during the irrigation season. Thus, the contract carefully regulated wintertime storage operations to gain greater flows in the summer months than existed under natural flow conditions.

---

362 Ibid., Section 24.

Primary Storage Capacity.[363] IPC was granted the right to perpetual use of 45,000 acre-feet of storage capacity in American Falls Reservoir to regulate its reserved water within the limitations of the contract. The company had the privilege of holding its primary water in storage after October 1 and until November 15 of each year. However, the right to any water which it had in its primary storage capacity of 45,000 acre-feet expired on November 15 and could not be carried over to the next year. Although called a "primary" storage right, the storage privilege provided to the company was not a first, or bottom, storage right in the reservoir, rather it was a storage right shared in common with all of the other reservoir spaceholders.

Secondary Storage Capacity.[364] IPC held a secondary right between October 1 and June 10 to use not to exceed 255,000 acre-feet of storage capacity in American Falls Reservoir, only when such capacity was "not then required and not being used" by the owners of irrigation storage rights. The secondary right of the company terminated when the capacity was needed for storage of water under the primary right for irrigation purposes. Therefore, this secondary right to storage capacity was merely a regulatory right and did not interfere with other rights of spaceholders who stored water for irrigation purposes. The right was granted to the company for the purpose of regulating and controlling the water reserved to the company under its right to the first 300,000 acre-feet of storable water arriving at the reservoir after October 1. Water accruing to this secondary storage right under provisions of the contract was meant for power use during the winter and spring months. However, the right to use any portion of this 255,000 acre-feet was relinquished as soon as the reservoir filled each year, and further, the company released on April 1 of each year all rights and claims to such capacity and water stored therein, except 60,000 acre-feet. All right and claim to the remainder, if any, was released on June 10 each year.

Summer Regulation of 5,000 acre-feet.[365] The company had the right to accumulate not more than 5,000 acre-feet in unused storage capacity

363 Ibid, Sections 16 and 20. In a review of the contract provisions in 1926, Interior Secretary Hubert Work using information provided by F. A. Banks, USRS engineer, wrote: 'The 45,000 acre feet of storage capacity that is part of the consideration is for the Power Company to run down the Snake River Canyon in the summer and offset in part, at the sites below Milner Dam, the curtailment of summer power at American Falls." Work to Wheeler, March 10, 1926. Folder: Reclamation Bureau, Box 1614, RG 48, NARA II.

364 Contract between the United States and the Idaho Power Company Relative to Power Rights at American Falls, Idaho, June 15, 1923, Section 18.

in the reservoir, out of the amounts of water to which it was entitled in order to reregulate its primary summer power right. Except that after the reservoir filled in any year, the company was entitled to only that portion of the 5,000 acre-feet accumulated after the reservoir filled. Upon a demand by the United States, the company was compelled to release any accumulated portion of the 5,000 acre-feet, whenever the available stored water in the American Falls Reservoir to which the Minidoka project and extensions were entitled, amounted to less than 5,000 acre-feet. This storage provision allowed the company to shape the water flows in the summertime to the company's load demand, and thus to more effectively use its power plant than under the unregulated conditions that prevailed prior to construction of the federal reservoir.

Limitations on Rights to Demand Secondary Storage.[366] The company agreed to limit its right to demand release of water from the reservoir for use at its power plants below Milner Dam. This right was limited to the 45,000 acre-feet of primary storage rights and to such portion of the 255,000 acre-feet of secondary storage rights as was needed for beneficial use for power development at Shoshone Falls or Twin Falls.[367] The releases of secondary storage were limited so as not to exceed the amount necessary to create the following total flows at the head of Shoshone Falls:

790 second-feet	Oct. 1st to November 30th
890 second-feet	December 1st to January 31st
790 second-feet	February 1st to February 12th
690 second-feet	February 13th to April 15th
790 second-feet	April 16th to June 10th

365 Ibid., Section 14. The use of 5,000 af of storage capacity in American Falls reservoir for reregulation of the company's summer power right allowed the company to adjust daily flows to meet peak load demands. The company did not possess this advantage before construction of American Falls dam when the river was unregulated.

366 Contract between the United States and the Idaho Power Company Relative to Power Rights at American Falls, Idaho, June 15, 1923, Section 21. Secretary Work later wrote that Articles 14 and 21 of the contract, according to information provided by USRS engineer Banks, was made with the intent of reducing the winter power use from an average of about 5,000 second-feet to an average of about 800 second feet. The 800 second-feet of winter flow that the company was retaining, Work asserted, was "the equivalent to the 255,000 acre feet of secondary storage right described in the contract." Work to Wheeler, March 10, 1926 Folder: Reclamation Bureau, Box 1614, RG 48, NARA II.

367 Section 22 of the contract provided for changing the place of measurement of the flow below Milner Dam if a powerhouse was constructed at Twin Falls. If so, the schedule of flow set out in Section 21 was to be reduced by 100 second-feet for each period.

These releases from American Falls Reservoir were to be controlled in such a way that the required flows at Shoshone Falls could be maintained while not using more than 3,000 acre-feet of storage capacity at Lake Walcott. These operations in no way affected the company's right to require releases at any time from storage at American Falls of any part of its 45,000 acre-feet of primary storage.

In consideration for the rights acquired from the power company in connection with the American Falls Reservoir under the June 15, 1923, contract, the United States agreed to pay to IPC $1,000,000 in cash, payable in four equal installments. The final payment was made on April 20, 1926.[368]

In addressing the terms of the agreement between the Bureau and IPC, W. G. Swendsen, Idaho Commissioner of Reclamation, summarized the effect of the agreement in a letter to Idaho Governor C. C. Moore, dated July 9, 1923, written only a few weeks following signing of the agreement:

> Briefly, it may be stated that the contract between the power company and the government provides for the regulation of Snake River in the interests of irrigation, adequate to the needs of a million and a half acre feet, reservoir capacity. If and when the reservoir is constructed beyond this capacity, it will be necessary, if the reservoir is to be filled each year, to acquire additional power rights. It is my belief that the contract as consummated is satisfactory, adequately protects the interests of the State, and is fair and equitable.[369]

State rights: Storage, Irrigation, Power

During negotiations for the purchase of certain rights and properties of Idaho Power Company at American Falls leading up to the signing of the 1923 contract, Idaho Power Company and the United States executed a memorandum of agreement regarding the filing of state water permits to protect the rights of the two contracting parties, whether or not the federal water storage project was built at American Falls.[370]

---

368 John H. Edwards, assistant secretary, to General Accounting Office, Claims Division, August 11, 1926. Folder: Reclamation Bureau, Box 1614, RG 48, NARA II.

369 Swendsen to Moore, July 9, 1923. Folder 6, Box 3, AR 2/13. ISHS.

370 Memorandum of Agreement, March 17, 1921, signed by F. F. Johnson, president IPC, Clarence T. Ward, Secretary of the Interior, and B. E. Stoutemyer, district counsel, U. S.

U.S. Permits at American Falls

District Counsel B. E. Stoutemyer filed the U. S. Reclamation Service's application for a permit to store 3,000,000 acre-feet at American Falls with the State of Idaho Commissioner of Reclamation W. G. Swendsen on March 30, 1921. On May 18, Swendsen approved the application as Permit R-269.[371] Construction on the reservoir began in 1925 and was completed in 1927. Acting through E. B. Darlington, superintendent, Bureau of Reclamation, the United States made proof of completion of works under its Permit R-269, with a priority date of March 30, 1921. The state Commissioner of Reclamation issued a certificate of completion of works for a storage capacity of 1,800,000 acre-feet per year (including 100,000 acre-feet of "bank storage"), being the full capacity of the reservoir, on July 12, 1932. On April 22, 1941, the United States made proof of beneficial use for a storage right of 1,800,000 acre-feet per annum. The Idaho Commissioner of Reclamation on July 9, 1942, issued state water right license 01-2064, granting to the United States storage rights to the full capacity of its reservoir.[372]

The place of use of the water stored and diverted was listed as lands included within the irrigation districts and water companies who held storage rights in the reservoir, the Minidoka project, and supplemental water to 1,000,000 acres of land throughout the Upper Snake River Valley. Domestic use was considered incidental to irrigation. A license was also issued on July 9, 1942, for Permit No. 15134 (File 01-02040), the diversion right for 8,000 second-feet for irrigation, and incidental domestic purposes, that had been filed with the application to store water in American Falls Reservoir on March 30, 1921.[373]

Reclamation Service. Cancelled Permit for American Falls, Permit 152001, File 01-2041. IDWR Back File 01-2041.

371 Application for a Permit to Construct the American Falls Reservoir, Application No. R-21513, Permit No. R-269, approved May 18, 1921. Idaho State Idaho State Water Right 01-2064 Back File.

372 The reservoir was constructed to only hold 1,700,000 acre-feet. The additional 100,000 acre-feet was considered part of the "bank storage." E. V. Berg, Commissioner of Reclamation to B. E. Stoutemyer, District Counsel USBR, July 13, 1942, enclosing License and Certificate of Idaho State Water Right, R-269; B. E. Stoutemyer to Commissioner, July 15, 1942. Folder: Minidoka; Box 718 Minidoka Project, 031 – 032, PCF, 1930-1945, RG 115, NARA-Denver.

373 E. V. Berg, Commissioner of Reclamation to B. E. Stoutemyer, District Counsel USBR, July 13, 1942, enclosing License and Certificate of Idaho State Water Right, R-269 and License and Certificate of Idaho State Water Right, 15134. Folder: Minidoka; Box 718 Minidoka Project, 031 – 032, PCF, 1930-1945, RG 115, NARA-Denver.

Idaho Power Company's filing

On March 17, 1921, nearly two weeks before the Reclamation Service filed its application for a permit to store 3,000,000 acre-feet of water in the American Falls Reservoir, IPC had applied as provided for in the memorandum of agreement with the United States for a permit to store 300,000 acre-feet of water at the reservoir site that would be inundated behind the Reclamation Service's proposed American Falls Dam.

In a companion permit the power company also requested a right to divert 1,200 cfs from the reservoir for power generation at American Falls, Twin Falls, Shoshone Falls, Auger Falls, Upper Salmon Falls, Lower Salmon Falls and Swan Falls. On June 21, Idaho's Commissioner of Reclamation, W. G. Swendsen, issued Permit R-273 (01-2065) for storage at American Falls with a priority date of March 17, 1921,[374] and for reservoir diversion for power generation purposes at multiple sites on the Snake River under Permit 15201 (01-2041) with the same priority date.[375] For both permits proof of completion was required by June 21, 1926, and proof of beneficial use five years later. The schedules for meeting these milestones were pushed back five years by an order granting extension issued to the company in September 1923.[376]

Idaho Power Company's Permits

As for IPC's water permits at American Falls, the company's special agent E. C. Kierstad filed proof of completion of works for Permits 15201 (01-2041) and R-273 (01-2065) on June 20, 1931.[377] This action

---

374 Application for a Permit to Construct the American Falls Reservoir and to Store the Unappropriated Waters of the State of Idaho, Application No. 21475, Permit No. R-273, approved June 28, 1921. Cancelled Idaho State Water Permit 01-2065 Back File.

375 Application for Permit to Appropriate the Public Waters of the State of Idaho, Application No. 21476, Permit No. 15201, approved June 21, 1921. Idaho State Cancelled Water Permit 01-2041 Back File.

376 Application for a Permit to Construct the American Falls Reservoir And to Store the Unappropriated Waters of the State of Idaho, Application No. 21475, Permit No. R-273, approved June 28, 1921; Cancelled Water Permit 01-2065 Back File; and Application for Permit to Appropriate the Public Waters of the State of Idaho, Application No. 21476, Permit No. 15201, approved June 21, 1921. Cancelled Idaho State Water Permit 01-2041 Back File.

377 Reference to this submission made in R. W. Faris, Commissioner of Reclamation, to Mr. E. C Kiersted, Special Agent, Idaho Power Company, Re: Permits Nos. 15201 & R-273, June 23, 1931. Cancelled Idaho State Water Permit 01-2041 Back File.

by IPC created some confusion among irrigation spaceholders – the American Falls Reservoir District, in particular. The district's president R. E. Shepherd had read about IPC's proof of completion for Permit 15201 (01-2041) in a public notice published in the local *North Side News*. Afterward, he wrote to R. W. Faris, the Idaho Commissioner of Reclamation (1931-38), and questioned the reason for IPC's 1,200 second-foot water right filing as the Bureau of Reclamation was responsible for construction of the works and "the rights of the Idaho Power Company in the American Falls Reservoir are all defined under contract between the Power Company and the United States Bureau of Reclamation."[378]

Faris and his department were equally at a loss to explain IPC's request for either permit. Faris sent a field agent to investigate and confer with IPC regarding their two permits. The proposed power plants at Twin Falls, Auger Falls, and Upper Salmon Falls mentioned in the permit had not been built and no additional capacity had been added at Shoshone Falls, Lower Salmon Falls or Swan Falls under Permit 15201. Therefore, no permit for completion of works could be issued for any of these facilities. At the East Side plant below American Falls an additional 2,000 second-foot unit had been added, but it was covered by Permit 16655 for which a Certificate of Completion was issued on October 7, 1927. This unit brought the total capacity of the East Side plant to 9,000 second-feet, an amount in excess of the capacity of the company's diversion works at American Falls. Thus, issuing another certificate under Permit 15201, reasoned Faris, would be "superfluous and a futile gesture." As for the reservoir permit, Faris noted, the reservoir had been constructed by the Bureau, not IPC, and a Certificate of Completion had already been issued by the State of Idaho to the Bureau on May 16, 1931. Further, under the 1929, *Woodville Decree*, the United States had been adjudicated all of the storage rights in American Falls Reservoir.

On September 29, 1932, the commissioner informed IPC his department understood that the company's rights to store water at American Falls had been waived and compensated for by its 1923 contractual agreement with the United States; therefore, unless

---

378 R. E. Shepard, President, American Falls Reservoir District, to Hon. R. W. Faris, Commissioner of Reclamation, May 7, 1931. Cancelled Idaho State Water Permit 01-2065 Back File.

otherwise persuaded by the company, he intended to deny both of the company's proofs of completion.[379]

Kiersted responded that the storage permit contemplated construction of a reservoir of 300,000 acre-feet for storage purposes (01-2065), the company already having continuous flow rights at American Falls by earlier appropriations, uses and permits. The use permit (1-2041) merely contemplated use of this water from the storage reservoir and "complimented the storage permit." These rights, argued Kiersted, were not additional to the already-claimed waters, but merely contemplated using them from the reservoir. As for the many power sites downstream, both developed and undeveloped, referred to in the permits, Kiersted suggested that IPC made no claim for these local developments, "excepting only as they related to the storage and use works, and the storage and use works to them." Each operating plant and undeveloped power site had its own independent water right, but also had "a certain relation to the reservoir." Any plan for storage of water in reservoirs upstream, noted Kiersted, had to take into consideration its impact on the run-of-the-river power plants on the Snake River located below. The company's own plans at American Falls had done so and these plans "became so recognized in the June 15, 1923 contract."

Moreover, the United States, IPC, and the State of Idaho had all agreed to the provisions of the 1923 contract, wrote Kiersted. The U. S. plans for American Falls Reservoir and those of IPC "were merged as neither could be carried on independently." Ultimately, the United States carried out the proposed development, but just on a more comprehensive scale than proposed by the power company. "By giving way" to the larger federal project, argued Kiersted, "the company did not waive any of its rights" – they were acquired independently on their own, had their own legal standing and were not merely contract rights.

---

379 Commissioner of Reclamation to Idaho Power Company, Attention: Mr. E. C. Kierstad, Special Agent, September 29, 1932; Cancelled Idaho State Water Permit 01-2065 Back File. The *Woodville Decree*, arising from the *Woodville Canal Company vs. Clark and Edwards Canal and Irrigation Company, et al.*, in the District Court of the United States for the District of Idaho, Eastern Division, reaffirmed both the *Rexburg* and *Foster* decrees in 1929. It awarded the Secretary of the Interior 1.7 million acre-feet of storage in the American Falls Reservoir and 8,000 cfs for direct diversion under a priority date of March 30, 1921 to irrigate the land described in the water application filed with the Idaho Commissioner of Reclamation. Idaho Power Company was not a party to the suit.

The certificates of proof of completion were valid under the statutes of Idaho, Kiersted concluded, and should be issued.[380]

In a follow-up letter, dated December 31, 1932, Faris reiterated his position citing the *Woodville Decree* that "adjudicated and decreed to the United States all the storage capacity of the American Falls Reservoir" as his "main reason" for opposing issuance of a license. The commissioner also rejected Kierstad's request "to withdraw proof under Permit No. 15201" and have "a certificate of completion of works under Permit No. R-273 be issued in the regular form" with "additional qualification and reservations" – namely,

> That the right to store waters under this permit is to be construed in relation to and in connection with Permit No. R-269 in the name of United States of America bear date of priority as of March 30, 1921, upon which permit certification of completion of reservoir and the right to store water was issued on July 12, 1932, and the right granted herein to store water is not in addition to the right to store water under said permit but is a part thereof.

Faris maintained that the power company had "abandoned all claim of right under Permit No. 15201," and in accordance with the decision rendered in *Quigg Irrigation Co., et al, vs. Dietz et. al.*, as Faris understood the court's ruling, IPC had "nothing upon which to predicate a certificate of completion of works." Although acknowledging his lack of legal expertise, the commissioner nevertheless thought "that Idaho Power Company must rely upon its contract with the United States, referred to in its proof of completion of works, for its right to store water in American Falls reservoir."[381]

Leo M. Bresnahan of the Idaho state Attorney General's office in a memorandum opinion dated February 10, 1933, confirmed Faris's opinion on the matter and concluded that the commissioner would be justified in acting to deny IPC a certificate of completion of work. His opinion concluded:

---

380 E. C. Kiersted, Special Agent, Idaho Power Company, to R. W. Faris, Commissioner of Reclamation, October 8, 1932. Cancelled Idaho State Water Permit 01-2041 Back File.

381 Farris to Idaho Power Company, Attention: E. C. Kierstad, Special Agent, Re; Permits No. 15201 and R-273, December 31, 1932. Cancelled Idaho State Water Permit 01-2065 Back File.

> We are, therefore, of the opinion that upon the facts submitted to us in this case, that to apply the work of the Federal Government in the construction of the American Falls Reservoir upon the permit in question of the Idaho Power Company, would in effect pivot two permits at the same time upon identical work and that such work should not be applied upon the permit in question for the reason that each must stand upon separate or additional construction, otherwise compliance with the terms of one permit would work a fictitious compliance with the other and such proof of completion of works is insufficient to authorize the issuance of a certificate of completion of works by the Department.[382]

The Bureau's counsel B.E. Stoutemyer discussed the matter of issuance of a certificate of completion with James L. Boone, an attorney for IPC, and Commissioner Faris. While not necessarily opposed to issuance of the certificate, Stoutemeyer argued for inclusion of a clause that recognized government ownership and control over the reservoir and that the right of IPC to store 300,000 acre-feet was limited by terms of the 1923 contract. Issuance of a certificate was in no way to impair the contract or the rights of the United States under its certificate of completion of works for 1,800,000 acre-feet of water under Reservoir Permit R-269. In his letter to Commissioner Faris, Stoutemyer specifically addressed and reiterated the subordinate status of the company's storage rights for power purposes to those of irrigation spaceholders.

> As you have probably noticed in reading the contract in question, the right of the company to store the 300,000 acre feet in the reservoir temporarily when the capacity of the reservoir is not needed for storing water for irrigation purposes, is a secondary right which does not interfere with the primary right to use the same capacity for irrigation purposes, – the secondary right of the company terminating when the capacity is needed for the storage of water under the primary right for irrigation purposes.

Under these circumstances, issuance of a certificate of completion and ultimately a license would not enlarge the rights of the power company, Stoutemyer believed, and therefore, would not be

---

382 Leo M. Bresnahan, Memorandum Re: Letter to R. W. Faris, Commissioner of Reclamation, under the Date of January 4, 1933, February 10, 1933. Cancelled Idaho State Water Permit 01-2065 Back File.

objectionable from the standpoint of the federal government, or other irrigation interests, on Snake River.[383]

In the three decades following Stoutemyer's letter, the issue of IPC's American Falls water permits lay dormant. Hearings were to be scheduled on the permits by the Idaho Commissioner of Reclamation, but apparently were never held.[384] Finally, in the summer of 1967, the issue re-emerged after the Idaho Department of Water Administration – successor to the Idaho Department of Reclamation – uncovered the unresolved permits during a review of its records. On July 19, 1967, State Reclamation Engineer R. Keith Higginson wrote to IPC to inform company officials that his department intended to cancel the permits.[385]

Once again, IPC objected and pressed for issuance of water rights licenses. Writing back to Higginson, in November 1967, IPC Vice President H. R. Moore argued that the company had "carefully checked its files concerning these permits and examined the contract between the Company and the United States Bureau of Reclamation," and had come to the conclusion there was nothing "in the contract which prevents or restrains the issuance of these permits; and in fact, the contract recognized the filed rights of Idaho Power Company." Noting that the company had "complied with the statutory requirements," Moore requested that the department "issue the required certifications and thus bring the matter to a conclusion."[386] In the wake of Moore's letter, Higginson met separately with the Bureau and the power company and elected in October 1968, "to take no further action until other issues are settled in the future."[387]

Two years later, in the fall of 1970, the power company reopened the matter. On September 18, James E. Bruce, now the IPC's vice president

383 B. E. Stoutemyer, District Counsel to R. W. Faris, State Commissioner of Reclamation, April 3, 1933. Cancelled Idaho State Water Permit 01-2065 Back File.

384 No records have yet come to light that indicates such proceedings were initiated. See Commissioner of Reclamation to Hon. James R. Bothwell, Permit R-273, June 9, 1933. Canceled Idaho State Water Permit 01-2065 Back File.

385 R. Keith Higginson, State Reclamation Engineer, to Idaho Power Company, Attn: Gale Farley, Re: Permit R-273; 15201, July 19, 1967. Canceled Idaho State Water Permit 01-2065 Back File.

386 H. R. Moore, Vice President, Idaho Power Company, to State of Idaho, Department of Reclamation, Attention: Mr. R. Keith Higginson, State Reclamation Engineer, Re: Permit R-273, 15201, November 9, 1967; and H. R. Moore, Vice President, Idaho Power Company, to Mr. R. Keith Higginson, State Reclamation Engineer, Re: Permits R-273 and 15201, July 25, 1967.Canceled Idaho State Water Permit 01-2065 Back File.

387 Memorandum – Permits R-273, 15201, October 25, 1968. Canceled Idaho State Water Permit 01-2065 Back File.

and secretary, wrote to Higginson to once again request that the Department of Water Administration "proceed to issue the required certificates."[388] Less than a month later, on October 1, Higginson issued an "Order to Show Cause" to the power company "as to why Proof of Completion of Works should not be rejected and Permits 15201/01-2041 and R-273/01-2065 cancelled and held for naught."[389]

On December 1, 1970, Bruce and IPC attorney R. P. Parry supplied the power company's answer. After briefly reviewing the history of IPC's involvement with American Falls, the two argued that with regard to "the 300,000 acre-feet covered by Permit No. R-273" the 1923 contract "preserved, defined and specifically provided for" that amount under paragraphs 16 and 18.[390] Furthermore, paragraphs 6, 9(d) and 30 "provided that the Company was not surrendering, or in any way disposing of its existing water rights but was only permitting the United States to use them in connection with its American Falls Reservoir." Those rights, moreover, Parry and Bruce noted quoting from the language of the contract, "were reserved and were '***not to be considered as mere contract rights***'"[391]

For the purposes of the operation of the American Falls Reservoir, however, Bruce and Parry observed, the Secretary of the Interior "held the water available in American Falls Reservoir under Permits R-269 and 15134 as trustee for the Company and all other contract holders." IPC had "continuously applied to a beneficial use its rights in said stored water…and the Company and all other such contract holders are the equitable owners of the right to so use these waters."[392] Given this

388 James E. Bruce, Vice President and Secretary, Idaho Power Company, to Mr. R. Keith Higginson, Idaho Department of Water Administration, Re: Permits R-273 and 15201, September 18, 1970. Canceled Idaho State Water Permit 01-2065 Back File.

389 Before the Director of the Department of Water Administration of the State of Idaho, In the Matter of Permits No. 15201/01-2041 and R-273/01-2065 in the name of Idaho Power Company, Order to Show Cause, October 1, 1970. Canceled Idaho State Water Permit 01-2065 Back File.

390 Before the Director of the Department of Water Administration of the State of Idaho, In the Matter of Permits No. 15201/01-2041 and R-273/01-2065 in the name of Idaho Power Company, Answer to Order to Show Cause, December 1, 1970, 4-5. Canceled Idaho State Water Permit 01-2065 Back File.

391 Before the Director of the Department of Water Administration of the State of Idaho, In the Matter of Permits No. 15201/01-2041 and R-273/01-2065 in the name of Idaho Power Company, Answer to Order to Show Cause, December 1, 1970, 6. Canceled Idaho State Water Permit 01-2065 Back File. Asterisks in original.

392 Before the Director of the Department of Water Administration of the State of Idaho, In the Matter of Permits No. 15201/01-2041 and R-273/01-2065 in the name of Idaho Power Company, Answer to Order to Show Cause, December 1, 1970, 7. Canceled Idaho State Water Permit 01-2065 Back File.

information, and citing the *Woodville Decree* of June 29, 1929 that confirmed the Secretary of Interior's water filings for the American Falls Reservoir, Parry and Bruce on behalf of IPC requested the department find,

> … (a) that all of the rights the Company covered by Permit Nos. R-273 and 15502 [*sic, 15201*] have been merged into rights granted to the Secretary of the Interior in Permit Nos. R-268 [*sic,* R-269] and 15134; (b) that the Secretary of the Interior holds and maintains the rights under said permits as trustee for the Company of all of the rights covered by Permit Nos. R-273 and 15502 [*sic, 15201*]; and (c) that in accordance therewith no further action need be taken with respect to Permit Nos. R-273 and 15502 [*sic, 15201*].[393]

More than four months later, on April 21, 1971, the Department of Water Administration issued a "Memorandum Decision." This ruling concluded that IPC had "failed to submit proof that it has constructed the necessary works to divert the water and apply it to beneficial use"; nevertheless, it adopted the power company's three-point request. The "rights established by the Idaho Power Company, by reason of Permit Nos. 01-2041 and 01-2065" were found to be merged with those "granted to the United States of America." Further, "by reason of contracts and agreements entered into between the Company and the United States of America, Bureau of Reclamation, those rights are held in trust for use by the Company under the water rights of the United States at American Falls Reservoir." As IPC's "rights to water at American Falls Dam and Reservoir" were "covered by reason of Licenses issued to the United States of America," Permits 01-2041 and 01-2065 were ruled as "duplicated" and ordered "cancelled."[394] The following day, an order of cancellation for the two permits was issued.[395]

---

393 Before the Director of the Department of Water Administration of the State of Idaho, In the Matter of Permits No. 15201/01-2041 and R-273/01-2065 in the name of Idaho Power Company, Answer to Order to Show Cause, December 1, 1970, 8. Canceled Idaho State Water Permit 01-2065 Back File.

394 Before the Director of the Idaho Department of Water Administration, In the Matter of Permits Nos. 01-2041 and 01-2065 in the name of Idaho Power Company, Memorandum Decision, April 21, 1971, 2. Canceled Idaho State Water Permit 01-2065 Back File.

395 State of Idaho, Department of Water Administration, In the Matter of Cancellation of Permit to Appropriate the Public Waters of the State of Idaho No. R-273/01-2065 in the name of Idaho Power Company, Order of Cancellation, April 22, 1971. Canceled Idaho State Water Permit 01-2065 Back File.

Upper Basin Water Management

The 1923 contract created a framework for the management of the water resources of the Upper Basin – but, as the back-and-forth regarding IPC's water permits at American Falls illustrates, it did not eradicate conflict between irrigation and hydropower interests. The contract was quite voluminous and complicated; its terms were subject to interpretation by both parties, and disagreements arose almost immediately.

For its part, IPC disagreed with how the Bureau accounted for excess water available for power purposes under Section 24 of the contract. Under the 1923 contract, the power company reserved a right to use winter waste water for power purposes at its American Falls plant. After a wet year in 1927 resulted in available "waste water," the Bureau argued that in calculating the storable run-off available for wasting, consideration should be given to unfilled capacity in all the reservoirs on Snake River rather than to unfilled capacity at American Falls Reservoir alone. This reading of the contract was favorable to the Bureau and its basin-wide storage operations. According to the Bureau's calculations, the amount of water available for waste under terms of the contract was less than the approximately 2,500 cfs required to be released for operation of the Minidoka power plant. Release for operation of the Minidoka plant would thus give IPC more water at American Falls than contractually required, but significantly less than the 7,000 to 8,000 cfs it could beneficially use at its recently-expanded American Falls plant. While the power company seemingly acquiesced to the Bureau's interpretation, by 1929, it took exception. That year, IPC vice-president and manager, M. L. Hibbard complained to U.S. Reclamation Commissioner Elwood Mead that the Bureau's construction of the terms "unfilled storage capacity" was untenable and Section 24 of the contract meant only "the existing or unfilled storage capacity of American Falls reservoir."[396]

Apparently, this issue was not resolved.

In subsequent years, the Bureau issued its report on unfilled storage capacity in compliance with Article 24 showing the cumulative unfilled

396 Stoutemyer to Darlington, October 1, 1927; Walter to Superintendent Darlington, October 5, 1927; Hibbard to Mead, December 14, 1929. Folder 301.1 MP, Dams & Reservoirs, American Falls Dam Jan. 1, 1926 - Dec. 31, 1929 (Folder 301.1), Box 729 Minidoka Project, 301 – 301.1, GAPR, 1919-1945, RG 115, NARA-Denver.

capacity of its various reservoirs and the power company responded with its protest against including any other than the unfilled capacity of American Falls Reservoir alone.[397] Neither side conceded the point, but as a practical matter of the administration of this provision of the contract, the United States operated the dam in close cooperation with the company and, in fact, a considerable amount of waste water in the winter time was made available to the company for power purposes.[398]

Dry years also brought controversy as to how the storage reservoir was operated. As far as the Bureau was concerned, in a water deficient year they had the right to hold essentially all the winter and spring flood flows in the reservoir – save IPC's 45,000 acre-foot primary power right – for the use and benefit of irrigators (spaceholders) who contracted to store water in American Falls. The Bureau remained determined to conserve as much irrigation water as needed to assure a reliable water supply to irrigators, while the power company endeavored to obtain as much water for power generation as possible under its rights specified in the contract.

The Bureau's district counsel B. E. Stoutemyer, for instance, maintained in a 1937 letter to the Minidoka project superintendent that the United States was not responsible for furnishing its *stored water* for the purpose of keeping up the specified flow at Shoshone Falls under Section 21 of the contract. All the government was obliged to do was bypass a limited amount of *natural flow* for that purpose in the wintertime period from October 1 through June 10, without drawing down the reservoir.

"If we are by-passing all of the natural flow that there is available at the reservoir," noted Stoutemyer, "we are doing all that can be required even though other parties having [superior] irrigation rights may divert such natural flow before it reaches the Shoshone Falls." He added: "If the Power Company wants any stored water turned out for the benefit of

---

397 For example, see Mead to Idaho Power Company, December 23, 1931; Hubbard to Mead, December 29, 1931 and Mead to Idaho Power Company, January 6, 1933; Hubbard to Mead, January 13, 1933. Folder: 301.1 Minidoka – Correspondence re: Dams and Reservoirs, American Falls Dam, 1930-1934 (Folder: 301.1 Minidoka -- Correspondence), Box 752 Minidoka Project, 301 – 301.1, PCF, 1930–1945, RG 115, NARA-Denver.

398 William Burpee, Interior Department Solicitor, "Informal Notes on the 1923 American Falls Contract with Idaho Power Company," (1962), 32. 05-14-2010 USBOR Response to IPC, AGO 06-17-2010-1 CO IAGO, AF006222- AF00681.

its power plant at Shoshone Falls, it would have to order such stored water released out of its primary right of 45,000 acre-feet."[399]

Prevailing water supply conditions in the upper basin – whether a year was wet or dry – further complicated water management, forcing the Bureau of Reclamation and IPC to re-evaluate on an almost yearly basis how contract provisions would be realized and applied. This became especially important as Idaho Power looked to develop the hydroelectric potential of the area below Milner Dam beginning in the 1920s and as groundwater pumping brought about an expansion of irrigated acreage beginning in the late 1940s.

Proposed IPC projects such as Twin Falls and Upper Salmon Falls and later Hells Canyon – all of which when fully developed would have relied in part on return flows past Milner – brought the power company, the federal government, and the State of Idaho into conflict a number of times.

These conflicts led both to additional agreements that modified portions of the 1923 contract and in some cases, to the imposition of limitations on the operations of IPC's facilities below Milner. Throughout this period, all sides nevertheless continued to look to the provisions of the 1923 contract to guide the allocation of Snake River water to irrigation and hydropower uses.

399 B. E. Stoutemyer to Minidoka Project Superintendent, May 7, 1937. Folder: 223.03 Minidoka – Lease of Power Contracts, Idaho Power Company, Jan. 1930 through March 1942, Box 740 Minidoka Project, 223.03, PCF, 1930-1945, RG 115, NARA-Denver.

PART TWO

IRRIGATION, HYDROPOWER AND SUBORDINATION

A careful consideration of the available water supply for arid lands leads to the conclusion that when the total flow of Snake River and its tributaries are completely controlled in the interest of irrigation, through the construction of storage reservoirs and otherwise, the water supply will be insufficient to accommodate all of the lands susceptible of irrigation....it is extremely important that any power permits granted in connection with these applications shall be conditioned upon the State's present and future right to divert, use and impound as much water as may be necessary for a complete development of its agricultural resources, both for irrigation, domestic and other consuming uses...

W. G. Swendsen, Idaho Commissioner of Reclamation, to Executive Secretary, Federal Power Commission, October 26, 1922

The right to divert and appropriate the unappropriated waters of any natural stream to beneficial uses, shall never be denied, except that the state may regulate and limit the use thereof for power purposes.

Idaho State Constitutional Amendment, 1928

...It seems to us that it would be contrary to sound public policy to allow any additional power rights to attach for use in the Snake River Canyon to such an extent as to require the waste of water over Milner Dam, since all the water available in Snake River above Milner Dam is needed for irrigation purposes even at the present time. . . the need to conserve all the available water above Milner Dam for irrigation purposes will become more and more evident as the years go on. On the other hand there are many places in the State of Idaho where additional power can be developed without infringing on the water supply that is needed for irrigation purposes.

Bureau of Reclamation District Counsel B. E. Stoutemyer to Idaho State Commissioner of Reclamation R.W. Faris, June 21, 1937

6

Water Rights, Power Licensing and the Amendment

The construction of American Falls Reservoir and expansion of Idaho Power in the Snake River Basin in the 1920s solved a series of problems but also created a sensitive issue for Idaho.

With those developments in place, the state had to weigh the needs of irrigation – a water use which it had, constitutionally and as a matter of public policy, declared the highest use of the Snake River – against the growing need of hydroelectric power development below American Falls. The Snake River was the life-blood of irrigators, a treasured, scarce resource to be harnessed and used to transform the arid environment. For nearly two-thirds of the state, water resources were a vital asset for agricultural production, and water requirements of arid lands capable of reclamation by irrigation far exceeded total available water supply. According to the Idaho Department of Reclamation, only about half of the water supply available for irrigation was being diverted in the mid-1920s. Much of the remaining supply ran to "waste" below Milner Dam as winter and spring "flood flow." As a matter of public policy, care needed to be taken to direct and control grants of water rights so as to safeguard the remaining available water supply for future agricultural purposes.

The Snake River and its tributary streams also possessed great potential for producing hydroelectric power – energy needed for pumping irrigation water into highline canals, as well as to provide heat and electricity to towns and to rural families and businesses. In fact, according to the U. S. Bureau of Water Resources, Idaho was blessed with abundant superb power sites, ranking fifth among all of the states of the union in available water power. This potential could be vastly increased by storage reservoirs regulating stream flow to the advantage of hydroelectric power production. Idaho's power industry then was still in its formative stage of development. Only about 7 percent of the state's estimated power potential had been developed and only an additional 1.1 percent was covered by existing rights; the remaining 92 percent of the power sites were still subject to future appropriation and development.[400]

The State of Idaho was obliged to protect the interests of its citizens by managing its water resources in the public interest, considering irrigation, power, mining, domestic, and other uses. It had, however, no water agency with the authority to construct water projects or to reserve water and water rights to meet the future needs of irrigated agriculture. Under the existing system of water appropriation, the state Department of Reclamation had little or nothing to say as to where or how its unappropriated water supply should be used – the state exercised no direction in the matter. To properly use its unappropriated water, state officials from time to time expressed interest in a state water plan that delegated to a department or agency of state government authority to manage the state's water resources among competing users for the long-term benefit of the public. No such power was delegated to the state Department of Reclamation when it was created in 1919. Instead, Idaho initially looked to the Federal Power Commission (created in 1920) and its licensing process for private hydropower facilities to safeguard the state's commitment to agriculture as the primary use of the Snake River's water.

Up to this time, Idaho Commissioner of Reclamation W. G. Swendsen wrote in 1922, "the diversion of water and its use in irrigation and power development … generally followed the lines of least resistance, and little or no consequence has been given to our ultimate future needs." Water utilization in the state had reached a stage which "demands a general, complete, and careful consideration of the economic problems involved, in order that the interests of the State and its present and future citizens may be safeguarded." The stage of development had been reached where large storage reservoirs soon would be constructed to conserve and make use of the remaining unappropriated water of the state. The question remained – who would control this surplus water supply; private companies, communities, the federal government, or other public agencies, and for what purposes? Idaho's water resources officials, like Commissioner Swendsen,

400 W. G. Swendsen, Commissioner, Department of Reclamation, State of Idaho, to Hon. C. C. Moore, Governor of Idaho, Re: Idaho Water Power, August 18, 1924, 2. Folder 10: Reclamation, Department of: Correspondence, 1923-1926, Box 3, AR 2/13, ISHS.

believed that the State of Idaho had to assume for itself the responsibility of limiting the potential impact of hydropower developments on future irrigation.[401]

This growing awareness of the public interest in controlling water resources emerged in the context of several clashes between irrigators and hydroelectric power companies in the early 1920s. The construction of American Falls Reservoir, discussed previously, was the major battleground, but several other skirmishes were fought that had significant long-term implications. IPC's proposed Twin Falls and Upper Salmon Falls facilities proposed for sites a short distance downstream of Milner Dam were two such conflicts that challenged the irrigators' right to dry up the Snake River at Milner during irrigation season. Finally, Pacific Power & Light's proposed development of ten power sites along the Snake River between Lewiston, Idaho and Homestead, Oregon highlighted the failings of the state's appropriative water laws and showcased certain weaknesses in the office of the state Commissioner of Reclamation.

In these instances, irrigators, Idaho reclamation authorities and the Bureau of Reclamation opposed hydropower projects they believed jeopardized the future of irrigated agriculture in the Snake River Basin. At a minimum, they sought to impose limitations on water use by power companies while providing legal protections for the future use of water for irrigation, as against its continuing development for hydroelectric power purposes. The power companies, for their part, resisted such proscriptions as they might undermine the companies' future ability to generate electricity and make federal-produced power more attractive. Irrigators, the State of Idaho, and the USBR, however, did not wish to deny all hydropower developments. Many local irrigators did support cheap public power; but while the operations of IPC's Twin Falls project were ultimately subordinated to irrigation and both its Upper Salmon Falls and Pacific Power & Light's developments were denied federal sanction, IPC's Swan Falls and Lower Salmon Falls plants were licensed by the Federal Power Commission with no restrictions placed on their operation. These plants, quite simply, were not seen as a threat to the broader public interest.

---

401 W. G. Swendsen, "Future Development," *Second Biennial Report of the Department of Reclamation State of Idaho, 1921-22* (Boise: 1922), 250-251.

Nevertheless, after the Pacific Power & Light's state water appropriation applications exposed the inability of the state to control its own water destiny, Idaho amended its constitution in 1928 to extend greater power to state officials to regulate hydropower developments.

Creating the Federal Power Commission

The Federal Power Commission (FPC) was created by the Federal Water Power Act of 1920, a measure aimed at regulating the use of public lands and navigable waters of the United States for power purposes. Many hydroelectric power developments in the early twentieth century used power sites on navigable streams and rights of way over public lands for the purpose of generating electrical energy from falling water. Several nineteenth century congressional acts governed granting rights of way over public land and federal reservations for the purposes of conveying water, but the Act of May 14, 1896 (29 Stat., 120) was the first act of Congress which specifically recognized the existence of hydroelectric power development and made specific provision for the use of public land for this purpose.[402] The Act of February 15, 1901 (31 Stat., 790) authorized the Secretary of the Interior to adopt regulations and issue revocable permits for use of public lands and forest reserves for hydroelectric power development and other purposes. In 1910 and 1912 the secretaries of Agriculture and Interior approved general regulations to govern applications for power sites.[403] None of IPC's (or its predecessors-in-interest) power developments were constructed under these regulations governing use of public lands for hydroelectric power developments, a fact that later in licensing hearings gave rise to claims by the Federal Power Commission and others that some of the company's Snake River power plants had been illegally sited and operated on public land.[404]

Passage of the Federal Water Power Act (FWPA) came at the end of a nearly two decades'-long struggle to define the jurisdiction of the

402 Prior congressional acts included Act of July 26, 1866 (Rev. Stat. Sec. 2339); Act of July 9, 1870 (Rev. Stat. Sec. 2340), and Act of March 3, 1891 (26 Stat., 1095).

403 O. C. Merrill, Chief Engineer, USFS, USDA, "The Relation of Legislation to Electric Power Development," December 20, 1915. Folder: Water Power – June to December 1916, General Correspondence, 1906-1975, Records of the Department of Agriculture, Record Group 16 (RG 16), NARA II.

404 This was particularly true of Lower Salmon Falls, Swan Falls, and Upper Salmon Falls. See relevant sections below.

federal government in the electrical industry, particularly where proposed or existing projects were located wholly, or in part, on federal reserved land or the public domain (the jurisdiction of the Department of Agriculture or Department of Interior), or where they affected navigable waterways (the jurisdiction of the war department). This interplay of jurisdictions is important in understanding the eventual composition of the Federal Power Commission created by the FWPA. The FWPA represented the combination of two separate lines of bills, one that governed development of dams on navigable waterways, and another that dealt with generation of power on the public domain and the construction of dams for irrigation and other public purposes.[405] As the federal court noted in *Alabama Power Company* v. *Gulf Power Company*, "these measures never became laws because the Senate and the House of Representatives could not agree on all of the forms and all the provisions. Finally the two pieces of proposed legislation were in material respects combined into one called the Federal Water Power Act."[406]

Various pieces of water power legislation were introduced to Congress by the Wilson administration beginning in 1916, that suggested organization of a commission to regulate power development on public lands, but each met with opposition by two principal groups: first, advocates of state's rights against federal control and, second, water power companies. In the view of conservationists and Progressive Era reformers interested in regulating the electrical industry, some 20 interlocking companies which had come into control of the bulk of the electrical generation capacity in the nation threatened to seize monopoly control over the industry. Although some of these powerful entities expressed a willingness to cooperate in securing federal water power legislation, others demanded practically perpetual free access to power sites on the public domain and loose federal regulation. On the other hand, the Progressives, and others in favor of publicly-owned and regulated power systems, believed that the public should derive broad benefits from development of water resources on the public domain and accused "the power trust" of systematically

---

405 Franklin K. Lane, Secretary of the Interior, to Henry L. Myers, chairman, Committee on Public Lands, U. S. Senate, July 31, 1918. Water Power Development, General, File #2-194, 65th Cong., Part 1 (File #2-194), General Land Office; and Merrill, "The Relation of Legislation to Electric Power Development," December 20, 1915. Folder: Water Power; General Correspondence, 1906-1975, RG 16, NARA II.

406 *Alabama Power Co. v. Gulf Power Co.*, 283 Fed. 606-09; Sen. Doc. 210, 68th Cong, 2nd Sess.

stifling efforts to apply a greater level of regulation or impose federal charges for the privilege of using the public domain or navigable streams.[407]

During World War I, Congress recognized a growing national demand for electricity and elected to streamline and standardize the authorization process for new hydroelectric facilities through passage of the Federal Water Power Act in 1920. The central tenets of the act, a lease not exceeding 50 years and revocable only by cause and reasonable recapture provisions, were aimed at providing enough certainty to stimulate capital investment in the development of hydroelectric power, while at the same time protecting the public interest in power sites by restoring the ability to secure public lands to future public uses. Passage of the act was aided by the direct intervention of President Woodrow Wilson (1912-1920), who instructed his secretaries of Interior, War, and Agriculture to prepare a water power bill that combined their previously separate jurisdictions into a commission structure. The bill was an administration measure which was assigned by the president for the personal attention of his Interior Secretary, Franklin K. Lane. Beginning in 1916 until its passage four years later, Wilson's cabinet officers – the secretaries of Interior, Agriculture, and War – worked with key congressional members to develop a compromise bill upon which all sides could agree.[408]

The increased need for electrical power resulting from U. S. participation in the war and crippling shortages in coal and oil supplies compelled immediate attention to water power legislation. In December 1917, President Woodrow Wilson laid the matter before the committees of the House of Representatives having jurisdiction over water power legislation and at the same time presented a draft of a bill that had the approval of his three cabinet secretaries. Hearings on the bill were undertaken by a special House Committee on Water Power, but failed to pass in 1918. A similar bill successfully made its way through the following congressional session.[409] The Federal Water Power Act (41 Stat., 1063), passed by Congress and signed into law by

---

407 Jerome G. Kerwin, *Federal Water Power Legislation* (New York: Columbia University Press, 1926), 172; Gifford Pinchot, "The Long Struggle for Effective Federal Water Power Legislation," *George Washington Law Review* 9(1945); Samuel P. Hays, *Conservation and the Gospel of Efficiency: The Progressive Conservation Movement, 1890-1920* (Cambridge: Harvard University Press, 1959).

408 Federal Water Powers and their Development: Laws Heretofore Enacted, General Land Office Water Power Legislation. File #2-194, 65th Cong. Pt. 3, RG 48, NARA II.

President Wilson on June 10, 1920, created the Federal Power Commission – comprised of the secretaries of War, Interior, and Agriculture – and charged the commission with responsibility for licensing (a limited 50-year lease) non-federal hydroelectric projects that not only occupied federal lands or made use of navigable waters, but also affected those waters, made use of water at a government dam, or otherwise affected interstate commerce. In connection with the water power legislation, no consideration was given to any act authorizing the permanent alienation of public lands valuable for water power. Congressional and administration policy in the years leading up to passage of the act had always been to grant nothing but revocable licenses or permits.[410]

During 1918 hearings on the federal power bill, western congressmen took an interest in how licensing of hydroelectric plants might present opportunities or pose problems to irrigation development in the arid western states. Western congressmen John E. Raker of California, William L. La Follette of Washington, and Carl Hayden of Colorado sat on the House Special Committee on Water Power that considered the bill. O. C. Merrill, chief engineer of the U. S. Forest Service, representing the Secretary of Agriculture, was the principal administration representative. Section 10(a) of the proposed bill authorized the FPC to consider uses other than merely hydroelectric power production in determining whether a proposed project was "best adapted to a comprehensive plan" for the waterway.

> Section 10. All Licenses issued under this Part shall be on the following conditions:
>
> (a) The project adopted, including the maps, plans, and specifications, shall be such as in the judgment of the Commission will be best adapted to a comprehensive plan for improving or developing a waterway or waterways for the use or benefit of interstate or foreign commerce, or the improvement and utilization of water-power development, and for other beneficial public uses, including recreational purposes; and if necessary in order to secure such plan the Commission

---

409 Senate Committee on Commerce, *Development of Water Power,* Senate Report No. 180, 66th Cong., 1st sess., September 12, 1919.

410 Franklin K. Lane, Secretary of the Interior, to President Wilson, February 1915 and Lane to Davis F. Houston, Secretary of Agriculture, November 24, 1917. File #2-194, General Land Office, RG 48, NARA II.

> shall have authority to require the modification of any project and of the plans and specifications of the project works before approval.[411]

As the legislative history of the Federal Power Act makes clear, irrigation was considered one of the "other beneficial public uses" to be considered in determining whether a project was "best adapted to a comprehensive plan for improving or developing a waterway." Rep. Raker and O. C. Merrill, in committee hearings on the bill, addressed the issue of conflicting and complimentary uses of water for power generation and irrigation within the context of the "comprehensive plan" required in Section 10(a). Raker wanted to extend the authority of the commission to require water power licensees, where appropriate, to make the water so developed available for irrigation use after it passed through the power plant. While development of electrical energy was the primary goal of the legislation, Merrill allowed that the commission's responsibilities under the act extended beyond power considerations alone.

> Mr. Merrill. ...This is what the commission should have authority to do and what the commission should do in considering any development on any stream or stream system. It should make a thorough examination of the stream and prepare, as the bill provides, for a scheme of development that will use to the full the resources of that stream for every purpose.
>
> Mr. Raker. That is just what I am asking about.
>
> Mr. Merrill. Irrigation, water power, water supply, navigation, and whatever there may be; and it should have a plan drafted which would comprehend that, and then it should, in considering applications for a license, approve the application of that applicant who will carry out to the best degree the plans prepared.[412]

In deciding on the best plan of development, Merrill, Raker, Secretary of Interior Franklin K. Lane, Secretary of Agriculture David F. Houston, and others agreed that the act required the FPC to consider potential irrigation uses.

---

411 41 Stat. 1063.

412 House Committee on Water Power, *Hearings*, March 18 to May 1918, Part 1, 65th Cong, 2d sess (1919), 93-94.

> Mr. Raker. It is the law to-day that where a man makes an application for hydroelectric plant and another one makes an application for the same site for an irrigation plant your department or the other departments will go in and see which [is] the highest use, and if the irrigation use is the higher use they will deny the hydroelectric application. Is that not right?
>
> Mr. Merrill. They can and they do.
>
> Mr Esch. That is settled by the last part of section 7, where, after speaking of the preference that may be given to municipalities, the language is:
>
> *The commission may likewise give preference to the applicant the plans of which, it finds and determines the best adapted to develop, conserve, and use in the public interests the navigation and water resources of the region.*
>
> Mr. Merrill. The same thing is also covered in section 10, subparagraph (a)...
>
> Mr Esch. So you get the utmost use, whether for irrigation or for power?
>
> Mr. Merrill. And that is the intention...that the plan for development that should be approved should take into consideration all uses that can be made of the water and of the site in that particular locality, as a unit.[413]

The phrase "other beneficial public uses" in section 10 (a) of the Federal Water Power Act, stated Merrill, included irrigation as a public use. Citing the situation on the Sacramento River in California where power reservoirs were situated in the mountains above the valley and irrigation was downstream, Raker wanted assurance that use of water for irrigation received protection from unreasonable interference from use of water for power developments. Merrill responded that a wide array of non-power uses should be considered in the issuance of licenses:

> Mr. Merrill: In my judgment it is covered by the provisions of subparagraph (a) of Section 10, as well as by the last part of section 7 in the preferences. Not only should the scheme of development be considered in relation to irrigation and to navigation and to water power, but also to flood control. There will be many projects in different parts of the United States where flood control is concerned, and where it should be within the power of the commission to require structures, reservoirs, etc., to be built in such manner or used in such

413 Ibid., 95.

> manner that they may be of benefit in flood control as well as in irrigation, navigation, and water-power development.[414]

Later the House Special Committee on Water Power herd from Secretary Franklin K. Lane, the principal administration spokesman for the Water Power Act. Again, Raker brought up the issue of irrigation and its protection under the act. He wanted "irrigation" as a "beneficial public use" to be specifically called out along with navigation and water power in Section 10(a) of the bill, "so there would be no possible doubt as to what the intention of Congress was in such legislation." Secretary Lane responded that he could "not see any objection" to including the word "irrigation."[415]

The complexities of the FPC considering power developments in relation to irrigation and other beneficial uses in the context of a comprehensive plan of river development and placing this authority in a distant federal Bureaucracy caused some western congressmen to challenge the wisdom of adopting Raker's proposed amendment to Section 10 (a). Chief among his critics was Rep. Frank W. Mondell of Wyoming, who wanted the commission's authority strictly limited to consideration of water power developments. Mondell viewed inclusion of irrigation in the bill as "utterly intolerable," an invitation to encroach on state water law, and "an attempt to put our irrigation development under the control of a new Bureau at Washington. ..." Rep. Taylor of Colorado expressed similar concerns.[416] Other congressmen noted that the arid western states had adopted the law of prior appropriation and other constitutional provisions that adequately protected irrigation interests in any comprehensive federal program of river development. In fact, development of a comprehensive plan for river development including irrigation guaranteed the effects on irrigation would be taken into consideration. Rep. Nicholas J. Sinnott of Oregon noted that the proposed bill required the FPC to follow state water law, and therefore, states could protect their irrigation interests before the commission. Mondell agreed.

---

414 Ibid, 96-97.

415 Ibid, 458-59.

416 *Congressional Record,* 58:2, June 30, 1919, 2134-35.

> Mr. Sinnott. Mr. Mondell, in most of the Western States, either by statute or enactment or constitutional provision, it is provided that the use of water for irrigation purposes is a higher use than the use of water for power purposes. Do you find anything in this bill that in any way conflicts with that or impinges upon that policy?
>
> Mr. Mondell. No; I should say not, because of the fact…no one can begin or carry out any of these power developments in the public-land states until and unless they have secured from the proper State authorities the right to divert the water. Of course the question as to the character of the use would be brought out and settled and determined at the time of the issuance of the water right or the water permit.
>
> Mr. Sinnott. If the State authorities thought it advisable they would require a certain use of that water for irrigation purposes as a condition precedent of granting the right to appropriate the water for power purposes.
>
> Mr. Mondell. They have power to do that in most of the States.[417]

One of the primary weaknesses of the 1920 act was that it placed the administration of the act in the hands of a commission of three persons who oversaw other departments of government and were too busy to give sufficient personal attention to conducting a major national program regulating use and development of water power on public lands and navigable waterways. Lack of continuity and staff support for the original commission's work presented difficulties in exercising the kind of comprehensive control over the nation's water resources envisioned under the FWPA of 1920. In 1930, the commission was reorganized. Under the Federal Power Act of 1930 (46 Stat., 797), the FPC was transformed into an independent regulatory agency headed by a full-time, five-member commission appointed by the president with the approval of the senate. Commissioners were not allowed to engage in any other business, vocation or employment while serving on the commission and the commission was authorized to employ a staff of its own. Five years later, Congress further strengthened the FPC mandate with passage of the Federal Power Act of 1935 (49 Stat., 838), which encouraged closer cooperation with state utilities commissions through joint conferences, investigations, hearings, boards, and exchange of information. The act reflected and reaffirmed the basin-wide and multiple-use tenants of New Deal national water policies and the commission's duty to approve hydroelectric projects "best adapted to a

417 Ibid, 621-23, 654

comprehensive plan for improving or developing a waterway or waterways…for beneficial public uses."[418]

The State of Idaho played a vital role in establishing the FPC as an agent for regulating use of Idaho's waters for power purposes in conformity with the state's constitutionally-established preference for irrigation. The federal licensing process also highlighted certain inherent limitations imposed by the Idaho state constitution on administration of water appropriations by the Idaho Commissioner of Reclamation. As both IPC and Oregon-based Pacific Power & Light Company (PP&L) moved to secure prime sites along the Snake River for the development of hydropower in the early 1920s, Idaho Governor D. W. Davis and his Commissioner of Reclamation W. G. Swendsen maneuvered to protect the state's water supply "for irrigation and domestic purposes" by appealing directly to the commission. Their efforts led to the FPC implementing "the policy of permitting the states affected [by a hydropower licensing application] to submit such data and requests as they desire," in addition to permitting the states "to make such appearances as they may deem advisable to or before the Commission in connection with any applications" – policies that subsequently became an integral part of licensing applications for the FPC and its successor, the Federal Energy Regulatory Commission (FERC).[419]

Idaho's Reclamation Commissioner W. G. Swendsen also served on the Board of Engineers appointed by the FPC to study development of the Columbia River and its tributaries for power, irrigation, and navigation purposes. Although the Snake River was excluded from the purview of their investigation, Swendsen later recounted that the board's report to the FPC endorsed "provisions therein which afford an adequate protection to the State's irrigation and other interests" – particularly to the waters of northern Idaho, Priest Lake, Pend Oreille Lake, and Clark's Fork. One specific provision, according to Swendsen, recommended

---

418 16 U.S.C., 791-823; and Dozier A. DeVane, "Highlights of Legislative History of the Federal Power Act of 1935 and the Natural Gas Act of 1938," *The George Washington Law Review* 14 (1945-46), 34-38.

419 W. G. Swendsen, Commissioner of Reclamation, *Second Biennial Report of the Department of Reclamation, State of Idaho, 1921-1922* (Boise, 1922), 25.

> That no permits be granted for power purposes down river [the Columbia] except under conditions which reserve to Idaho the right to use such water as may be required for irrigation or other consuming purposes at all times in the future, together with other recommendations regarding the protection of resorts, summer homes, and other property rights that might be affected.[420]

In subsequent hearings dealing with licensing of commercial hydroelectric power plants in Idaho on the Snake River, the state's on-going interest in protecting its future agricultural development was the frequent basis for protests. When conditions warranted, the state, as well as the Bureau of Reclamation, protested against licensing power plants until provisions were included that subordinated the power company's water rights to those of irrigated agriculture, both in the present and into the future.[421]

Lower Salmon Falls and Swan Falls, 1923-1937

Although occurring about the same time, federal licensing of IPC's Lower Salmon Falls and Swan Falls plants contrasted sharply with hotly contested licensing of the company's upstream Twin Falls facility and its failed attempt to secure an FPC license for its proposed Upper Salmon Falls plant. Both Lower Salmon Falls and Swan Falls were existing facilities, acquired by IPC in 1916 and 1915 respectively, that occupied public lands of the United States.[422] Neither generated controversy during the FPC licensing process, however, and both were licensed relatively quickly. Indeed, while FPC applications for Lower Salmon and Swan Falls were submitted a few years after the application for Twin Falls, both were licensed in 1928 – several years before Twin Falls.

420 Ibid., 25-26.

421

Research at the Federal Energy Commission (FERC) in Washington, D. C. – the federal agency that succeeded the FPC – seeking this Board of Engineers report failed to uncover the document. According to the FERC records managers, most of the materials of the Commission are on a 50-year retention schedule and the report was likely destroyed.

422 George Smith, Director, United States Geological Survey, Memorandum for the Secretary, December 8, 1928 and E. O. Patterson, Solicitor, to the Honorable, The Secretary of the Interior, February 21, 1929. Folder: General Land Office, Misc. Rights Of Way, Idaho Power Company, May 13, 1926 to May 14, 1935, Box 631 2-11, Henrylyn, Lon, CCF 1907-1936, RG 48, NARA II.

Since they were licensed by the FPC, neither power plant was made subordinate to upstream irrigation interests. Even after IPC began expanding generating capacity at its Lower Salmon Falls site in the early 1930s, and had to obtain amendments to its FPC license for the facility, no subordination clause was introduced. Located far below Milner and Twin Falls, lacking storage capacity (as in the case of Lower Salmon Falls) or having only regulatory storage (as in the case of Swan Falls), and operating largely on return flow and groundwater recharge as run-of-the-river plants, neither the federal government nor the State of Idaho considered Lower Salmon Falls and Swan Falls as significant, future impediments to irrigation development.

Lower Salmon Falls

The Lower Salmon Falls facility had been constructed by Great Shoshone & Twin Falls Water Power Company in 1909 with IPC acquiring the plant in 1916. With the passage of the FWPA in 1920, the power plant – which occupied public lands – fell under purview of the newly-created FPC. On November 22, 1923, IPC's vice-president W. R. Putnam made an application before the FPC "for a license for a minor part of a constructed project" at Lower Salmon Falls.[423]

The power company's application described the Lower Salmon Falls power site as essentially a run-of-the-river facility, although it did not designate the site as such. The application indicated that the facility consisted of an "overflow diversion dam…timber-faced on concrete piers about 960 feet in length and about 25 feet in height" and a single reinforced concrete powerhouse measuring approximately "95 feet in height, 110 feet in width, and 74 feet in length." The powerhouse and its forebay formed part of the dam. The dam possessed seventeen headgates – all of which were 11 feet in height, with two having a width of 22.5 feet, the other fifteen being 10.5 feet. Under the project's "scheme of development," no water was stored. Instead, river flow was "diverted to forebay, whence it passes directly through three turbines

---

423 Application for License (Act of June 10, 1920, 41 Stat., 1063), Lower Salmon Falls Power Project (Idaho), Idaho Power Company, November 22, 1923, 1, and Application for License (Act of June 10, 1920, 41 Stat., 1063), Lower Salmon Falls Power Project (Idaho), Idaho Power Company, May 1, 1926, 1. File Folder 100-2, Project No. 457, Formal, Part 1, December 18, 1923 to December 1946 (File Folder 100-2, Project No. 457, Formal, Part 1), Box 2, Accession No. 138-62A-0938 (Box 2); Idaho Power Company (Lower Salmon Hydro-Electric plant), Snake River, Federal Power Commission, FERC.

with a combined rating of 9,350 hp under a head of about 35 feet, to excavated tailrace about 100 feet in length."[424]

The three Westinghouse generating units within the powerhouse totaled 7,600 kva of installed capacity.[425] The power was intended for "lighting, general power, irrigation, and other purposes, in conjunction with the interconnected power system" of IPC.[426] The application specifically listed 11 "Large Irrigation Pumping plants Served" by the power facility in the states of Idaho and Oregon, including the Gem Irrigation District, the Snake River Canal Company, the Ontario-Nyssa Irrigation Company, and the North Side Pumping Company, Ltd.[427]

The power company's Lower Salmon Falls application also offered a statement of the company's water rights at the site, a statement provided by Idaho's Commissioner of Reclamation, W. G. Swendsen. According to Swendsen, at the Lower Salmon Falls site, the power company held two water rights:

> (1) Water License issued by the Commissioner of Reclamation of the State of Idaho on October 31, 1919, confirming to the Idaho Power Company a perpetual right to the use of 1700 cubic feet per second of the waters of Snake River, (2) Water License issued by the Commissioner of Reclamation of the State of Idaho on October 31, 1919, confirming to the Idaho Power Company a perpetual right to the use of 1300 cubic feet per second of the waters of Snake River in addition to the 1700 cubic feet per second above described, both licenses being issued under Permit to Appropriate the Public Waters of the State of Idaho, No. 3685

Swendsen further stated, "no protests or contests are pending in this office against said water rights and that the holder is not delinquent in the matter of proof required by the laws of the State of Idaho."[428]

Unlike Idaho Power's Twin Falls FPC application, its Lower Salmon Falls application generated no protest or controversy. The Bureau of Reclamation, which initially had opposed the Twin Falls project, felt –

---

424 Application for License, November 22, 1923, 1. File Folder 100-2, Project No. 457, Formal, Part 1, Box 2, FERC.

425 Exhibit "M," attached to Application for License, November 22, 1923. File Folder 100-2, Project No. 457, Formal, Part 1, Box 2, FERC.

426 Application for License, November 22, 1923, 2. File Folder 100-2, Project No. 457, Formal, Part 1, Box 2, FERC.

427 Ibid., 3.

428 Exhibit "E," attached to Application for License, November 22, 1923. File Folder 100-2, Project No. 457, Formal, Part 1, Box 2, FERC.

as Acting Commissioner Ottamar Hamele expressed in a letter to the Federal Power Commission on January 22, 1924 – that the recently executed June 15, 1923 contract with Idaho Power Company at American Falls "fully protected" the Bureau's "interests" and thus it did not have "any objection to the granting of the application for license."[429] Five months later, in May 1924, FPC Executive Secretary O. C. Merrill contacted the governor of Idaho, Charles C. Moore, as well as the mayors of Gooding and Twin Falls, Idaho and the board of supervisors for the counties of Gooding and Twin Falls, seeking their "objections to or comments on this proposed project" by July 20.[430] Additionally, under provisions of the FWPA, the power company was required to publish notice of its filing in local newspapers for a period of eight weeks and directing protesters to contact the FPC. Judging from the FPC file on the Lower Salmon Falls application, neither these public officials nor members of the public lodged a protest with the commission.[431]

In July 1925, after receiving applications from IPC for Lower Salmon Falls, Swan Falls, Horseshoe Bend, and a transmission line from the Horseshoe Bend plant, the FPC determined that both Lower Salmon Falls and Swan Falls be treated as "major" projects.[432] This ruling impelled Idaho Power to submit an amended application for its Lower Salmon Falls facility on May 1, 1926.[433]

---

429 Ottamar Hamele, Acting Commissioner, to Executive Secretary, Federal Power Commission, January 22, 1924. Folder: 412.12 – (A-Z) Idaho, Applications for Power Permits, Thru 1929 (Folder 412.12), Box 341 412.06-412.12 (Box 341), GAPR 1919-1945, RG 115, NARA-Denver. Nearly three weeks prior to the Hamele submitting his letter, the Bureau's Acting Chief Engineer, R. F. Walter expressed substantially the same opinion in a memorandum to the Commissioner: "The interests of the Reclamation Bureau in this matter and probably also those of irrigation generally on Snake River are believed to be fully protected in the American Falls contract of June 15, 1923 and especially paragraph 21 thereof. It is therefore recommended that the application be granted." R. F. Walter, Acting Chief Engineer, to Commissioner, Application before Federal Power Commission by Idaho Power Co., Serial No. 457, Idaho, January 3, 1924. Folder: 412.12, Box 341, GAPR, 1919-1945, RG 115, NARA-Denver.

430 O. C. Merrill, Executive Secretary to the Honorable Charles C. Moore, Governor of Idaho, May 24, 1924. File Folder 100-2, Project No. 457, Formal, Part 1, Box 2, FERC. Merrill's letter indicates to whom copies were sent.

431 For one such example, see Publisher's Affidavit, undated. File Folder 100-2, Project No. 457, Formal, Part 1, Box 2, FERC.

432 Chief Engineer to Mr. M. O. Leighton, July 20, 1925. File Series P-503, Project No. 503, Correspondence, Part 1, May 9, 1924 to September 30, 1946, Box 78, Accession No. 138-65A-1792, Swan Falls, Federal Power Commission, FERC.

433 Exhibit "I," Application for License, May 1, 1926. File Folder 100-2, Project No. 457, Formal, Part 1, Box 2, FERC.

IPC's amended application not only changed the license sought from minor to major but also revised the description of the project works. The headgates of the dam were identified as "two each 15 feet wide by 12 feet high, and three each 12 feet wide by 9 feet high."[434] The amended application also referenced Idaho Power's proposed plants at Twin Falls and Upper Salmon Falls (FPC Projects 18 and 19, respectively), as elements of the power company's system, and provided additional data on the power capacity of Lower Salmon Falls. According to the amended application, the facility made use of average static head of 36 feet, and while the total flow available to the plant 90 percent of the time was 5,500 cfs, the hydraulic capacity of the works totaled only 2815 cfs, only one-half of the quantity of water available to it 90 percent of the time. Consequently, the power capacity of the project was 8,107 hp (nearly 6,050 kW), far below its potential generating capacity. Thus, the hydraulic capacity of the generating works was the controlling factor in its power production, not lack of a water supply.[435]

Two years later, on June 25, 1928, the FPC ruled that the Lower Salmon Falls facility was "best adapted to a comprehensive scheme of improvement and utilization for the purposes of water-power development and of other beneficial uses" and it issued a license for Idaho Power's Lower Salmon Falls facility, designating it Project 457.[436] The license was made effective retroactive to January 1, 1928 and was set to expire on June 30, 1970. Unlike the power company's Twin Falls facility – which by virtue of its proximity to the major public and private irrigation projects above Milner appeared so threatening to irrigators – no subordination provision was made a part of the license for Lower Salmon Falls.[437]

In the early 1930s, not long after IPC secured a license for its Twin Falls facility, the company itself moved to amend its FPC license for Lower Salmon Falls so as to make greater use of the available flow of the Snake River to generate power. On December 15, 1934, IPC's

434 Application for License, May 1, 1926, 1. File Folder 100-2, Project No. 457, Formal, Part 1, Box 2, FERC.

435 Exhibit "I," Application for License, May 1, 1926. File Folder 100-2, Project No. 457, Formal, Part 1, Box 2, FERC.

436 The Federal Power Commission, License on Government Lands, Issued June 25, 1928, Effective Jan. 1, 1928, Project No. 457, Idaho, Idaho Power Company (Lower Salmon Falls Project), 1-2. File Folder 100-2, Project No. 457, Formal, Part 1, Box 2, FERC.

437 Ibid., 3. File Folder 100-2, Project No. 457, Formal, Part 1, Box 2, FERC.

president K. M. Robinson filed an application for amendment, received by the commission on February 9, 1935, to replace two of the plant's three turbines with more efficient units. As Robinson explained,

> Existing turbines Nos. 1 and 2 were designed for a head of 50 feet. The site as developed provides a head of only about 37 feet. The turbines have been operated with specially constructed runners, but they have never driven their connecting electric generators to full capacity. Both the turbines and their governing equipment are in very poor condition, making it necessary to replace them with up to date equipment. Accordingly, it is proposed to replace the two old turbines with new ones, each of 3,000 hp capacity.[438]

Even with the proposed new turbines, the facility would still be unable to make use of the "flow available 90 percent of the time" as that remained "in excess of the capacity of the installation." Under a static head of 37 feet and making use of 2,820 second-feet of water, the power capacity of the generators would now total 8,350 hp (approximately 6,227 kW) – an incremental increase of about 250 hp over the facility's existing power capacity.[439]

According to Robinson, the installation of new turbines was not expected to have any effect on his company's existing water rights or upon the flow of the Snake River. IPC's president noted that the power company currently had a right to develop and use 3,000 cfs, and "[i]t is believed that the alterations contemplated will not require additional water rights." In keeping with IPC's business philosophy to develop power sites with incremental increases to meet demand, Robinson noted, "if it is found later that more water is required, rights thereto will be secured, and the Commission will be notified." As to the impact of the "proposed change of turbines," Robinson stated unequivocally that there would be "no effect on the normal flow of Snake River."[440]

Within months of filing its application, Idaho Power determined that more water was indeed required. On March 20, 1935, IPC filed an

438 Idaho Power Company, Lower Salmon Falls Power Project – No. 457, Snake River Idaho, December 15, 1934. File Folder 100-2, Project No. 457, Formal, Part 1, Box 2, FERC.

439 Exhibit I, Lower Salmon Falls Project, Power Capacity of Project, Idaho Power Company, Lower Salmon Falls Power Project – No. 457, Snake River, Idaho, December 15, 1934. File Folder 100-2, Project No. 457, Formal, Part 1, Box 2, FERC.

440 Exhibit H, Lower Salmon Falls Project, Effect of the Proposed Operation of the Project Works on Normal Stream Flow, December 15, 1934. File Folder 100-2, Project No. 457, Formal, Part 1, Box 2, FERC.

application for permit with the State of Idaho for an additional 300 cfs for its Lower Salmon facility. In the "Remarks" section of its water rights application, the power company explained that it intended to replace two existing generating units with two new ones, the capacities of which required a larger volume of water. On May 23, 1935, Idaho Commissioner of Reclamation R. W. Faris issued Permit No. 18107, which required proof of completion of works by May 23, 1940 and proof of beneficial use by May 23, 1946.[441]

Armed with a right to a greater amount of water and approval of the FPC to install new turbines, the power company went to work to boost the power capacity of their facility. A little more than one year later, the it succeed in increasing the overall capacity of its plant to 9,265 hp (nearly 7,200 kW) – nearly 1,000 hp greater than what the company had initially proposed to the FPC in its December 1934 application. The commission accepted the change, and ordered at its February 18, 1936 meeting, "[t]hat for the purposes of computing annual charges and for other purposes of the license, the power capacity of the project No. 457 be fixed at 9,265 horsepower, effective as of May 13, 1935."[442]

Building ahead of schedule, IPC made good on its proofs of completion and beneficial use, and on August 20, 1937, Commissioner Faris issued the power company State Water License No. 18107 (later designated 02-2059), in the amount of 250 cfs with a priority date of March 20, 1935.[443] Two days later, the FPC ruled in favor the company's amendment license application. Noting that the proposed changes were "desirable" and "not incompatible with the public interest," the commission ordered that IPC's Lower Salmon Falls license be amended to provide for substituting new turbines with modern type gates for those now existing in power units Nos. 1 and 2; for making the requisite changes in screens; and such changes in the governor and other apparatus as may be incident to the installation of the two new turbines.[444]

---

441 Application for Permit to Appropriate the Public Waters of the State of Idaho, Application No. 25299, Permit No. 18107, approved May 23, 1935. Idaho State Water Right 02-2059 Back File.

442 Extract from the minutes of a meeting of the Federal Power Commission held February 18, 1936: Readjustment of power capacity (Project No. 457). File Folder 100-2, Project No. 457, Formal, Part 1, Box 2, FERC.

443 State of Idaho, License and Certificate of Idaho State Water Right, Water License No. 18107, Amount 250 Sec. Ft., Priority March 20, 1935, August 20, 1937. Idaho State Water Right 02-2059 Back File.

IPC made no further additions to Lower Salmon Falls and sought no other amendments to its FPC license until after World War II.

Swan Falls

Much like Idaho Power's Lower Salmon plant, its Swan Falls facility had been acquired from another company in the late 1910s and because it too occupied government lands, it fell under the purview of the FPC in the early 1920s.

On January 26, 1924, a little more than a month after IPC filed its Lower Salmon application, the power company applied for its "Swan Falls power project" as "a major project, already constructed."[445] M. O. Leighton, on behalf of the power company, submitted the engineering data in support of the application to the FPC on May 6, 1924.[446]

In its application, the power company described the Swan Falls facility as essentially a run-of-the-river plant consisting of a dam and power house.[447] The Swan Falls dam was a combination rock-filled crib and concrete gravity dam. On the left bank of the Snake River, the rock-fill crib portion extended 158 feet from a concrete *abutment*. At the northern end of the crib section, the dam became a concrete gravity type with 21 tainter gates each measuring 7.5 x 12 feet and extended 326 feet to a rocky outcrop. The Swan Falls power house had three

---

444 Extract from the minutes of a meeting of the Federal Power Commission held March 22, 1935: Amendment of license (Project No. 457). File Folder 100-2, Project No. 457, Formal, Part 1, Box 2, FERC.

445 Application for License (Act of June 10, 1920, 41 Stat., 1063), Swan Falls Power Project (Idaho), Idaho Power Company, January 26, 1924. Folder: 412.12, Box 341, GAPR, 1919-1945, RG 115, NARA-Denver.

446 M. O. Leighton to Federal Power Commission, May 6, 1924. File Series P-503, Project No. 503, Correspondence, Part 1, May 9, 1924 to September 30, 1946 (File Series P-503), Box 78, Accession No. 138-65A-1792 (Box 78), Swan Falls, Federal Power Commission, FERC.

447 The available historic record suggests that the Swan Falls facility was not operated as a strictly run-of-the-river facility. USGS District Engineer, Thomas Newell, in September 1933 disclosed that "[i]n order to make the most of the available water supply it is the practice of the company to store water in the forebay above the plant during...off load hours and to increase the flow as needed during the period of greater power requirements." These "diurnal fluctuations," which occurred between 12 midnight and 7 AM, were sufficiently noticeable to cause difficulty in setting water stage recorder at USGS Murphy gaging station and affected the Weiser staff record station adversely. Although Idaho Power Company was willing to work with the USGS and the FPC in the "necessary collection of stream flow records," Newell's letter suggests that IPC would continue to operate the facility in this manner. Thomas Newell, District Engineer, to The Chief Hydraulic Engineer, U. S. Geological Survey, September 21, 1933. File Series P-503, Box 78, FERC.

concrete sections measuring 290 feet in length and formed part of the dam at this rocky outcrop. Situated between the northerly end of the plant and the north shore abutment, a distance of about 160 feet, were a set of by-pass gates and a concrete fish ladder. Although the dam created a backwater pond, "[n]o storage reservoir" existed as part of the project works. Water for power generation was directed from the forebay through openings in the concrete substructure of the power plant to the turbine pits, and from there expelled through the turbines and draft tubes to the tailrace in the river channel.[448]

As noted earlier, the Swan Falls facility had ten power generating units capable of producing between 11,000 and 20,000 hp and from 8,200 to nearly 15,000 kW of energy. This variance related to the flows at the plant: 90 percent of the time, the flow at the facility was 6,900 cfs or more, while 50 percent of the time, the flow was at least 12,500 cfs. The absolute minimum flow in Snake River at Swan Falls, according to USGS data, was about 5,400 second-feet corresponding to an output at the Swan Falls plant of 8,600 hp, or a little more than 6,400 kW.[449]

As for the water rights that the power company owned at the site, Idaho's Commissioner of Reclamation W. G. Swendsen filed a "Statement of Nature, Extent, and Ownership of Water Rights" for IPC's application to the FPC. Swendsen stated that IPC's rights consisted of:

> First: Judgment and Decree were made and entered into in the Circuit Court of the United States for the District of Idaho on the second day of April, 1907, in an action wherein the Trade Dollar Consolidated Mining Company was plaintiff and A. R. Cruzen and A. A. Fraser were defendants, wherein the honorable Court decreed to said Trade Dollar Consolidated Mining Company 2150 cubic feet per second of time of the waters of Snake River;
>
> Second: Water License issued by the State Engineer of the State of Idaho on the 4th day of August, A. D. 1911, confirming to the Swan Falls Power Company the right to the use of 1840 cubic feet per second of the time of the waters of Snake River, in addition to 2150 cubic feet per second decreed by court as aforesaid;

448 Application for License (Act of June 10, 1920, 41 Stat., 1063), Swan Falls Power Project (Idaho), Idaho Power Company, January 26, 1924, 1. Folder: 412.12, Box 341, GAPR, 1919-1945, RG 115, NARA-Denver.

449 Hoyt, "Report Relative to Water Power Resources of Snake River," August 1922, 69-70. Folder: D-100, 12-G-51, WPB, 1894-1961, RG 57, NARA II.

> Third: Water License issued by the State Engineer of the State of Idaho on the 16th day of August, A. D. 1911, confirming to the Swan Falls Power Company the right to the use of 1460 cubic feet per second of time of the waters of Snake River, in addition to 2150 cubic feet per second decreed by court as aforesaid and 1840 cubic feet per second confirmed by License as aforesaid and,
>
> Fourth: Water License No. 14362 issued by the Commissioner of Reclamation of the State of Idaho, confirming to Idaho Power Company the right to the use of not to exceed 4000 cubic feet per second of time of the waters of Snake River, in addition to the decreed and licensed rights as aforesaid.[450]

The first three water rights had been acquired by IPC along with the Swan Falls plant, and the fourth was obtained by the power company itself on May 26, 1920.[451]

The "rock filled timber crib" dam, according to IPC's application, was,

> ... completed long prior to the spring of 1918 and the water claimed...was applied to beneficial use on April 27, 1918, and the use thereof gradually increased until June 14, 1918, when the entire amount of water claimed was and ever has been applied to beneficial use, viz., to generating electric energy for lighting, power and other purposes...[452]

The FPC licensing for Swan Falls, in contrast to Twin Falls and even Lower Salmon Falls, proceeded without complications. On June 5, 1924, nearly six months after IPC filed its federal application, the Bureau of Reclamation weighed in on the Swan Falls licensing. Writing to O. C. Merrill, FPC executive secretary, Commissioner of Reclamation Elwood Mead, declared, "The approval of this permit will

---

450 Copy of "Statement of Nature, Extent, and Ownership of Idaho State Water Rights," Application for License, Swan Falls Power Project (Idaho), Idaho Power Company, January 26, 1924. Idaho State Water Right 02-2032 Back File. The rights enumerated in Swendsen's statement were respectively, 02-0100, 02-4000, 02-4001, and 02-2032.

451 State of Idaho, License and Certificate of Idaho State Water Right, Water License No. 14362, Water District No. 50, Amount 4000 sec. ft., Priority 7-29-19, May 26, 1920; Idaho State Water Right 02-2032 Back File.

452 Application for Permit to Appropriate the Public Waters of the State of Idaho, Application No. 20149, Permit No. 14362, District No. 18, July 29, 1919; Idaho State Water Right 02-2032 Back File.

not interfere with any existing project of the Bureau of Reclamation, nor with any development contemplated at the present time."[453]

Four years later, on June 25, 1928, the same day that the FPC licensed IPC's Lower Salmon Falls facility, it also licensed the company's Swan Falls facility as Project 503. Like the Lower Salmon Falls license, the Swan Falls license was made effective retroactive to January 1, 1928 and was set to expire on June 30, 1970; and like Lower Salmon Falls – but unlike Twin Falls further upstream – no explicit subordination provision was made a part of the license.[454]

The power company moved to amend its Swan Falls license in August 1936.[455] As with the amendment filed for Lower Salmon Falls in December 1934, the power company did so with an eye to making improvements to its existing Swan Falls power site. Specifically, its August 4, 1936, filing sought FPC permission,

> … for the construction of a rock-fill embankment faced with steel sheet piling superimposed on the present crib dam, a concrete retaining wall, a concrete spillway and stop-log structure between the existing dam and power house, and the addition of one 6,600 kVA booster transformer.

On November 24, 1936, the commission ruled that, "the proposed changes are for the purposes of lessening maintenance costs and securing water tightness in the present rock-fill and timber dam, and for the purpose of improving voltage conditions in the plant." The FPC found that the amendment, as offered, would not affect annual charges or alter the basic conditions upon which the license originally was issued. The commission ordered the license amended to reflect these physical alterations to the Swan Falls power site.[456]

---

453 Elwood Mead, Commissioner, to Executive Secretary, Federal Power Commission, June 5, 1924. File Series P-503, Box 78, FERC. The Bureau's Superintendent in Boise likewise felt that this power plant "does not affect any present or proposed development of the Bureau of Reclamation, and it is therefore recommended that it be granted." See J. B. Bond, Superintendent, Boise, Idaho, to Chief Engineer, Application before Federal Power Commission by Idaho Power Company – Serial No. 503, Idaho. File Series P-503, Box 78, FERC.

454 The Federal Power Commission, License on Government Lands, Project No. 503 – Idaho, Idaho Power Company (Swan Falls Project), effective on January 1, 1928. Folder: Licenses for Idaho Power Projects by Federal Power Commission, 1928-1933, Box R 2/4, AR 20, ISHS.

455 Amendments of licenses (major), (Projects Nos. 457 and 503), December 23, 1933. File Folder 100-2, Project No. 457, Formal, Part 1, Box 2, FERC.

456 From the Minutes of November 24, 1936, Application, Idaho Power Company, Project, No. 503, Authorization for Amendment of license (major), enclosed with a letter from Leon M. Fuquay,

While the power company sought and obtained another amendment to its license in order to replace Turbine Units 1 and 2 in 1946, the plant did not attract much further government attention during the first three decades of its licensing.[457] In the early 1960s, when the Bureau of Reclamation moved to develop its long-awaited Mountain Home project – an irrigation project that would have destroyed IPC's entire Swan Falls power site – Swan Falls would become a major focus of federal, state, and private interests.[458]

IPC's Struggle to License Twin Falls

Twin Falls and Upper Salmon Falls power plants were the first two power facilities that IPC sought to license under the Federal Power Act of 1920, and overall only the eighteenth and nineteenth applications to be considered by the commission nation-wide.[459] The Twin Falls plant, deriving its name from the pair of waterfalls in the Snake River at its proposed location, was the fifth power site developed by IPC on the Snake River and was situated approximately four river miles above an old plant at Shoshone Falls acquired by IPC in 1916. The proposed Twin Falls plant was sited, in part, on land that had been withdrawn from entry by the Reclamation Service in 1904, for possible use as a power site in support of pumping units for the Minidoka project.[460] It

Acting Secretary, Federal Power Commission to Bureau of Water Resources, Department of Reclamation, March 5, 1937. Folder: Federal Power Commission Prior To 1941, Box R 2/4, AR 20, ISHS.

457 Federal Power Commission, Amendment No. 4 of License, Project No. 503-Idaho, Idaho Power Company, September 18, 1946, enclosed in a letter from J. H. Gutride, Acting Secretary, Federal Power Commission, to Department of Reclamation, Bureau of Water Resources, September 26, 194. Idaho State Water Right 02-2032 Back File.

458 J. H. Gutride, Secretary, Federal Power Commission, to Mr. Harold T. Nelson, Regional Director, Bureau of Reclamation, Region 1, September 4, 1963. Correspondence File, Part 2, Federal Power Commission No. P-503, November 1, 1946 to __ (Correspondence File, Part 2, FPC No. P-503), FERC.

459 When originally sited, IPC's East Side American Falls power plant did not fall under the purview of federal regulations governing power sites. The company owned the site, the plant was erected in 1913 before passage of the Federal Power Act, and the upper Snake River, at the time, was not considered a navigable water of the United States. plant expansion in 1923-1924 and 1926-1927 also did not trigger an application for a FPC license, although the plant, through provisions of the June 15, 1923 contract with the United States, was designed to make use of water stored at a government-owned reservoir.

460 Minutes of Public Hearing, Rupert, Idaho, October 7, 1921, 26. Formal File 100-2, Part 1, Federal Power Commission No. 18, July 7, 1920 to December 31, 1927 (Formal File 100-2, Part 1), FERC; George Hoffman, Report on Application Before Federal Power Commission by Idaho Power Company, Serial No. 18, for Preliminary Permit covering Twin Falls site, on Snake River, 5, enclosed in a memorandum from Project Manager, Minidoka Project to Chief Engineer, Denver,

was part of IPC's larger plan to develop a total of seven power sites on the reach below Milner Dam where the river plunges into its canyon and remains unmolested by irrigation diversions for some 200 miles.[461]

In 1920, the upper Snake River still carried large excess flood flows that "wasted" down the river in winter, but it was a highly regulated stream the remainder of the year, and was largely depleted of water by irrigation at Milner Dam in the summer months. Nevertheless within the rock-walled canyon below Milner, there were no irrigation diversions and the river's flow was replenished year-round by groundwater, springs, drainage, canal seepage, and return irrigation flows. In this reach of Snake River canyon below Milner Dam, IPC sought to capitalize on the perennial river flow to generate hydroelectric power at several major falls located on this stretch of the river.[462]

Between the USGS gauging station at Milner and the one nearest to Twin Falls, located about seven miles below the proposed site of the Twin Falls Powerhouse, there was an inflow of approximately 550 cfs made up of seepage and canal losses from the North and South Side Twin Falls irrigation tracts and groundwater flow coming off the Snake River Plain. Small springs also contributed water to the stream at frequent intervals. During irrigation season, surface flow in the river past Milner Dam frequently consisted of nothing more than leakage through the dam. W. G. Hoyt, a hydraulic engineer for the USGS, estimated inflow between the Twin Falls plant site and the Twin Falls gauging station at about 140 cfs, making the inflow from groundwater, seepage and springs between Milner and the proposed power plant about 410 cfs. Therefore, regardless of irrigation diversions, some 500

Colorado, Subject: Inclosing report on the application before the Federal Power Commission by the Idaho Power Company, Serial No. 18, for Preliminary Permit covering Twin Falls Site, on Snake River, March 7, 1921, enclosed in a memorandum from Chief Engineer to Director, Washington, D. C., Subject: Report on the application before the Federal Power Commission by the Idaho Power Company for preliminary permit covering Twin Falls site on Snake River, March 12, 1921. Folder: 412.12 Idaho, Federal Power Applications (Application #18-Twin Falls Site on the Snake River), Idaho Power Company, 1930 thru (412.12 Idaho), Box 809 412.06_412.12 - Colorado – Idaho, General Correspondence File, 1930-1945 (GCF, 1930-1945), RG 115, NARA-Denver.

461 Minutes of Public Hearing, October 7, 1921, 13-14, and 55-56. Formal File 100-2, Part 1, FERC.

462 W. G. Hoyt, "Report Relative to Application of Idaho Power Company For a Preliminary Permit to Develop Twin Falls Site on Snake River, Idaho: Permit No. 18 Federal Power Commission, Idaho 26642," July 1921, 9-11. Folder: D-100, 12-G-43, WPB, 1894-1961, RG 57, NARA II.

to 600 cfs could be anticipated at the power site chosen by IPC about "90 percent of the time," estimated Hoyt.[463]

The application for a state water right license for the Twin Falls plant, initially filed in 1918, proved controversial. It was the closest power plant sited below Milner Dam (about twenty-two miles); the final point of diversion for irrigation projects served by storage releases from the proposed government reservoir at American Falls.[464] That meant it also was the most vulnerable and sensitive to upstream irrigation depletions. Water rights for the Twin Falls power plant were acquired under State Application No. 24073 and Permit No. 17070, approved April 2, 1929, for a power plant with a planned maximum capacity of 1,500 cfs. The actual License and Certificate of Water Right No. 17070 (02-2056) was issued by the Idaho commissioner of reclamation on August 20, 1937, to IPC upon submission of proof of the company making beneficial use of a maximum flow of 1050 cfs. The power plant was designed to use the average flow in the river of approximately 600 cfs at a fall of 205 feet to develop 9,500 horsepower.[465]

IPC filed its supporting papers for a preliminary permit for Twin Falls before the FPC on July 7, 1920, only 27 days subsequent to the approval of the FWPA, and while the company was in the middle of contract negotiations with the Reclamation Service over the American Falls reservoir site. Through this contract, the Reclamation Service sought to limit the company's power rights on Snake River and subordinate its American Falls power rights in deference to future irrigation storage requirements in the upper basin. The Reclamation Service and local water users believed construction of a Twin Falls hydroelectric plant had the potential to threaten the water supply of existing and proposed irrigation projects located upstream that used (or

463 Ibid., 10.

464 See Application for a Permit, To Construct the American Falls Reservoir, And to store the Unappropriated Waters of the State of Idaho, Application No. R. 21475, Permit No. R. 273, District 70, 2, enclosed with Exhibit "D," Project No. 18, Application for License (Act of June 10, 1920, 41 Stat., 1063), Twin Falls Project, Snake River, Idaho, Idaho Power Company, enclosed with memorandum from Assistant Director Morris Bien to Chief Engineer, Denver Colo., Subject: Application before Federal Power Commission by Idaho Power Co., Serial No. 18, Idaho, February 10, 1922. Folder: 412.12 Idaho, Box 809, GCF, 1930-1945. RG 115, NARA-Denver.

465 State of Idaho, License and Certificate of Idaho State Water Right, Water License No. 17070, Amount 1,050 Sec. Ft., Priority April 2,1929, issued August 20, 1937; Idaho State Water Right 02-2056 Back File; and "Application for License," no. 18, Twin Falls Project, Snake River, Idaho, Idaho Power Company, Exhibit D. Folder: 412.12 Idaho, Box 809, GCF, 1930-1945. RG 115, NARA-Denver.

might in the future) waters of the Snake River for irrigation. For several months following the company's filing for a preliminary power permit, the Reclamation Service considered making a formal protest against the company's application, citing its ongoing American Falls Reservoir contract negotiations with IPC, effects of a potential loss of winter power generating capabilities at Minidoka Dam, and the possible need for a Bureau hydroelectric power facility at the same Twin Falls site to serve future irrigation pumping requirements and winter power demands on the Minidoka project.[466]

The Bureau asked for a delay by the FPC in making any findings until these outstanding issues could be resolved.[467] Meanwhile, the cities of Burley and Rupert, as well as the Minidoka and Burley irrigation districts, filed a formal protest, collectively asking, "that action on the said application…be with-held until a full and complete investigation can be made."[468] Director Arthur P. Davis of the Reclamation Service forwarded a report by George L. Hoffman, an engineer for the service, to the FPC on April 5, 1921.[469] The report repeated some of the conclusions reached by the joint federal-state Board of Engineers on the proposed American Falls Reservoir a year earlier. Along the Snake River from its headwaters to Milner, wrote Hoffman, an additional irrigable area of 1,439,000 acres needed a water supply from the Snake River. Plans for storage reservoirs being considered jointly by the Reclamation Service and the State of Idaho would ultimately conserve all water in the Snake River "susceptible of economic diversion for irrigation purposes." "Considerable time," however, "[would] be necessary to complete" these projects, and

466 B. E. Stoutemyer, USRS district counsel, to Chief Counsel, Washington, DC, January 17, 1921; Barry Dibble, Minidoka Project manager, to Chief Engineer, Denver, CO, March 7, 1921; and Hoffman, Report. Folder: 412.12 Idaho, Box 809, GCF, 1930-1945. RG 115, NARA-Denver. USBR chief engineer Fred Weymouth did not believe it was necessary to obtain control of the Twin Falls site as long as some satisfactory arrangement was made with the power company to meet its commercial power needs at commercial rates at American Falls. If not, he favored getting control of the Twin Falls site and any other sites along the Snake River that the Service needed. Weymouth to USRS Director, March 12, 1921. Folder: 412.12 Idaho, Box 809, GCF, 1930-1945. RG 115, NARA-Denver.

467 Authorization for preliminary permit, excerpt from the minutes of the meeting of the Commission held March 24, 1924. Formal File 100-2, Part 1, FERC.

468 Protest against issuing preliminary permit to the Idaho Power Company for power development at the Twin Falls, on Snake River Idaho, between Sections 4 and 12 in Town. 10 So. Range 18 E. B. M., June 10, 1921. Formal File 100-2, Part 1, FERC.

469 A. P. Davis, Director to O. C. Merrill, Executive Secretary, Federal Power Commission, April 5, 1921. Folder: 412.12 Idaho, Box 809, GCF, 1930-1945, RG 115, NARA-Denver.

sometime in the future, flows currently available for power use would need to be curtailed.[470]

Hoffman also maintained that the Bureau itself needed to retain the Twin Falls power site in the interest of conserving water and making maximum use of the waters of Snake River above Milner Dam for irrigation. When irrigation development in the upper basin reached a stage where irrigation water stored in American Falls Reservoir could not be safely released from the dam in the winter months to meet downstream needs, winter power requirements would need to be met from a source other than the Minidoka or American Falls power plants. Located nearest to the Minidoka project and supplied by seepage, groundwater, irrigation return flows, and springs below Milner Dam, Twin Falls was considered by local irrigators to be the most logical undeveloped power site for a small plant to replace the reduced winter power supply at Minidoka Dam or American Falls. Hoffman recommended that IPC's application be rejected "until such time as some satisfactory arrangement can be made for winter power for the Minidoka project." Ultimately Reclamation Director A. P. Davis recommended to the FPC on April 5, 1921, that a permit for Twin Falls not be granted until it could be determined whether the site was needed by irrigation developments being undertaken by the Reclamation Service in the Snake River Valley.[471]

Public hearings on the issue extended into 1922. The FPC investigation entailed public hearings in October 1921 at Rupert that foreshadowed subsequent debates among IPC, local water users, and the Reclamation Service (and later, the Bureau of Reclamation) over the issues of power development and subordination on the Snake River. At these hearings, the company faced a largely hostile crowd. Minidoka water users through their spokesperson W. R. Griswold expressed concern that the power company's proposed Twin Falls facility would adversely affect power development for irrigators on Snake River. Twin Falls power, Griswold noted, could replace Minidoka power at a reasonable cost without sacrificing any of the stored water supply available above Milner Dam for power generation.[472] Griswold made

470 Hoffman, Report, 2-3. Folder: 412.12 Idaho, Box 809, GCF, 1930-1945. RG 115, NARA-Denver.

471 Hoffman, Report, 5-6. Folder: 412.12 Idaho, Box 809, GCF, 1930-1945. RG 115, NARA-Denver.

472 Minutes of Public Hearing, October 7, 1921, 13. Formal File 100-2, Part 1, FERC.

further reference to the "McConnell Report," the 1920 report of the joint federal-state Board of Engineers, which pointed out that construction of American Falls Reservoir would significantly limit winter power production. Water users maintained that what power could be developed during the winter – through, as stated in the McConnell Report, "accretions from springs and return flow" to the area "between Milner Dam and Twin Falls" – should be developed either by "themselves" or the "Reclamation Service."[473]

IPC noted that the proposed facilities were wholly sited on private land purchased by the company or its predecessors-in-interest. As a power proposition, Twin Falls was crucial to the company's development and its continued service to its customers. Company officials considered it the most economically beneficial site for consumers as well as the company. Through its spokesperson W. R. Putnam, vice-president and general manager, the company noted that its "predecessors in interest" had filed water rights at the site in 1903, a year before withdrawal of the land for reclamation purposes.[474] IPC was now "the holder of valid water permits from the State of Idaho to use the normal flow of Snake River at the Twin Falls site" as well as stored water at American Falls, "such permits being in priority of right, no other claimants appearing" in the records.[475]

According to the power company, development of the Twin Falls site was immediately necessary to unite IPC's bifurcated and unbalanced electric power system. According to Putnam, a facility at Twin Falls would link the company's fledgling eastern wing surrounding American Falls with its larger, more stable and mature interconnected system in the west.[476] The Reclamation Service's designs on American Falls, he maintained, made construction of a Twin Falls power site more imperative as the effect of a government dam and power plant at American Falls would be to limit IPC's expansion potential at that site, thus complicating the company's problem of supply to its eastern territory. Putnam stressed the continuing growth of the company's own demand in the service area, the company's prior investment at the site, and the obligation of the company as a public utility to meet the electrical needs of its customers at the lowest rates possible. Minidoka

473 Ibid., 25. Formal File 100-2, Part 1, FERC.

474 Ibid., 53-54. Formal File 100-2, Part 1, FERC.

475 Ibid., 66. Formal File 100-2, Part 1, FERC.

476 Ibid., 56. Formal File 100-2, Part 1, FERC.

water users' arguments for equity were "especially unreasonable," Putnam said, when "the welfare of a much wider territory and a larger body of the public are involved in the Company territory" than in the area served by the Minidoka project.[477] The federal reclamation project itself would be more than accommodated by hydroelectric power produced by the proposed government plant at American Falls Dam. From IPC's perspective, Putnam concluded:

> There can be no sound reason, and certainly no merit, in the contention that the Twin Falls site should be withheld from early development, to the betterment and well-being of a large number of persons the year around, merely at some time in the future, more or less remote, it might be used by a smaller number of persons only a few months in the year.[478]

IPC forecasted demand for power to double on its system by 1930. This growth required that, "the system of the company so expand and enlarge with it that it not only keep step with the progress of the country but be in advance of it." The company's legal rights and demand projections aside, Putnam argued that Idaho Power alone was "in logical position to develop the Twin Falls site most economically and beneficially," as it held priority in its water right, owned property at the site, and possessed storage space for power uses in American Falls Reservoir.[479]

> Since the utilization of both the natural flow of the river and of the stored water can be effected at lesser cost, and more efficiently, through its one plant than through two separate plants, one utilizing stored water and the other natural flow, as must be the case were the Minidoka project to undertake development; and not only in the plant installation itself, but in the transmission and distribution systems as well, because the project site is within easy touch of the presently existing wire system of the Company, to which connection would be but a simple matter, while necessitating many miles of line to reach the Minidoka project. The greater economy would appear to be in the development of the site by the Company.[480]

477 Ibid., 58-59; and Before the FPC, In the Matter of the Application of IPC for Preliminary Permit, Twin Falls Power Site, Answering Brief of Applicant to Brief of Protestants on Minidoka Project, November 7, 1921. Formal File 100-2, Part 1, FERC.

478 Minutes of Public Hearing, October 7, 1921, 60. Formal File 100-2, Part 1, FERC.

479 Ibid., 61-63. Formal File 100-2, Part 1, FERC.

480 Ibid., 63. Formal File 100-2, Part 1, FERC.

While the power company insisted on the need for the Twin Falls project, Griswold and others nevertheless questioned why, with Snake River's many potential power sites possessing nearly 400,000 undeveloped horsepower, did the company put such a high premium on this specific site which would return such a small amount of power compared to other sites further downstream?

Putnam replied that the Twin Falls project was modest out of necessity. Idaho Power already had made an investment there, the local market for power was growing, and the plant required less money to produce sufficient power to meet demand more so than any other proposed or actual plants.[481] Moreover, the company's vice president and general manager pointed out, Idaho Power's development of the plant under the conditions outlined in the American Falls contract was consistent with the mandate of the FPC, "to so regulate and control, so far as possible, the use of water so that the most economical results from the use of that water would be secured." The contract, should it be executed, "results in all points below possibly American Falls, certainly below Milner, in the Idaho Power Company being able to handle the generation of power on the Snake River with less wastage of power and in a more economical manner than any other – than any group of users of water could handle it."[482]

The FPC ultimately agreed to withhold action on the preliminary permit until the American Falls contract negotiations had concluded. On February 1, 1922, after the company had executed a tentative agreement calling for the Reclamation Service to take over a part of their power generating facilities at American Falls and regulating winter releases for power generation at Twin Falls and Shoshone Falls, IPC filed another application for a FPC license for the Twin Falls facility under the same project number. In the application, E. W. Hill, IPC vice-president, noted operation of Twin Falls power plant during irrigation season would be greatly affected by diversions of water at Milner into the North Side and South Side Twin Falls irrigation districts and by canal diversions at Minidoka.

The effect of these diversions was to dry the natural flow of the river at Milner for about two months during the irrigation season in a normal

481 Ibid., 76-77. Formal File 100-2, Part 1, FERC.
482 Ibid., 77-78. Formal File 100-2, Part 1, FERC.

year and up to double that time in years of lesser flow. According to Hill, during irrigation season about 400 cfs entered the stream as side-flow between Milner and the Twin Falls site. During non-irrigating season, there was currently sufficient water to operate the proposed powerhouse to its capacity. However, construction of the proposed USBR dam at American Falls and creation of a reservoir with an initial capacity of at least 1,500,000 acre-feet would affect these flows in the future. IPC anticipated that,

> … upon completion of the reservoir to its initial stage the only natural flow available at the site [Twin Falls] from the beginning of the irrigation season in April until the reservoir begins to spill, which in an average year will be about December 1st, and in a minimum year not until March or April, will be make-up from Milner to Twin Falls.

On account of this situation, in the pending 1923 contract, IPC intended to insert language that allowed the company to retain a right to certain waters in American Falls Reservoir to guarantee proper operation of its American Falls power plant, and other power plants located below Milner Dam.[483]

As IPC pressed forward with its preliminary licensing proceedings before the FPC, it conducted studies during the summer of 1923 to determine for itself the amount of summer inflow into the Snake River at various points between the USGS stations at Milner Dam and Shoshone Falls to demonstrate the dependability and limitations of this flow. The power company set up two gaging stations: a temporary station at the highway bridge near Murtaugh, above the mouth of Dry Creek; and a permanent USGS station, known as Kimberly Station, located about one-half mile downstream from the Twin Falls power site.

Measurements were also obtained at the headworks of Shoshone Falls plant, where the river's entire flow could be diverted through the powerhouse for measurement. The company also recorded the larger streams of wastewater flowing over the rim of the canyon and located points of subterranean flow along this stretch of the river. Data from the government gaging stations at Milner Dam (established in 1909) and four miles downstream from Shoshone Falls (established in 1911) was additionally collected. C. E. Tappan, the engineer who conducted

---

483 Exhibit H, "Statement of the Effect of the Proposed Operation of the Project Works on the Normal Flow of the Stream", E. W. Hill, vice-president IPC, Application No. 18 for License Twin Falls Project, Snake River Idaho, February 1, 1922. Folder: 412.12 Idaho, Box 809, GCF, 1930-1945, RG 115, NARA-Denver.

the study for IPC, noted that little water from the river above Milner would be available to furnish the powerhouses except when winter and spring floods spilled at American Falls. In most years, the river at Milner, wrote Tappan, was totally used during irrigation season to satisfy decreed rights to the normal flow of the river:

> After the flood season the river is made practically dry at Milner. All the water possible is shut off at the gates and diverted to the Twin Falls canals. During the irrigation season the operation and control of Snake River is subject to the irrigation requirements of the projects along Snake River above the Milner Dam. ... As long as the flow of flood water continues, Snake River does not require close regulation and it is only after the flood water has passed and the irrigation demand requires all of the water in the river that the gates at Milner are shut and the river practically made dry at that point.[484]

Tappan summarized in general terms how the upper river was managed for the benefit of irrigation. From July through September the average daily flow diverted at Milner Dam into the Twin Falls North Side Canal was 2,376 cfs and through the Twin Falls South Side Canal 3,132 cfs, or a total average daily diversion of 5,508 cfs. Thus, the plains on both sides of the Snake River below Milner Dam were in a high state of cultivation and in many places they were farmed right up to the edge of the rim of the canyon. With the large volume of water placed on these lands for irrigation, some of this water diverted at Milner found its way back into the river through waste ditches or subterranean channels. During flood season, water in excess of reservoir storage capacity and domestic and stock watering requirements was allowed to pass Milner; however, during the irrigation season only enough water was delivered at Milner to fulfill the irrigation rights of these two Twin Falls canals. There was therefore some leakage through the dam. In the summer of 1923, for example, the gates were closed for most of the period from early July through the end of September, but flows at Milner gaging station never fell below 9.2 cfs. Over the period from 1912 to 1923, as shown in **Table 6.1**, average daily flows for the Snake River at Milner Station for the critical irrigation months of July through September indicate that while flows

484 C. E. Tappan, "Report Covering Water Measurements and Studies of Stream Flow of Snake River between Milner and Shoshone Falls." (Idaho Power, Company, 1923), 3-4. D-100-12-G-54; WPB, 1894-1961, RG 57, NARA II.

were sharply curtailed past the dam, some water in excess of the diversion requirements spilled past the dam. During irrigation season in the extreme dry years of 1915 and 1919, little more than leakage passed through Milner Dam.

Table 6.1. Low Water Flow in Second-Feet Measured at Milner Dam Gauging Station, 1912-1923[485]

Year	Month *June*	*July*	*August*	*September*
1912	24,500	9,310	3,610	8,040
1913	21,500	8,240	1,540	5,050
1914	16,900	2,190	28	1,680
1915	3,820	11	12	43
1916	10,700	4,540	652	737
1917	20,800	12,700	318	1,680
1918	----	----	----	----
1919	363	22	14	37
1920	5,400	360	18	391
1921	16,000	137	55	1,370
1922	7,580	106	117	16
1923	7,660	981	84	10

Based on Tappan's 1923 study, it appeared that there was a daily average inflow of approximately 587 cfs in Snake River Canyon between Milner and Shoshone Falls. Over the same eleven-year period from 1912 through 1923, the average gain per month in the flow of Snake River between gaging stations at Milner and Perrine Bridge (four miles below Shoshone Falls) is shown (in cfs) in **Table 6.2**. These figures showing the average gain in the river between Milner and Perrine Bridge exceeded the minimum flow standards for October 1 through June 10 for the river at Shoshone Falls established in Section 21 of the June 15, 1923, contract between the U. S. and IPC. [486]

Table 6.2. Average Gain in Flow of the Snake River by Month from Milner Dam to Perrine's Bridge, 1912-1923 (in cfs)[487]

January	1064	May	1077	September	719
February	814	June	772	October	916
March	832	July	543	November	899

485 See USGS, *Compilation of Records of Surface Waters of the United States through September 1950, Part 13, Snake River Basin*, Water-Supply Paper 1317 (GPO, 1950).

486 Section 21 of the June 15, 1923 Contract between the U. S. and IPC.

487 USGS, *Compilation of Records*.

April	889	August	562	December	1010

The greater part of the increase or gain came from the south side of the canyon and the evidence from the 1923 study by Tappan for IPC suggested that this increased flow was the direct result of irrigation. The study also suggested that "the inflow between Milner and Shoshone Falls is considerably affected by short water conditions on the irrigation tracts and in the discharge of Snake River above Milner."[488]

It was assumed that the government would build a new 30,000-kW government power plant when it constructed the federal dam at American Falls. During negotiations for the American Falls power plant, it was informally understood that if a satisfactory contract was entered into for the government's power needs at American Falls, the Reclamation Service would not object to Idaho Power receiving a permit to develop the power site at Twin Falls to be operated under stipulations set out in Section 22 of the contract.[489] With the adoption of the June 15, 1923, contract, the Burea of Reclamation in a letter dated March 1, 1924, formally dropped its opposition to Idaho Power's preliminary permit for development of the Twin Falls site.[490]

Protests by local water users, however, continued unabated – in part because they desired lower-cost electricity provided by a public entity or rural cooperative, and because they desired to maximize the irrigation potential of the upper basin. As a result of their protests, USBR district counsel Stoutemyer intervened and once again pointed out the problems associated with construction of any plant of a size sufficient to demand passing water over Milner Dam for downstream power production "as against future use of the water at Milner and above for the development of irrigation." Once again, the FPC delayed taking action on the Twin Falls application.[491]

In 1924, IPC sought a three-year extension of time for completing its pre-construction studies under the preliminary permit from the FPC so as to revise its plans in light of the restrictions placed on winter water

488 Tappan, "Report," 3-4. D-100-12-G-54, WPB, 1894-1961, RG 57, NARA II.

489 O. C. Merrill to USBR Director Davis, February 10, 1922. Folder: 412.12 Idaho, Box 809, GCF, 1930-1945, RG 115, NARA-Denver; and *Minidoka County News*, April 24, 1924.

490 Davis to Executive Secretary, FPC, March 1, 1924. Folder: 412.12 Idaho, Box 809, GCF, 1930-1945, RG 115, NARA-Denver.

491 Redacted memorandum, "Information about Contract with the Idaho Power Company – American Falls, 1962, 76-77 Boise Field Solicitor's Office, Bureau of Reclamation; Minutes of Public Hearing, 13-14 and 17-18. Formal File 100-2, Part 1, FERC.

releases under the 1923 American Falls contract.[492] The FPC compromised by granting a two-year extension. The city of Burley and other municipalities continued their protests and now sharpened their arguments about protecting future agriculture – if IPC were granted a permit to build its proposed plant, the company stood to secure a superior power right at the Twin Falls site before the Burea of Reclamation had even decided on its final plans at American Falls.[493] On April 26, 1926, four days before that permit was set to expire, Idaho Power requested another one-year extension. Noting a "somewhat complicated situation at American Falls" had "raised some difficult questions with respect to power production all along the river below American Falls," the company felt it "necessary to make detailed studies of the change that will take place in river regimen."[494] This explanation proved sufficient for the commission: It ruled that an extension was "reasonable and necessary," in order for Idaho Power "to continue investigation work and other preliminary arrangements."[495]

Finally, on April 1, 1927, Idaho Power submitted its formal license application to the FPC proposing construction of a reinforced concrete dam nearly 900 feet long, with a spillway of almost 740 feet and maximum height of 47 feet. The Twin Falls dam, as contemplated by IPC, would have backed up water for three miles, thereby creating a reservoir with a storage capacity of about 2,960 acre-feet. The powerhouse was to contain a 17,000 hp vertical turbine capable of generating 12,000 kW while making use of a maximum of 1,140 cfs. The power generated was to be dedicated "for light, general power and traction purposes," as well as "for irrigation pumping, eleven large irrigation districts now being served under the applicant's system" – the Gem Irrigation District, the Snake River Canal Company, and the Twin Falls North Side Pumping Company among them. In addition, the company noted that 56 towns, including American Falls, Aberdeen, Boise, Caldwell, Gooding, Jerome, King Hill, Milner, Nampa, Pocatello, Shoshone, Twin Falls, and Weiser were currently served, and would continue to be served by IPC and its new Twin Falls plant.[496]

492 Authorization for preliminary permit, March 24, 1924. Formal File 100-2, Part 1, FERC.

493 Ibid. Formal File 100-2, Part 1, FERC.

494 Idaho Power Company to Executive Secretary, Federal Power Commission, April 26, 1926. Formal File 100-2, Part 1, FERC.

495 Amendment of preliminary permit, excerpt from the minutes of the meeting of the Commission held June 23, 1926 Formal File 100-2, Part 1, FERC.

496 Application for License, Twin Falls, 4-5. Formal File 100-2, Part 1, FERC.

While Idaho Power emphasized that all facets of demand would be met, the development of irrigation agriculture and its water power demand for pumping purposes was largely responsible for the growth in electric load.[497]

The company went further in arguing that it was uniquely situated to make the most efficient use of Snake River water for power generation at its multiple electrical generating plants on the river. The Twin Falls plant itself, operating on an average flow of about 430 cfs, would produce 5,620 hp, but would materially increase power production at other IPC power plants in the system. Unlike most of these other plants, the proposed facility at Twin Falls contained some water storage capacity. Generating power at full plant capacity, using the maximum flow of 1,140 cfs (the capacity of the proposed penstock), the reservoir at Twin Falls would provide slightly more than daily regulation of the stream. E. W. Hill, vice-president of IPC, pointed out the resulting benefits:

> With its storage capacity of 2900 ac ft it will be possible for the applicant to operate the plant at a low load factor, while Shoshone, the next station below on the Snake River can be operated at a high factor by means of its own storage capacity. Except for the two plants…and the Swan Falls plant, the applicant does not own any station at the present time possessing daily or weekly capacity, except the American Falls plant, which during a small portion of the year will have a regulating effect on the stream.
>
> The Company has in use several hydro-electric plants which, on account of natural conditions, had to be developed for producing power of higher load factor than that of the power market. The addition of the Twin Falls power development to the Idaho Power Company's system will conserve the use of water now wasted on account of lack of storage capacity at the present hydro-electric plants of the Company. The effectiveness of the Twin Falls site in conserving the water power resources of Snake River must consequently be much greater if added to the Idaho Power Company's system, than if used separately or added to any other power system within practicable transmission distance.[498]

At one time in 1924, with the American Falls contract recently signed, the Bureau of Reclamation had expressed willingness to drop its

497 Application for License, Twin Falls, 5. Formal File 100-2, Part 1, FERC.

498 E. W. Hill, "No. 18. Application for License, Twin Falls Project," April 1, 1927. Folder: 412.12 Idaho, Box 809, GCF, 1930-1945, RG 115, NARA-Denver.

objection to granting a Federal Power Commission license to Idaho Power Company for its Twin Falls development. However, the 1927 IPC application for FPC Project No. 18, once forwarded to the Bureau for comment, provoked a strong negative response. The size of the proposed plant, and its theoretical water demand, appeared objectionable. S. O. Harper, acting chief engineer, forwarded a copy of the Twin Falls license application to E. B. Darlington, superintendent of the Minidoka project, for his comments. The Minidoka project "as at present constructed," responded Darlington, would not be adversely affected, but if the power company perfected a right for its state water filing of 1500 cfs, or even 1140 cfs, the maximum diversion capacity of the proposed facility, future irrigation projects could be placed in jeopardy.

Even the proposed future enlargement of the American Falls Reservoir to 3,000,000 acre-feet would be placed at a disadvantage by reason of the prior rights that might be established by the company at the Twin Falls site. The Twin Falls project would carry a state water right priority date of June 17, 1920, if the company proceeded with construction under their license, and that would be nearly a year senior to the American Falls Reservoir filing, which was stamped May 18, 1921.[499] Since the Twin Falls development was likely to affect the storage and natural flow rights of projects such as the proposed 40,000-acre Gooding Division of the Minidoka project, Darlington recommended that the Bureau of Reclamation demand a stipulation from Idaho Power Company "providing that the irrigation water rights have a preferential status over the power water rights which they seek to establish."[500]

After reviewing Darlington's letter, R. F. Walter, chief engineer for the Bureau, also suggested that no license be granted unless conditions were added that prevented the company from interfering with competing uses of water for irrigation purposes. Walter objected because the application contemplated use of far more water than what

499 Darlington did allow that Article 9(e) of the American Falls contract of June 15, 1923 might restrict the Power company's Idaho State Water Right at Twin Falls, but the effect of establishing the early priority date needed to "be examined from a legal standpoint."

500 E. B. Darlington to Chief Engineer [R. F. Walter], May 13, 1927. Folder: 412.12 Idaho, Box 809, GCF, 1930-1945, RG 115, NARA-Denver. Size of Gooding Division indicated in Department of the Interior, Bureau of Reclamation, Annual Project History, Report of Construction and Maintenance, Minidoka Project, Idaho, Volume XXV, Calendar Year 1931, 16. Box 301 Minidoka, PHFHR, 1902-1932, RG 115, NARA-Denver.

entered the Snake River between Milner Dam and the power site at Twin Falls. The excess could only be supplied by wasting water over Milner Dam, which meant it would enter into the deep canyon below and be lost for irrigating purposes. If this water was ever to be recaptured for irrigation uses by future settlers or for the authorized Gooding Division or other divisions of the Minidoka project, it would be burdened with a heavy and perhaps prohibitive cost of condemning water rights and the power plant. Furthermore, Walter noted that state law supported his case. In Idaho, wrote Walter, "irrigation is considered a superior public use to power development and is designated as a superior public use by the Constitution of the State of Idaho." Walter instructed district counsel B. E. Stoutemyer to draft a suggested form of stipulation protecting future irrigation interests to be embodied into the application before any license was issued by the FPC.[501] The stipulation drawn up by Stoutemyer was forwarded to USBR Commissioner Elwood Mead for approval. In a letter dated June 17, 1927, Mead requested that the FPC in granting a license to include the following stipulation obligating the company:

> ...for itself, its successors and assigns...hereby stipulates and agrees that all power rights and all rights to the use of water for power purposes heretofore or hereafter acquired for use at or in the vicinity of those certain falls in the Snake River known as Twin Falls, or at or in connection with any power plant or power works at or in the vicinity of said falls, shall be held and considered at all times to be subject, inferior and subordinate to all rights heretofore and hereafter acquired by the United States or other parties for irrigation purposes...; that no rights in excess of those authorized and provided for under the contract of June 15, 1923...will be claimed by the company at or in connection with the company's power plant or plants at or in the vicinity of Shoshone Falls as against parties desiring to use the water of Snake River for irrigation purposes.

Mead further urged that such a "stipulation and agreement ... be considered a covenant running with the title to the said power plants at Twin Falls and Shoshone Falls and all rights in connection therewith," and that "a recordable contract or deed embodying the provisions, conditions, and limitations" laid out in the stipulation "may be recorded in the counties in which said power sites are located." Finally, the

501 Stoutemyer to Chief Engineer, May 20, 1927, with preamble and proposed stipulation attached. Folder: 412.12 Idaho, Box 809, GCF, 1930-1945, RG 115, NARA-Denver.

commissioner insisted that the "stipulation shall be effective to bind the applicant and its successors and assigns without signature or execution by or on behalf of the United States,"... effective ... "upon the granting of the foregoing application or the issuance of a license thereunder."[502] As he explained in his letter to FPC Executive Secretary O. C. Merrill, the stipulation was needed "in order to fully protect the interest of the projects under construction and operation by this Bureau on the Snake River, including the Minidoka project, and the American Falls reservoir."[503]

For three years following that Mead letter, the issue laid dormant. According to Merrill's successor F. E. Bonner, the last that the commission had heard from IPC or its representative, M. O. Leighton, was on August 16, 1927, when Leighton acknowledged receipt of Mead's proposed stipulation and indicated "that upon the completion of certain studies, a further presentation would be made." That further presentation proved to be a revised application for the Twin Falls plant, forwarded to the Federal Power Commission by Leighton on June 21, 1930.[504]

Idaho Power's revised application was substantially the same as its previous one – albeit with some adjustments to enlarge the storage capacity of the reservoir, to slightly increase the power generating capacity of the plant, and to alter the maps and legal description of the lands upon which the facility was to be sited. Nothing in the application addressed the subordination stipulation proposed by Mead three years earlier. Upon learning of IPC's new application, the Bureau insisted that Mead's stipulation be included in any license issued for the Twin Falls facility.[505]

This time the power company responded, submitting a substitute stipulation, drafted by IPC attorneys sometime in the late summer or early fall of 1930.[506] The preamble to the company's substitute

---

502 Enclosed with Mead to Executive Secretary, Federal Power Commission, June 17, 1927. Folder: 412.12 Idaho, Box 809, GCF, 1930-1945, RG 115, NARA-Denver.

503 Mead to Executive Secretary, June 17, 1927. Folder: 412.12 Idaho, Box 809, GCF, 1930-1945, RG 115, NARA-Denver.

504 F. E. Bonner, Executive Secretary, Federal Power Commission, to M. O. Leighton, August 27, 1930. Folder: 412.12 Idaho, Box 809, GCF, 1930-1945, RG 115, NARA-Denver.

505 See Bonner to Leighton, August 27, 1930. Folder: 412.12 Idaho, Box 809, GCF, 1930-1945, RG 115, NARA-Denver.

506 The existing, publicly-available documentation on the issue held by the National Archives-Rocky Mountain Region (Denver) furnishes no clear indication of when Idaho Power offered its

stipulation struck all references to future water supply shortages for irrigation in the Snake River Basin as well as the reference to the superiority of irrigation as a public water use under the Idaho constitution. Instead it simply tied the upper limit of water use to the rights of the company at Shoshone Falls. The power company's stipulation, like the Bureau's, held that Idaho Power's water rights for power development at the Twin Falls site were "inferior and subordinate" to all water rights acquired by the United States or other parties "heretofore or hereafter" for irrigation purposes. However, IPC's substitute subordination clause did include one caveat: a provision for use at Twin Falls of the water rights the power company already possessed and used at its Shoshone Falls plant. Specifically, the company's "substitute" stipulation read as follows:

> NOW, THEREFORE, saving, reserving and excepting the right to use for power purposes in any power plant or plants to be constructed under this license or any extension or modification thereof, the amounts of water and water rights equivalent to the amounts of water and water rights now owned by the application for use at its Shoshone Falls power plant less the natural and uncontrolled inflow into the Snake River between Twin Falls and Shoshone Falls....

The company's attorneys struck from the subordination provision the language proposed by the Bureau which expressly limited the power company and its successors and assigns to the rights authorized and provided for in the contract of June 15, 1923, in connection with the company's power plant or plants in the vicinity of Shoshone Falls as against Snake River irrigators. Instead, the company agreed to subordinate its Twin Falls power right to upstream irrigation uses – so long as the company was guaranteed to receive the same water rights owned at its Shoshone Falls power plant minus the amount contributed by uncontrolled side flows that enhanced the flow of Snake River between Twin Falls and Shoshone Falls.[507]

---

alternative stipulation. It is included in a piece of correspondence dated October 25, 1930 between the Bureau Commissioner and D. F. McGee, IPC's Vice President. See Commissioner to D. F. McGee, October, 25, 1930. Folder: 412.12 Idaho, Box 809, GCF, 1930-1945, RG 115, NARA-Denver.

507 Enclosed with Commissioner to D. F. McGee, October, 25, 1930. Folder: 412.12 Idaho, Box 809, GCF, 1930-1945, RG 115, NARA-Denver.

The substitute stipulation proposed by IPC was forwarded to Darlington and Stoutemyer for comment. Darlington recommended retention of the stipulation proposed by Commissioner Mead as it provided more protection to irrigation rights than the ambiguous language proposed by the IPC attorneys. The power company's proposal ran contrary to the 1923 contract and left the water rights issue open to potential dispute, alleged Darlington:

> Presumably the Idaho Power Company was granted certain licenses for use of water at Shoshone Falls, with priorities senior to the Government's American Falls filings, but it is our understanding that the rights acquired by these licenses from the State Engineer's office are modified and restricted by the contract of June 15, 1923, between the company and the United States. The substitute stipulation proposed by the company ignores the effect of the contract of June 15, 1923, on these permits or license rights, and it is the view of this office that the language proposed by the company leaves considerable room for dispute in the determination of "the amounts of water and water rights now owned by the applicant."[508]

Bureau district counsel B. E. Stoutemyer agreed with Darlington, and moreover, advocated limiting any Twin Falls power facility built by IPC to a size justified by the amount of inflow to Snake River below Milner Dam. This restriction in plant size would prevent IPC from establishing a right to waste-water passing over Milner Dam through beneficial use, "so that it will never be necessary to sacrifice water which will be required for irrigation purposes by wasting it over the Milner Dam for the purpose of operating the proposed power plant at Twin Falls..." There was a number of available power sites located further down Snake River canyon which did not require wasting water at Milner Dam. The power company preferred the Twin Falls site, noted Stoutemyer, only because of its higher head and slightly cheaper cost per horsepower.[509]

After he received Stoutemyer's report, Commissioner Elwood Mead informed D. P. McGee, vice-president of IPC, that the company's stipulation was objectionable in that it "failed fully to subordinate the

---

508 Darlington to Commissioner USBR (through Chief Engineer), October 29, 1930. Folder: 412.12 Idaho, Box 809, GCF, 1930-1945, RG 115, NARA-Denver.

509 Stoutemyer to USBR Commissioner, November 1, 1930. Folder: 412.12 Idaho, Box 809, GCF, 1930-1945, RG 115, NARA-Denver.

power right to the irrigation interests of the Snake River Valley." In fact, he added, it might be construed to grant the power company more water rights than it possessed under the terms of the 1923 contract.[510] McGee expressed surprise at Mead's interpretation of the company's proposed stipulation, noting it was not the company's intention to "modify the provisions of the contract of June 15, 1923."

McGee promised to produce a revised stipulation that would meet the Bureau of Reclamation's objections while also "protecting the legitimate interests of the Idaho Power Company"[511], and early in 1931 delivered it. The Bureau's previous stipulation, argued McGee, made it impossible under the company's mortgage and deed of trust to obtain financing for the Twin Falls project as it "had the effect of enabling the Bureau of Reclamation to condemn our Shoshone Falls plant in the interest of an ambitious irrigation development, and leave our Twin Falls plant without water rights, and therefore valueless." McGee's new proposed stipulation subordinated the Twin Falls plant to upstream irrigation uses, excepting and reserving the right to use for power purposes at Twin Falls "that water, derived from sources lying upstream from said Twin Falls, which applicant now owns the right to use for power purposes at its Shoshone Falls plant, as such water becomes available for power use at said Twin Falls during its passage down the channel of Snake River to and for use at said Shoshone Falls plant..."[512]

Once again, Mead referred McGee's revised stipulation to the Bureau district counsel B. E. Stoutemyer in Portland, Oregon for comment. Once again, Stoutemyer objected, using the same arguments as before, and this time elaborated on his reasoning. In the 1923 contract between the United States and IPC, the power company had reserved 300,000 acre-feet of water in American Falls Reservoir for power purposes. The water was being used at American Falls and the Shoshone Falls plants. All, or a part of, the 300,000 acre-feet would eventually become necessary to meet growing irrigation demands in the Snake River Basin, wrote Stoutemyer. Settlers on future reclamation projects would be required to pay the cost of condemning a part of the

510 Mead to D. P. McGee, vice-president IPC, November 11, 1930. Folder: 412.12 Idaho, Box 809, GCF, 1930-1945, RG 115, NARA-Denver.

511 McGee to Mead, November 19, 1930. Folder: 412.12 Idaho, Box 809, GCF, 1930-1945, RG 115, NARA-Denver.

512 McGee to Mead, January 13, 1931. Folder: 412.12 Idaho, Box 809, GCF, 1930-1945, RG 115, NARA-Denver.

company's power rights at the American Falls and Shoshone plants to secure the necessary water. If IPC were allowed to build another power plant at Twin Falls under the conditions proposed by the power company, a third large power plant and its substantial water rights would also need to be condemned at a possibly prohibitive cost to future irrigators. Under the form of stipulation proposed by the Bureau, noted Stoutemyer, the irrigators would not need to condemn the plant at all. Furthermore, if the Twin Falls plant was built at a size that could be fully supplied with the water coming into the canyon below Milner Dam it could never be affected by upstream irrigation and there would be no problem with financing its construction. If additional power was needed by the power company, it could be developed at any number of available power sites further down the Snake River in southern Idaho without interfering with irrigation.[513] Stoutemyer, Chief Engineer R. F. Walter, and Commissioner Elwood Mead all agreed that the form of stipulation suggested by the power company should be rejected and the Bureau's own stipulation insisted upon before the FPC.[514]

McGee was so informed on February 4, 1931.[515]

In May 1931, James L. Boone, an attorney for the power company, launched negotiations with the USBR and FPC to develop yet another compromise stipulation that might satisfy all parties. Six months later, Boone met personally with Stoutemyer in Portland, Oregon leaving with Stoutemyer a draft stipulation intended to accompany IPC's new application to the FPC. The stipulation, marked "Draft A," was a proposed substitute for the prior stipulations prepared by the company and the Bureau. Stoutemyer found Draft A to be a "reasonable compromise" of the dispute, except as to the stored water from the company's 45,000 acre-feet of primary storage capacity in American Falls Reservoir, which was described as "a prior right" to that of other spaceholders. IPC reserved a secondary right to store and use 300,000 acre-feet of water in the 1923 contract. The 45,000 acre-feet referred to in paragraph 16 of the contract related to storage capacity reserved for the company which was to be put on an "equal basis" to other primary

513 Stoutemyer to Mead, January 20, 1931. Folder: 412.12 Idaho, Box 809, GCF, 1930-1945, RG 115, NARA-Denver.

514 R. F. Walter to Mead, January 26, 1931 and Mead to Walter, February 11, 1931. Folder: 412.12 Idaho, Box 809, GCF, 1930-1945, RG 115, NARA-Denver.

515 Acting Commissioner P. W. Dent to McGee, February 4, 1931. Folder: 412.12 Idaho, Box 809, GCF, 1930-1945, RG 115, NARA-Denver.

storage rights of irrigation spaceholders. Thus, Stoutemyer objected to the 45,000 acre-feet being described as a "prior right" and redrafted the stipulation, calling his "Draft B." It read:

> ...except (a) the right to use for power development at this project the water from the 45,000 acre feet of primary storage capacity which the licensee holds in American Falls reservoir under paragraph 16 of that certain contract between the United States of America and Idaho Power Company, dated June 15, 1923, and the discharge of which it may control under the terms of the contract.[516]

Commissioner Mead found the form of stipulation as drafted by the district counsel generally satisfactory in protecting upstream irrigation development. With the addition of inserting that the company's right to waste entering the channel below Milner Dam was a valid right only "as against the United States," Mead approved modified "Draft B."[517] Boone's reply was not favorable, noting in a telegram to Stoutemyer that Mead's suggestions "practically nullify compromise such as I presented to you and which had approval of our people."[518] Boone met with USBR representatives Messrs. Mead, Dent, Debler and Stoutemyer in Denver in mid-April 1932, to discuss and agree upon a new stipulation. The parties agreed to revert to the form of stipulation (i.e., "Draft B") formerly agreed upon by Stoutemyer and Boone. If it

---

516 Stoutemyer to Mead, January 20, 1932. Folder: 412.12 Idaho, Box 809, GCF, 1930-1945, RG 115, NARA-Denver, As American Falls Reservoir was filling in 1926, Banks and Stoutemyer of the Bureau and IPC vice-president and general manager, W. R. Putnam, had disagreed over this issue of whether the 45,000 acre-feet of primary storage *capacity* was a "prior right" to the first water flowing into the reservoir or whether it was to be put on an "equal basis' with all other irrigation water with the power company being entitled "to the use and benefit of its proportionate share of the capacity of the reservoir as rapidly as it is created." Stoutemyer maintained that under the contracts between the U. S. and the irrigation districts and companies (the contents of which IPC was fully aware as copies were attached to the Power Company contract) a provision was included to the effect that the power company's 45,000 acre-feet of storage was to be put "on the same basis as the rights of irrigation companies and irrigation districts under the contract..." F. A. Banks to W. R. Putnam, May 26, 1926; Putnam to Banks, May 27, 1926; and B. E. Stoutemyer to Banks, June 5, 1926; USBR-Boise. This issue was never settled in 1926 as the reservoir filled the following season, but in 1973 after the Bureau placed a storage limitation on the deteriorating American Falls Reservoir, the same controversy erupted between IPC and the USBR. James E. Bruce, vice-president and secretary IPC, to E. F. Sullivan, Regional Director, USDI USBR, July 3, 1973. USBR-Boise.

517 Mead to Stoutemyer, January 25, 1932. Folder: 412.12 Idaho, Box 809, GCF, 1930-1945, RG 115, NARA-Denver.

518 Stoutemyer to Mead, February 11, 1932.Folder: 412.12 Idaho, Box 809, GCF, 1930-1945, RG 115, NARA-Denver.

was found satisfactory by the Secretary of the Interior and IPC executives, the stipulation was to be filed with the Federal Power Commission.[519] It was filed with the FPC in the form agreed upon in Denver on May 6, 1932.[520]

The Bureau thus obtained from the FPC a subordination clause in the federal license for the Twin Falls project. The stipulation was inserted into the license as Article 14, and was accepted by Idaho Power's Board of Directors by resolution at their meeting on May 11, 1934.[521] The license for Project No. 18, Twin Falls, issued to Idaho Power Company, was executed by the chairman of the FPC on June 11, 1934. The stipulation confined the company's entitlements to: a) its contractual right under the 1923 contract to the release of 45,000 acre-feet from the American Falls Reservoir for power development, as its system operations required; and b) the right to use for power development "the seepage, percolation, drainage, spring or springs, waste, and/or other influent waters which do not flow or spill over Milner Dam, but which enter in, arise in, and flow in and along the channel of Snake River between the downstream toe of Milner Dam in Snake River, Idaho, and the site of the project covered by this license."

IPC further agreed,

> ...for itself, and its successors and assigns, to forego any and all claims which it might otherwise have for damages or other relief, in connection with any power rights at the site of this project, other than those mentioned in (a) and (b) above and which are necessary for irrigation purposes....

IPC expressly reserved all rights it had under the June 15, 1923, contract, including the company's secondary contractual right to release water, when available, for use below Milner Dam which was limited to certain flows at the head of Shoshone Falls from October 1 to March 15, each year. The Twin Falls plant, therefore, was dependent on its

519 Boone to Stoutemyer, April 25, 1932 and Stoutemyer to Mead, April 27, 1932. Folder: 412.12 Idaho, Box 809, GCF, 1930-1945, RG 115, NARA-Denver.

520 Mead to Boone, May 6, 1932 and Mead to D. C. Green, vice-president, Electric Bond and Share Co., May 6, 1932. Folder: 412.12 Idaho, Box 809, GCF, 1930-1945, RG 115, NARA-Denver.

521 Idaho Power Company, Minutes of Special Meeting of Board of Directors, May 11, 1934; Idaho State Water Right 02-2056 Back File.

water supply to the return flows arising in the river bed between Milner Dam and the Twin Falls plant, plus the limited amounts of water permitted under the 1923 contract to flow past Milner Dam.[522]

The power company's original water right application had been filed in 1918, but it had lapsed during the FPC application process in 1924. The Idaho Commissioner of Reclamation accepted a new filing (Application No. 24073) made by IPC for 1500 cfs to be diverted from the Snake River at or near Twin Falls with a priority date of April 2, 1929. The application was approved by the Commissioner of Reclamation as Permit No. 17070.[523] The permit provided that one-fifth of the work be completed before October 2, 1931, and the whole of the work before April 2, 1934, and full beneficial use be made on or before April 2, 1940. Because of delays in securing a permit from the FPC, Idaho Power Company filed an application in the Idaho Department of Reclamation for an extension of time for compliance with the terms of the provisions of its permit to appropriate water. The extension was granted. The FPC license was finally issued in May 1934, thereby allowing the company to proceed with development of its power project. As previously mentioned, IPC's board of directors at a special meeting on May 11, 1934 agreed to accept the license "embodying the provisions, conditions and limitations set out in the stipulation and agreement contained in Article 14 of the Federal Power Commission license." Time for completion of the project works under State Water Rights Permit No. 17070 was extended to June 11, 1939, and time for making beneficial use was extended to June 11, 1945.[524]

IPC proceeded with construction of its project ahead of schedule. The Twin Falls plant went into commercial operation on December 3, 1935. The project works had sufficient capacity to convey at least 1,250

522 Federal Power Commission, License on Government Lands, Project No. 18, Idaho, Idaho Power Company, June 11, 1934. File Folder 100-2, Project No. 18, Formal Pt. 2, January 1, 1928 thru December 31, 1948 (File Folder 100-2, Formal, Pt. 2), Box 13, Accession No. 138-88-006, FERC.

523 Tyler to Mead, December 1, 1930. Folder: 412.12 Idaho, Box 809, GCF, 1930-1945, RG 115, NARA-Denver.

524 Application for Permit, Application No. 24073, Permit 17070, Idaho Power Company, April 2, 1929. File Folder 100-2, Formal Pt. 2, Box 13, Accession No. 138-88-006, FERC; Idaho Power Company, Minutes of Special Meeting of Board of Directors, May 11, 1934; State of Idaho Department of Reclamation, Order Extending Time, In the Matter of the Application of the Idaho Power Company to Extend the Time for Requirements Under Permit No. 17070, August 11, 1934. Idaho State Water Right 02-2056 Back File. In part, IPC moved forward on its Twin Falls facility because the severe downturn in the American economy in the 1930s prompted the company to delay its plans to develop a power site at Upper Salmon Falls.

cfs from the point of diversion to the place of use. Over the next year and a half, an average of about 902 cfs and a peak of 1,030 cfs had been used to develop electric energy at the Twin Falls power plant. IPC submitted proof of completion of its works to the Idaho Department of Reclamation on June 29, 1937.[525]

After reviewing published notices of the company's proof of completion and its statement regarding application of water to beneficial use, Lynn Crandall, District No. 36 watermaster, posted letters to Idaho Commissioner of Reclamation R. W. Faris and to USBR district counsel B. E. Stoutemyer inquiring about the stated priority date of April 2, 1929, for the Twin Falls development. Crandall recalled that the FPC license for the Twin Falls project contained a reservation clause, Article 14, subordinating the power company's water right to "future storage rights above Milner." Thus, the priority date for Twin Falls power plant water rights should only apply against future rights downstream from Milner. Crandall inquired specifically whether Stoutemyer intended to ask Faris to include a similar subordination clause in the state water right license for Twin Falls as "it would be desirable to have it reaffirmed in the State water right license so that it will be clearly evident that the date of priority applies only to rights below Milner and is not adverse to subsequent developments above Milner." Crandall also inquired whether a similar subordination provision had been attached to the FPC application for a license involving expansion of generating facilities at Lower Salmon Falls and a 300 cfs increase in water use. If so, Crandall opined, a similar restriction should be included in the Lower Salmon Falls water right license.[526]

On June 21, 1937, Stoutemyer wrote to Commissioner Faris regarding the Twin Falls water right permit but said nothing of Lower Salmon Falls. Recalling the Bureau's insistence of a stipulation protecting present and future irrigators in the Snake River Valley in the FPC license for Twin Falls, Stoutemyer suggested that the State of Idaho require a similar provision in its water right license.

---

525 State of Idaho Department of Reclamation, Report of Engineer, Permit No. 17070, "Proof of Completion of Works," June 29, 1937 and "Proof of Application of Water to Benerficial Use" June 29, 1937 Idaho State Idaho State Water Right 02-2057 Back File.

526 Lynn Crandall to R. W. Faris, June 16, 1937 and Crandall to B. E. Stoutemyer, June 18, 1937. Idaho State Idaho State Water Right 02-2057 Back File.

> ...It seems to us that it would be contrary to sound public policy to allow any additional power rights to attach for use in the Snake River Canyon to such an extent as to require the waste of water over Milner Dam, since all the water available in Snake River above Milner Dam is needed for irrigation purposes even at the present time. . . the need to conserve all the available water above Milner Dam for irrigation purposes will become more and more evident as the years go on. On the other hand there are many places in the State of Idaho where additional power can be developed without infringing on the water supply that is needed for irrigation purposes.[527]

Since IPC had already agreed to this limitation in its federal license, Stoutemyer did not see why there would be any objection to provide additional protection by inclusion of a similar subordination clause by the state engineer in any state license. The Idaho Commissioner of Reclamation issued a Certificate of Completion of Works and a License and Certificate of Water Right under Permit 17070 on August 20, 1937 without requiring the subordination clause requested by Crandall and Stoutemyer.[528]

Failure at Upper Salmon Falls

The value of Upper Salmon Falls as a hydroelectric power site had long been recognized.

In 1907 the Great Shoshone & Twin Falls Water Power Company, under control of the Kuhn interests, began development work at the site with a canal being dug along the south bank of the river. After the financial crash of 1913 and the failure of the Kuhn interests in Idaho the following year, the Great Shoshone & Twin Falls Power Company fell into the hands of a receiver, and the bond holders of the company entered into negotiations with the Electric Bond & Share Company of New York City. Out of those talks the Electric Investment Company was organized, and at a judicial sale on February 1, 1916 it acquired the Great Shoshone & Twin Falls Power Company and other power properties in southern Idaho. Idaho Power Company, organized in May 1915, took over the properties acquired by the Electric Investment

527 Stoutemyer to Faris, June 21, 1937. Idaho State Water Right 02-2057 Back File.

528 Deputy Commissioner of Reclamation to A. C. Inman, attorney IPC, August 20, 1937. Idaho State Idaho State Water Right 02-2057 Back File. If Faris ever responded to Crandall and Stoutemyer, no record of it appears in the water right files for the Twin Falls plant held by the Idaho Department of Water Resources.

Company on August 1, 1916. IPC then began upgrading its power plants and co-ordinating the various plants and systems it had acquired into a more unified, efficient organization.

In the early 1920s, when IPC proposed to construct a power plant at Twin Falls, it simultaneously sought to develop the power site at Upper Salmon Falls. Since the Upper Salmon Falls site – like Twin Falls – occupied some public lands, the company filed for a preliminary permit with the FPC on June 17, 1920, in order to establish a priority date with respect to a license under the FWPA "while securing data and performing the acts necessary to perfect an application for such a license."[529]

According to its preliminary application (FPC No. 19), Idaho Power sought to construct a low concrete dam across Snake River at the head of Upper Salmon Falls about 1,000 feet in length and eight feet above low water. The dam site would be located roughly 60 miles below Milner Dam and operated for the proposes of generating electrical power for "light, general power and traction purposes; also for irrigation pumping."[530] The proposed plan of development called for construction of a canal running from the south end of the dam, and ultimately from both ends, to a power house site or sites situated about 1.5 miles below the head of the falls. The company estimated 80 feet as the "average effective head" at the site. Assuming an average flow of 5,000 cfs, IPC projected generating "approximately 33,000 horsepower" at 70 percent efficiency. Below the power plant the river turned abruptly to the south, then bent gently to the north, flowing into the pond water at IPC's existing Lower Salmon Falls power development. The new Upper Salmon Falls power facility would be connected by a 5.5-mile transmission line to the existing power plant at Lower Salmon Falls. The company anticipated that its dam would create a pond covering 875 acres with a storage capacity of 2,200 acre-feet.[531]

---

529 Application for Preliminary Permit (Act of June 10, 1920), Upper Salmon Falls Site, Idaho Power Company, June 17, 1920, 1. Folder 412.12, Idaho Applications for Power Permits under Act 6-10-20 by the Idaho Power Company, 1930 thru (Folder 412.12), Box 809, GCF, 1930-1945, RG 115, NARA-Denver.

530 Ibid., 1. Folder 412.12, Box 809, GCF, 1930-1945, RG 115, NARA-Denver.

531 Ibid., 2; and Report on Application Before Federal Power Commission by Idaho Power Company, Serial No. 19, 1, Folder 412.12, Box 809, GCF, 1930-1945, RG 115, NARA-Denver.

In anticipation of developing the Upper Salmon Falls site, Idaho Power had applied for and secured a water permit from the State of Idaho (Permit No. 13,948) for 5,000 cfs, but the company's proposed development required a "right to use lands held and controlled by the United States." Permission for this use was a necessary precondition to perfect its state water right and required an FPC license. Additionally, the proposed project was situated on private lands owned in part by IPC and in part by others. Negotiations for the purchase of those lands, the company explained in its application, would be initiated "on granting of the present application" by the FPC.[532]

None of these anticipated dealings transpired. As with the Twin Falls site, the Reclamation Service objected to the company's proposed development of Upper Salmon Falls, as did the Wendall and Gooding chambers of commerce and the county commissioners of Gooding County. The latter groups protested on many issues, but were concerned most with the large area to be impacted by the proposed development at Upper Salmon Falls and by the general business strategy of IPC to tie up power sites by partially developing so many large sites on the river.[533]

The construction of American Falls reservoir would open large tracts of new land to irrigation where there had existed only desert wastelands. Reclamation Service engineer George L. Hoffman prepared a report on the company's Upper Salmon Falls application, which explained the service's plans for the full development of irrigable land above Milner, construction of storage reservoirs at American Falls and other sites, and turning the river dry at Milner Dam during the summer irrigation season. During portions of the storage and irrigation seasons, Hoffman wrote, the proposed power plant would be dependent entirely upon the return flow between Milner and Upper Salmon Falls. According to USGS reports on groundwater conditions, the inflow to the river from springs and underground flow in this reach amounted "to

---

532 Exhibit "D," Responding to Regulation 3-E, Application for Upper Salmon Falls Site, Idaho Power Company, enclosed with Application for Preliminary Permit (Act of June 10, 1920), Upper Salmon Falls Site, Idaho Power Company, June 17, 1920. Folder 412.12, Box 809, GCF, 1930-45, RG 115, NARA-Denver.

533 W. G. Hoyt, Report Relative to Application of Idaho Power Company for A Preliminary Permit to Develop Upper Salmon Falls Site on Snake River, Idaho. Permit No. 19, Federal Power Commission, Idaho 26643 (July , 1921), 12-19. Folder: D-100, 12-G-44, WPB, 1894-1961, RG 57, NARA II..

approximately 5,000 second-feet," noted Hoffman.[534] W. G. Hoyt of the USGS noted in 1921 that the flow of the Snake River past the proposed power site for 90 percent of the time was 5,000 second-feet, and the flow for 50 percent of the time was 10,000 second-feet.[535] Frank B. Weymouth, Reclamation's chief engineer in Denver, counseled Reclamation Service Director A. P. Davis to oppose Idaho Power's proposed project at least until the government secured all of the water and power rights that it desired from the power company at American Falls. Otherwise, the service might need Upper Salmon Falls for its own use to meet project irrigation pumping requirements.[536] On March 26, 1921, Reclamation Service Director A. P. Davis forwarded Hoffman's report to FPC executive secretary O. C. Merrill, adding:

> From this report, you will note that the Upper Salmon Falls development is intimately tied up with the Minidoka project of the Reclamation Service, comprising 121,500 acres now under ditch with a 115,000 acre extension [the North Side Pumping Division] planned for the immediate future. It is also intimately concerned with our proposed American Falls reservoir for which preliminary work is now in progress.
>
> This reservoir [American Falls] will completely control and use Snake River above Milner dam and it would manifestly be opposed to the public interest to grant a private company rights which might enable them to handicap seriously, if not prohibit, the otherwise feasible complete development of the power and irrigation resources of Snake River Basin.

Noting that the company's "dam and a portion of both canals proposed to be constructed" were also "located on land that had been withdrawn by the Reclamation Service for the past 12 or 13 years," for potential power development purposes, Director Davis "earnestly recommended that no permit be granted for this site [Upper Salmon

534 Report on Application Before Federal Power Commission by Idaho Power Company, Serial No. 19, 1. Folder 412.12, Box 809, GCF, 1930-45, RG 115, NARA-Denver.

535 W. G. Hoyt, Report Relative to Application of Idaho Power Company for A Preliminary Permit to Develop Upper Salmon Falls Site on Snake River, Idaho. Permit No. 19, Federal Power Commission, Idaho 26643 (July, 1921), 11-12. Folder: D-100, 12-G-44, WPB, 1894-1961, RG 57, NARA II. Hoyt estimated that after American Falls Reservoir was completed, the flow for 90 percent of the time would be approximately the same, while the 50 percent flow rate would be significantly reduced to about 6,000 second-feet.

536 Weymouth to Director, March 14, 1921. Folder 412.12, Box 809, GCF, 1930-45, RG 115, NARA-Denver.

Falls] at the present time, pending the complete study of the American Falls development."[537]

The FPC apparently heeded the director's recommendation, and the commission did not grant the company a preliminary permit – a move that effectively killed the company's plans for Upper Salmon Falls in the early 1920s.

A few years later, in a letter dated January 7, 1924, FPC executive secretary O. C. Merrill requested a "final report" from the Bureau of Reclamation on the company's application.[538]

Three months later, on March 1, 1924, David W. Davis – former governor of Idaho and newly appointed commissioner of the re-organized Bureau of Reclamation – indicated in a letter back to Merrill that "it is noted that the Federal Power Commission is contemplating asking the applicant to withdraw his preliminary permit or to show cause why the application should not be rejected." Before mid-April 1924, the commission had decided to reject the application.[539]

The hydropower amendment of 1928

During the 1920s, hearings before the newly-organized Federal Power Commission regarding licensing proposed hydroelectric power developments provided a venue for individuals and state, federal or local agencies to enter protests against power projects that adversely affected irrigators in the Snake River Basin. The state water right application process in Idaho did not provide any such opportunity for public hearings. State officials frequently articulated their preferences for irrigation development on the Snake River, but lacking a state official with clear discretionary authority to deny water applications, or

---

537 A. P. Davis, Director, to O. C. Merrill, Executive Secretary, Federal Power Commission, March 26, 1921. Folder 412.12, Box 809, GCF, 1930-45, RG 115, NARA-Denver.

538 Letter referenced in a memorandum from Ottamar Hamele, Acting Commissioner, to Chief Engineer, regarding Report on Federal Power application No. 19, Idaho Power Co., Idaho, January 14, 1924. Folder 412.12, Box 809, GCF, 1930-45, RG 115, NARA-Denver.

539 O. C. Merrill, Executive Secretary, Federal Power Commission to Honorable William E. Borah, United States Senate, April 15, 1924. Land Utilization File, Box 561, Borah Papers, LOC Available historic records fail to disclose when Idaho Power's application for the Upper Salmon Falls site was rejected and why. No Application File for Project No. 19 could be found within the Federal Power Commission Records in Washington, D. C. In his letter to Borah, Merrill simply informed the senator of the rejection but did not give a specific date or reason.

any state water resources agency with authority over water resources planning and project development, the state had insufficient constitutional or legislative mandates to control and manage its own water resources.

The State of Idaho did act in the 1920s to limit and regulate the water rights of power interests through an amendment to Article XV, Section 3 of the Idaho constitution. The amendment was approved in 1928.

The proposed projects of Idaho Power Company at American Falls, Twin Falls, and elsewhere on the Snake River provided only part of the backdrop for the constitutional amendment. Another motivating force came from the actions of Pacific Power & Light (PP&L). In April 1922, the Portland, Oregon-based power company filed an application for a permit with the Idaho Department of Reclamation for use of the water of Snake River at ten power sites covering that part of the river lying between Homestead, Oregon and Lewiston, Idaho, a distance of approximately seventy miles. The portion of the stream where the power sites existed formed the boundary between the states of Idaho and Oregon. The company proposed to build ten dams and power plants to generate approximately 1,000,000 horsepower with about 85 percent of the water to be used in the plants originating in Idaho. The power sites were all located on the Snake River from 35 to 105 miles below IPC's lowest power plant at Oxbow and running downstream past the confluence with Salmon River.[540] Under the Idaho constitution, Idaho Department of Reclamation Commissioner W. G. Swendsen had little discretionary alternative but to grant the requested permit to appropriate available unappropriated water in the stream. Because the company proposed to build some of its plants on public land and the Snake River in this stretch was considered navigable, PP&L needed to seek a license from the FPC. Consequently, on or about April 24, 1922, PP&L also filed for a preliminary power permit (No. 293) before the FPC.[541]

540 O. C. Merrill, Executive Secretary FPC, to Mayor of Asotin, Washington, Governors of Washington, Oregon and Idaho, et al., April 24, 1922 and A. P. Davis, Director, USRS, to Executive Secretary, FPC, July 6, 1922. Pacific Power& Light, File Folder 100-2, Project No. 293, Formal, Part 1, April 22, 1922 to September 28, 1923 (File Folder 100-2, Project No. 293, Formal, Part 1), Box 36, Acc. No. 138-835-A-85 (Box 36), FERC; W. G. Swendsen, Commissioner, Department of Reclamation, State of Idaho, to Hon. C. C. Moore, Governor of Idaho, Re: Idaho Water Power, August 18, 1924, 2. Folder 10, Box 3, AR 2/13, ISHS.

The FPC application elicited strong objections from both the U. S. Reclamation Service and the State of Idaho. Arthur P. Davis, director of the Reclamation Service, noted in a letter to the Federal Power Commission that the proposed power project threatened the viability of certain upstream irrigation projects in which his agency had already made considerable investments, such as the Minidoka, King Hill and Boise irrigation projects. It also endangered proposed irrigation developments by the Reclamation Service in Oregon at Malheur, Owyhee, and Baker. In view of these irrigation investments and others in the Snake River Valley, Davis recommended that if the FPC gave favorable consideration to PP&L's application, the company be obligated "to recognize the prior right of all irrigation development, present or future, to the use of the water of Snake River and its tributaries."[542] On July 16, 1922, Commissioner Swendsen also addressed a letter to the FPC calling attention to the state's interests in "contemplated future developments in its [Snake River] upper reaches" and requesting an opportunity to be heard before any action was taken on the application. The commission's executive secretary responded that no action would be initiated on the application until October of 1922. In the meantime, Swendsen began assembling data for the State of Idaho to make its case before the FPC.[543]

Since power developments proposed by PP&L, like other proposed power developments on the Snake River made by IPC, depended upon guaranteed instream flows at certain points on the river, they had the potential to severely curtail future irrigation diversions as well as water storage on the Snake River and its tributary streams in Idaho, such as the Boise, Payette and Salmon rivers. And since irrigated agriculture was the mainstay of the southern Idaho economy, the state was alarmed over PP&L's application for state water rights for such a large power development. Idaho Governor David W. Davis requested Swendsen to prepare a memorandum with recommendations on how Idaho's interests

541 W. G. Swendsen, Commissioner, Department of Reclamation, State of Idaho, to Executive Secretary, Federal Power Commission, October 26, 1922. Folder: Reclamation (Folder: Reclamation); Folder Divider: Department of Reclamation, 1919-1922 (Folder Divider: Department of Reclamation). No Box, AR 2/12, ISHS.

542 A. P. Davis, Director, USRS, to Executive Secretary, FPC, July 6, 1922. File Folder 100-2, Project No. 293, Formal, Part 1, Box 36, FERC.

543 Swendsen to Executive Secretary, Federal Water Power Commission, July 18, 1922 and September 20, 1922. Pacific Power & Light, File Folder 100-2, Project No. 293, Correspondence, Part1, April 3, 1922 to March 4, 1935 (File Folder 100-2, Project No. 293, Correspondence, Part 1), Box 36, FERC.

in the Snake River might be protected, given the state constitutional provision governing issuance of state permits for water rights which held: "the right to divert and appropriate the unappropriated waters of any natural stream to beneficial use shall never be denied."

One of the responses recommended by Swendsen was to file a complaint and seek assistance from the FPC; the other was to amend Article XV, Section 3 of the Idaho constitution to give the state Commissioner of Reclamation more discretion to consider the broader public interest before issuing water right permits and licenses.

Swendsen's report, drafted at the urging of the governor, was sent directly to the executive secretary of the FPC on October 26, 1922 in the hope that in the commission's deliberations on PP&L's application, its members would act to protect the state's interests. The Snake River, Swendsen wrote in his letter, was Idaho's greatest economic asset and its waters needed to be conserved and dedicated to future irrigation uses:

> The Snake River drainage basin comprises 108,000 square miles, approximately three-fourths of which is located in the State of Idaho. In fact, approximately nine-tenths of the State of Idaho is given over to Snake River drainage basin. It is safe to say that 85% of the water available for power at the sites in question originates in Idaho.
>
> The basic resource throughout this part of the state is agriculture, and because of the arid climate and soil conditions the artificial application of water for irrigation, is essential to the growth of profitable agricultural crops. Any development that relates to the use of Snake River water is therefore of supreme and vital importance to the state. Indeed, if it were not for the agricultural development that has heretofore been had through a use of Snake River water for irrigation, approximately three-fourths of Idaho, which is, under present existing conditions, partly developed, would still be given over to non-inhabited desert wastes ...
>
> A careful consideration of the available water supply for arid lands leads to the conclusion that when the total flow of Snake River and its tributaries are completely controlled in the interest of irrigation, through the construction of storage reservoirs and otherwise, the water supply will be insufficient to accommodate all of the lands susceptible of irrigation.[544]

544 Swendsen to Executive Secretary, October 26, 1922. Folder: Reclamation; Folder Divider: Department of Reclamation, AR 2/12, ISHS.

As Swendsen made clear in his appeal to the FPC, protection of future Snake River Basin irrigation was the primary concern of the State of Idaho. With a careful plan for managing water resources in place, Swendsen argued, it would be possible to meet all of the state's projected electrical power needs while realizing maximum irrigation development. No such water resources management plan was currently in place; therefore, Swendsen cautioned the commission that power uses needed to be subordinated to future agricultural and domestic uses:

> … it is extremely important that any power permits granted in connection with these applications shall be conditioned upon the State's present and future right to divert, use and impound as much water as may be necessary for a complete development of its agricultural resources, both for irrigation, domestic and other consuming uses.[545]

Swendsen's concerns were not limited to PP&L's proposed ten-plant Middle Snake River development.

Abundant flows present in the stretch of Middle Snake River below Thousand Springs and the lower end of Hagerman Valley where canyon side-flows replenished Snake River with 6,000 cfs of more or less continuous discharge could not be counted upon for power purposes indefinitely, wrote Swendsen. A considerable part of this water, he theorized, came from irrigation return flows from "deep percolation on the Twin Falls North Side and Twin Falls South Side projects, where a total of 400,000 acres are irrigated." The exact amount contributed to the return flow by irrigation on these projects was not known, but Swendsen believed it constituted a significant portion of the total inflow. Thus, in considering the amount of water available for power purposes in the Snake River below Shoshone Falls, Swendsen advised the FPC that future conservation measures, such as technological innovation, improved farming methods, storage reservoirs, canal lining and other measures to prevent canal losses, needed to be taken into consideration. These water conservation actions would result in more uniform flows as well as a decrease in total return flows in the future.[546]

Below the lower end of Hagerman Valley and the proposed point of use for the PP&L projects there was only limited opportunity for irrigation development because of the high cost of pumping to boost

545 Ibid.
546 Ibid.

water from the streambed to the irrigable bench lands. Nevertheless, Swendsen estimated that eventually some 150,000 acres in this region might be developed by pumping and should be figured into irrigation water demand in the future and reservations made for this use in any permit issued for power purposes. Tributaries of Snake River, such as Boise River, Payette River, Weiser River and Salmon River all contained irrigable acreage in their watershed that might someday control practically the entire stream discharge for irrigation uses, when fully developed. Although the State of Idaho did not possess sufficient data to fully analyze the problem of future irrigation needs, Swendsen suggested that his summary of the situation might serve as a guideline in limiting preliminary FPC permits with the state being able to present more complete data before issuance of any final power permit. In this way, appropriate reservations could be incorporated in power permits dedicating the water supply of these streams for their ultimate irrigation uses.

This meant: If a preliminary permit was to be issued to PP&L, Swendsen urged the FPC to include appropriate reservations dedicating and reserving nearly the entire supply of the Snake River above Hagerman Valley, as well as the entire discharges of the Boise, Payette, Weiser and Salmon rivers for ultimate irrigation uses.[547]

The FPC forwarded a copy of Swendsen's report to M. O. Leighton, the consulting engineer to both PP&L and IPC on November 22, 1922. A week later, the commission announced that it would withhold action on the PP&L application until after September 1, 1923.[548] No hearings were held.

Citing changes in its power needs and the power market, PP&L vice-president F. G. Sykes informed the FPC on September 17, 1923, that the company would withdraw its application.[549] Because of that, the FPC did not need to rule on restricting the water rights of the power company as a condition of its permit. The larger issue of how the State of Idaho and its commissioner of reclamation would respond to similar future water power proposals on Snake River, however, remained.

547 Ibid. See also Swendsen, *Second Biennial Report*, 26.

548 William Kelly, chief engineer FPC, to M. O. Leighton, November 22, 1922 and Swendsen to Executive Secretary, FPC, December 9, 1922. File Folder 100-2, Project No. 293, Correspondence, Part 1), Box 36, FERC.

549 F. G. Sykes to FPC, September 17, 1923. File Folder 100-2, Project No. 293, Correspondence, Part 1), Box 36, FERC.

Under the statutes and constitution of the State of Idaho, state water rights for power purposes, such as the project proposed by PP&L, were initiated by making application for and obtaining permits from the state Department of Reclamation to use unappropriated water for the proposed development. Under the provisions of Article XV, Section 3 of the state constitution, the state administrative officer vested with authority to issue permits could not deny any permit to use available water, if made in the proper form, accompanied by the necessary data and fees.[550] With no constitutional provision empowering the state to regulate use of water for power purposes, and thus no legal means of preventing the company from appropriating Snake River water, Swendsen had been obliged to issue a state water permit to PP&L.[551]

The early 1920s applications for water rights for power developments at American Falls, Upper Salmon Falls and Twin Falls by IPC and by PP&L lower down on the Snake River compelled Idaho public officials to seek some means to allow the state greater control over management and protection of its water resources. Commissioner Swendsen addressed this shortcoming in his *Biennial Report for 1923-24*: "If the greatest future benefit is to be had from its water resource, the State must put itself in a position to control appropriations, particularly for power."[552]

At newly-elected Governor Charles Calvin Moore's urging, the Idaho state legislature in 1923 sought to enact legislation to provide this authority to the state's chief manager of its water resources. On March 31, the Idaho Senate passed Senate Joint Resolution No. 7, which submitted to the voters a proposed constitutional amendment to Article XV, Section 3 that empowered the legislature to vest authority in an appropriate administrative officer to deny permits and regulate water use "for power purposes," such use being "subject to such limitations as may be prescribed by law."[553] Passage of the resolution by the

---

550 Article XV, Section 3, Idaho State Constitution (1890).

551 Swendsen to Hon. C. C. Moore, Governor of Idaho, Re: Idaho Water Power, August 18, 1924, 2. Folder 10, Box 3, AR 2/13, ISHS.

552 W. G. Swendsen, Commissioner of Reclamation, *Third Biennial Report of the Department of Reclamation, 1923-1924* (Boise, 1924), 9.

553 *Journal of the State Senate of the Idaho Legislature, Seventeenth Session, Convened on the Eighth Day of January, A.D. 1923, and Adjourned on the Ninth Day of March, A.D. 1923, at Boise, Idaho, Published by the Authority of the Secretary of State*, 419, ISHS. Swendsen also made reference to this measure in his correspondence with Moore. Swendsen to Moore, August 18, 1924, 2. Folder 10, Box 3, AR 2/13, ISHS.

legislature in 1923, noted Swendsen, "was prompted by the absolute necessity of providing some legislative machinery whereby the use of water for power purposes may be safe-guarded in the interests of the public, to the end that ultimately the greatest good may be had from the water supply, considering irrigation, power, domestic and other uses." [554] As noted by Swendsen the power he sought not only provided for regulation of water use for power purposes, but also an expansion of the powers of his office to authorize management of the state's water resources according to a comprehensive plan designed to safeguard the public interest.

The following year, Governor Moore instructed Swendsen to make a general study of the power resources of the State of Idaho and the problem of coordinating the various uses of the water supply in the public interest. Swendsen's survey noted that Idaho ranked fifth among all of the states in available power sites, but only about 8 percent of these sites had been developed. While it had a surplus of excellent power sites, not all of them *would* be developed because the area of arid, but irrigable, land in Idaho far exceeded the acreage that could ultimately be reclaimed with the state's total surface water supply. Not more than half of the water flowing in streams and available for irrigation had been diverted. Much of this development had been achieved by federal reclamation and Carey Act projects since the early 1900s. How the remaining supply would be used was unknown. The federal government had laid claim to some unused irrigation water for expansion of its Boise and Minidoka reclamation projects, but the state had no overall plan on how to best use its remaining water supplies.

The voters had a chance to approve the constitutional amendment in 1924 and thereby provide the state the tools to control to some extent where, when, and to what use this surplus water would be put. State officials warned that failure to pass the constitutional amendment could have grave consequences, as the PP&L case had shown. Commissioner Swendsen summarized the situation in a pre-election letter dated August 18, 1924, to Governor Moore:

> The State was not in a position to protect its own interests … in the event these rights to the use of Snake River for power matured, the State would have been prevented from making any future consuming use in

554 Swendsen, *Third Biennial Report*, 9.

> irrigation or otherwise, of the waters of Snake River and its tributaries, including the Salmon … [T]hese power rights, if granted without restrictions, would have seriously affected the growth, and the prosperity of approximately nine-tenths of the area of the State.

In order to perfect its state water right application the power company had to obtain a license from the FPC to construct its proposed project. The commission, at the urging of former Idaho Governor David Davis and state Commissioner of Reclamation W.G. Swendsen, had acted to protect the state's interest. It granted no permit and ultimately PP&L withdrew its application. The intercession of the FPC notwithstanding, Swendsen believed it was essential that "the State … should be in a position to protect itself," instead of relying on the federal government.[555]

The voters, however, on November 4, 1924, rejected the constitutional amendment by a decisive margin, 32,682 to 18,163.[556] There is little discussion in the historic record about the campaign leading up to the vote on the constitutional amendment in 1924, and contemporary commentary leaves little evidence or reasons for the amendment's overwhelming defeat in 1924, or for its turn-around victory four years later. Swendsen, for one, speculated that "the lack of funds and proper facilities for presenting the merits" to voters who were protective of their private property rights regarding water were the causes for the 1924 defeat.[557]

Despite its poor showing in 1924, three years later another amendment was drafted and re-submitted to the voters – this time by the Idaho House of Representatives. The house approved Joint Resolution No. 13 on February 28, 1927, sending it to the voters for the general election in November 1928.[558] According to "Cato the Censor," in an editorial published in *The Idaho Statesman*, the renewed push for a constitutional amendment was a direct result of PP&L's earlier attempt to appropriate the waters of the Snake River:

---

555 Swendsen to Moore, August 18, 1924, 2. Folder 10, Box 3, AR 2/13, ISHS.

556 G. C. Hobson, Richard A. Lake, and Clara Bays, eds, *The Idaho Digest and Bluebook: A Compendium of Social, Industrial, Financial, and Governmental Statistics, Together with a "Who's Who" of Idaho People* (Caldwell, Idaho: The Caxton Printers, Ltd., 1935), Abstract Table No. 18.

557 Swendsen, *Third Biennial Report*, 9.

558 George N. Carter, Commissioner of Reclamation, *Fifth Biennial Report of the Department of Reclamation, State of Idaho, 1927-1928* (Boise, 1928), 29; and Colson, *Idaho's Constitution*, 177.

> It seems the Pacific Power & Light Company decided it wanted to appropriate some of those vacant power sites in the Seven Devils canyon of Snake river. Now the reclamation department has no authority to refuse the permit, but it successfully fought the question before the federal power commission ... Then it occurred to the powers that be that an amendment giving specific control over power sites to the state might be in order.[559]

Water users on the Minidoka project questioned whether the proposed amendment would affect the status of their water power rights or state water rights in general. George W. Lyle, secretary of the Burley Irrigation District, consulted B. E. Stoutemyer for his legal opinion on the matter. On the matter of affecting Minidoka project water rights, Stoutemyer wrote: "I would say that the proposed constitutional amendment gives the legislature a power which the legislature does not have under the present constitution. What the effect of this would be would depend largely upon what use the legislature made of the power after it was given." As for water rights in general, Stoutemyer continued, the constitution prescribed that appropriations of water for beneficial use could not be denied by the legislature. If the amendment passed, appropriations for power development could be limited by the legislature, or commissioner of reclamation, under authority of an act passed by the state legislature. The legislature might, for example, wrote Stoutemyer, disallow any appropriation for power development that interfered with irrigation or domestic needs. However, he believed existing vested water rights could not be taken away, whether for power or other purposes, as a completed water appropriation was a property right protected under the federal constitution and could not be taken without due process.[560]

In 1926, Swendsen, who had been one of the strongest voices for subordination of power rights to irrigation since his appointment as commissioner in 1919, left Idaho's Department of Reclamation, and was replaced by George N. Carter. The new commissioner was also in

---

559 "Un-Championed; No Tom Toms Beat for Them," *The Idaho* (Boise) *Statesman*, October 20, 1928.

560 Lyle to Stoutemyer, October 20, 1928 and Stoutemyer to Lyle, October 22, 1928. Folder: 143.12 Idaho: State Irrigation Laws, Water Codes, Etc., 1920 thru 1929 (Folder 143.12 Idaho); Box 107 143.12 – 143.31 (Box 107), General Files, 1919-1929, GAPR, 1919-1945, RG 115, NARA-Denver.

favor of the constitutional amendment as he believed it would give more discretionary power to the state to protect and conserve its unappropriated public waters; it might even change the water development strategies of power companies. For example, in a letter to Republican candidate for attorney general, W. D. Gillis, Commissioner Carter wrote that "this constitutional amendment … would give the state authority to deny granting a power right that interfered with the use of water for future irrigation projects" and might "encourage power companies to use the numerous supplies of water unsusceptible to use for irrigation."[561]

In a few short years the tide of public opinion had changed. Whereas in 1924, the constitutional amendment had been soundly defeated, in the general election held on November 6, 1928, the voters concurred with the commissioner of reclamation, the Bureau's general counsel, and the water users on the Minidoka project. The people of Idaho ratified the measure convincingly by a vote of 56,624 to 38,001.[562]

As amended in 1928, Section 3 read:

> The right to divert and appropriate the unappropriated waters of any natural stream to beneficial uses, shall never be denied, *except that the state may regulate and limit the use thereof for power purposes.* Priority of appropriation shall give the better right as between those using the water; but when the waters of any natural stream are not sufficient for the service of all those desiring the use of the same, those using the water for domestic purposes shall (subject to such limitations as may be prescribed by law) have the preference over those claiming for any other purpose; and those using the water for agricultural purposes shall have preference over those using the same for manufacturing purposes. And in any organized mining district those using the water for mining purposes or milling purposes connected with mining, shall have preference over those using the same for manufacturing or agricultural purposes. But the usage of such subsequent appropriators shall be subject to such provisions of law regulating the taking of private property for public and private use, as referred to in section 14 of article I of this Constitution.[563]

561 *Idaho Daily Statesman,* October 16, 1928. Folder: 143.12 Idaho, Box 107, GAPR, 1919-1945, RG 115, NARA-Denver.

562 Hobson, Lake, and Bays, *The Idaho Digest and Bluebook,* Abstract Table No. 20.

563 Idaho Constitution, Article XV, Section 3, as amended (1928).

By the late 1920s, the citizens of the State of Idaho had for the second time since 1889 seemingly laid to rest the issue of priority in the use of their state's water. Ratification of the 1928 amendment gave the state the ability to regulate and limit the development of its unappropriated water for hydroelectric power uses and maintained the constitutional preference for agricultural water uses over manufacturing and power generation.

The measure, however, had not gone as far as Swendson had hoped in 1924 in conferring broad powers on the state Commissioner of Reclamation to manage the water resources of the state in the public interest and the state legislature did little to seize upon its new constitutional authority to "regulate and limit" the use of water for "power purposes."

Subsequent expansion of both irrigated agriculture and hydropower development would continue to complicate the issue of water rights priorities on the Snake River and its tributaries.

7

A Failed Promise of Abundance at American Falls

During the water shortages of the 1910s the Reclamation Service and private irrigators, operating through Warren Act contracts, had cooperated to increase the water supply available for irrigation by expanding storage at Jackson Lake Reservoir. Their efforts were not enough to avert disastrous crop losses during the great drought of 1919. In the early 1920s federal reclamation authorities, canal companies, reservoir and irrigation districts, and IPC had negotiated and worked together to try harder, to build more storage capacity at American Falls Reservoir.

With the completion and filling of American Falls Reservoir in 1927, irrigators believed they had solved their near-term water shortage problems. In the first few years after 1928, however, during an unprecedented series of dry years, the new storage facility at American Falls failed to fill. During the first seven years of its operation, American Falls Reservoir failed to fill in every year except two.

A prolonged drought lasting through the first half of the 1930s tested once again the ability of water users to craft temporary legal agreements apportioning the waters of the basin to meet emergency needs or to join together as a community to allocate, share, or conserve scarce water resources and to expand the water supply of the region by constructing additional reservoirs on the Upper Snake River and its tributaries. A heightened awareness and concern for water conservation ultimately drove water users, the federal government, and IPC to implement yet another major agreement in 1934 that, while modifying portions of the 1923 contract, nevertheless reaffirmed that the upper basin was devoted to irrigation. The drought also prompted the State of Idaho to experiment by creating a new state water agency whose goal was to assert greater control over the state's public water not only as a way of safeguarding the constitutional preference for agriculture in water use, but also as a means of forging future water policy.

Operating American Falls

The three years leading up to the completion of American Falls Reservoir confirmed the urgent need for such a storage facility in the upper basin. The year 1924, the year before construction began on American Falls Dam, turned out to be the most extreme drought year on record up to that time. Run-off was less than 60 percent of normal and throughout the entire summer, spillage past Milner was limited to what leaked through the dam.[564] The following year precipitation was above normal, Jackson Lake filled early and from the standpoint of irrigation water supply 1925 was one of the most satisfactory years on record. Water flow in the Snake River was adequate to afford large hold-over storage at Jackson Lake, which helped avert total disaster the following irrigation season.[565]

In 1926 temperatures during the spring months in the upper Snake River Valley were above normal and the season was unusually dry. In early February, precipitation registered just 6.75 inches, half of the 1925 figure for the same time period and one-half inch less than in the record drought of 1924. Thus, as construction on American Falls Dam progressed in 1926 and the facility began to store water, stream flow on the Snake River was far below normal. On February 6, the board of directors of the American Falls Reservoir District called a meeting at American Falls of all the canal companies and irrigation districts who were spaceholders, to discuss making additional guaranteed payments to the dam contractor in order to speed up construction to allow storage of the maximum amount of water possible that year.

Because snow was so thin at the high elevations in the watershed, G. Clyde Baldwin, the Upper Snake River Basin watermaster, was predicting that Jackson Lake Reservoir would not fill; in fact, unless weather conditions changed dramatically, he doubted that it would come within 100,000 acre-feet of filling. Under such conditions, peak run-off on the Snake River might be expected in late May or early June. In fact, many canals already were unable to divert full heads of water at the low river stages which prevailed in late spring.

564 G. Clyde Baldwin, "Water Distribution and Hydrometric Work, District No. 36, 1924," (typescript, 1954), 13-16 and 31. Unsorted Water Distribution Reports; District 12A: 1944; District 36: 1924, 1933, 1940, Accession no. 20071910, AR 20, ISHS.

565 Baldwin, "Water Distribution and Hydrometric Work, District No. 36, 1925," (typescript, 1925), 6-11 Unsorted Water Distribution Reports; District 12A: 1944; District 36: 1924, 1933, 1940, Accession no. 20071910, AR 20, ISHS.

Baldwin ordered diversion to some of the youngest decreed rights on the river stopped as early as June 6. Anxious farmers on the Minidoka and Twin Falls projects, whose natural flow rights were junior to the majority of upstream diverters, looked to American Falls Reservoir to provide them with some measure of relief. It had been anticipated that with an aggressive construction schedule on the river portion of the structure that the reservoir might be capable of making 200,000 acre-feet of water available that season. As it turned out, a total of only 106,570 acre-feet of regulated water from American Falls Reservoir was delivered to the Minidoka project, North Side Canal Company, Twin Falls Canal Company, and Milner Low-lift Irrigation District during the 1926 irrigation season. As a consequence, the flow past Milner Dam was limited at all times during the summer to leakage past the dam.[566]

Under the terms of the United States' June 15, 1923, contract with Idaho Power Company, the company was entitled to approximately 800 cfs (255,000 acre-feet) under its secondary power right during the winter. The Bureau of Reclamation assumed that IPC would order the release of this water at a fairly uniform rate during the winter months and that this release would be available for government use at Minidoka. IPC insisted it had no obligation to make uniform releases, but that it could make use of whatever water was released by the United States.[567] At the time the contract was negotiated, the Bureau had contemplated that these more or less uniform releases by the company, together with limited releases of water passing through the government power plant at American Falls would be sufficient to take care of all the power requirements at the Minidoka power plant and maximize stored water for summer irrigation purposes. IPC attorney James L. Boone advised Putnam that if there was a limitation on how the secondary winter storage right was released it arose by implication. It was written

566 Baldwin, "Water Distribution and Hydrometric Work, District No. 36, 1926," (typescript, 1926), 10-14, 36, and 41. Unsorted Water Distribution Reports; District 5: 1936, 1937, 1938, 1939; District 12A: 1915, 1919-1920, 1927; District 36: 1926, 1927, 1938, Accession no. 20071914, AR 20, ISHS; R. E. Shepherd, General Manager, North Side Canal Company, Ltd. To Elwood Mead, commissioner of USBR, January 28, 1926 and R. E. Shepherd, president Board of Directors of the American Falls Reservoir District, to Elwood Mead, commissioner of USBR, February 7, 1926. Folder: 301.1 Minidoka – Correspondence re: Dams and Reservoirs, American Falls Dam, 1926-1929 (Folder: 301.1); Box 729 Minidoka Project, 301 – 301.1, General Files 1919-1929, GAPR, 1919-1945, RG 115, NARA-Denver.

567 Putnam to McGee, November 3, 1926. AFF001778-81. American Falls Power Situation, List of Various Reports and Letters, Idaho Power Company Records.

into the contract, noted Boone, that IPC's use of stored water must be consistent with "a perpetual program of necessary economy and conservation of water in the reservoir in the interest of summer irrigation." A program of extraordinary fluctuation in releases would be inconsistent with this purpose. Nevertheless, Boone concluded, the company was free and unencumbered in its right to fluctuate its releases at American Falls to meet its own power purposes at that plant.[568]

Before the first partial year of reservoir operation was over, Idaho Power began seeking ways to safeguard its interests at American Falls. Although the potential to produce additional power was markedly increased by the higher head at American Falls Dam, Putnam noted that the commercial value of the site was lessened because of the necessity of storing water in the winter for release during irrigation season. Therefore, the bulk of the power generated was seasonal, not firm power. Under the contract between the United States and IPC, the government was entitled to develop only a small amount of power during the winter months to meet the needs of the Minidoka project, not to exceed 10,000 kW. During the summer months, the government had plenty of water, but was limited to a maximum 30,000 kW power plant at American Falls. Putnam questioned whether the substantial $3,300,000 investment in this large power plant could be economically justified if it was working only a few months out of the year and was lying virtually idle the remainder of the time. Furthermore, the power plant was to provide electricity for pumping plants on the North Minidoka Extension, whose pumping units would not be finished until at least 1935, and consequently not require any power for at least eight years. Since it took only two years to construct a power plant, there was no need to begin construction. The money would be better spent by the Bureau of Reclamation elsewhere.

Alternatively, Putnam proposed that his company enter into a "cooperative program of power plant construction" with the United States in order to make: 1) the most economical use of the water available at American Falls; and, 2) the least investment for the greatest utilization of the power resources. He suggested that the power generated be used by the Bureau, IPC, and Utah Power & Light for multiple purposes in the region surrounding American Falls and that the

568 Boone to Putnam, November 8, 1926. AFF001785-92. American Falls Power Situation, List of Various Reports and Letters, Idaho Power Company Records.

government and IPC enter into a "joint program covering the development and utilization of this power," without interfering with irrigation.[569]

With the decision to build the so-called Gooding Division before the Minidoka Pumping Extension in 1927, the Bureau decided not to proceed with construction of the American Falls power plant and its first of four 7,500 kW generating units. The Minidoka Extension would be deferred until at least the completion of the Gooding project, scheduled for 1932. The full development of the Minidoka Extension would require at least ten years from the time construction started, consequently the Bureau anticipated that it would not reach full operation until at least 1942. F. J. Rankin, former chief engineer for IPC and now with the engineering department of the Electric Bond & Share Company, traveled to Denver to confer with Mr. Walters, chief engineer for the Bureau, about leasing any surplus power available from the proposed American Falls installation.[570] The Bureau's engineering and construction departments in Denver, however, decided to do nothing with the construction of the power plant in 1927.

The *Idaho Statesman* on May 10, 1927, published an article advising of the postponement of the construction of the proposed government hydroelectric plant for an indefinite period. The Bureau of Reclamation issued a statement carried in the *United States Daily* on May 26, 1927, relating to the deferral of the American Falls power development:

> The estimated cost of the installation of this power development, in four units, is about $3,000,000. The cost of the installation of one unit of 10,000 horse power, together with the tailrace excavation and other joint construction required for all four units, is about half this amount. As but $700,000 is available in 1928 appropriation, no power can be made available without an additional appropriation. Under the circumstances, unless the power development can be contracted in advance, at a fair return during the 10-year period prior to its

569 W. R. Putnam, "Memorandum: American Falls Power Situation," 3-240-2, November 22, 1926. AFF001800-03. American Falls Power Situation, List of Various Reports and Letters, Idaho Power Company Records; F. J. Rankin, "Memorandum: Proposed Joint Study of the American Falls Power Situation by Committee Representing the Bureau of Reclamation, the Idaho Power Company and the Utah Power & Light Company," December 15, 1926. AFF001804-06. American Falls Power Situation, List of Various Reports and Letters, Idaho Power Company Records

570 F. J. Rankin to Putnam, December 21, 1926. AFF001815-18. American Falls Power Situation, List of Various Reports and Letters, Idaho Power Company Records.

> requirements for the Minidoka north side pumping division, it will not be good business to make this improvement until just before the power is needed.

In the absence of a government dam at American Falls, winter power requirements for the Minidoka project were met by the Bureau's powerhouse at Minidoka Dam. It required an average release at American Falls of about 1,500 cfs to supply project towns and farms with light, power, and heat during the winter months. The irrigation storage rights of spaceholders – the rights of the United States notwithstanding – on the other hand entailed storing up to 800,000 acre-feet at the dam. To assure filling of the reservoir, approximately 4,000 acre-feet per day needed to be stored until there was no doubt of a mountain snowpack sufficient for spring runoff. Walter calculated that if American Falls received its normal, or average, winter runoff of 5,000 cfs, approximately 3,500 cfs of this flow would be needed to care for project power requirements, IPC's contract power rights, and spaceholder irrigation storage requirements, leaving 1,500 cfs for "further disposition, either for filling the storage rights of the United States or for release through the winter for power purposes." Because of a series of drought years in the Great Salt Lake Basin, Utah Power & Light Company could not meet its load without securing power from IPC, which had operated its American Falls power plant to the limits of its capacity and water supply the previous summer to meet this demand. As the United States had no use for its part of the storage capacity at American Falls until its irrigation distribution systems were extended to additional acreage on the Minidoka project, it had no reason to store more than 800,000 acre-feet in the reservoir. This situation presented the United States with an opportunity to lease up to 860,000 acre-feet of surplus water to derive an income from its investment instead of filling American Falls Reservoir in the winter. [571]

Walter requested authority to advertise to obtain bids from the power companies and irrigation interests for surplus inflow in American Falls Reservoir. The surplus water was to be stored or released, as the successful bidders designated, from time to time. The Bureau's plan to advertise to obtain bids for leasing the United States' own surplus water in American Falls Reservoir was approved by Commissioner Mead on

---

571 Mead to Walters, September 18, 1926. Folder: 301.1. Box 729, GAPR, 1919-1945, RG 115, NARA-Denver.

September 18, 1926.[572] W. R. Putman, vice-president and general manager of IPC, informed D. F. McGee of the Electric Bond Company of the pending sale of surplus winter water stored at American Falls and asked that he travel to Washington, D.C. to talk with Commissioner Mead about the amount of water to be made available and its unit cost. Meanwhile Putnam arranged to meet with F. A. Banks and District Counsel Stoutemyer in Boise to discuss the possible purchase of water owned by the government and reserved by them for use on the deferred Minidoka Extension.[573]

In dry years like 1924 and 1926, there was little flood water for storage as a result of spring snow melt and whatever storage was secured had to be obtained by saving winter flow. The Bureau of Reclamation loathed risking another year of drought. With the soil already dry, even a year of normal rainfall would result in subnormal run-off. As the 1926 irrigation season drew to a close, the Bureau looked ahead to its first full year of storing water behind American Falls Dam. Another low water year like 1926 could spell disaster to Snake River irrigators. Stoutemyer cautioned the Bureau's chief engineer, R. F. Walter, to take special measures to store all water possible for irrigators in the new reservoir during the winter months.

> … if the year 1927 proves to be another low water year like that of 1924 and 1926, the only thing which can save the irrigation interests of the Snake River valley from a great disaster, involving crop losses aggregating possibly as much as the entire cost of the American Falls reservoir, is the storage this winter in the American Falls reservoir of all the winter flow which can be saved. To gamble on sacrificing irrigation projects of the valley to this extent for the purpose of getting a few thousand dollars from the Power Company by selling to them the winter water which we have secured the right to store, would seem to me very unwise.

In the event of another low water year, American Falls stored water would prove much more valuable to irrigators than to the power

---

572 Walter to Mead, September 14, 1926; Mead to Walters, September 18, 1926 Folder: 301.1. Box 729, GAPR, 1919-1945, RG 115, NARA-Denver.

573 Putnam to McGee, October 28, 1926, AFF001777 and Putnam to McGee, November 1, 1926, AFF001755. American Falls Power Situation, List of Various Reports and Letters, Idaho Power Company Records.

company. If the water supply was above normal, there would be excess water owned by the Bureau in the reservoir.

Although IPC had offered to purchase surplus space in the reservoir, Stoutemyer said that the government project was primarily an irrigation enterprise and that farmers should have first opportunity to rent any stored water not needed by the government reclamation project. He counseled the Bureau to store enough water to take care of any and all irrigators' needs "then consider dickering with the Power Company to release some water for power development, but not before that time." Instead of contracting to sell surplus stored water to IPC as the company preferred, Stoutemyer favored its sale to junior natural flow appropriators who depended on supplemental storage water such as the North Side Canal Company and Twin Falls Canal Company during the irrigation season. In addition to being used for irrigation, this released water could also be used in the summer months at the American Falls and Minidoka power plants to generate electrical power to operate irrigation pumps.[574]

When apprised of the upcoming sale, irrigators in the Snake River Valley, led by R. E. Shepherd, board president of the American Falls Reservoir District, and the Committee of Nine, protested against the sale of excess water available at American Falls. More than fifty water users, said to represent more than one million acres of farm land in the valley, assembled at a convention at American Falls on November 10, 1926, where they passed a resolution condemning the proposed sale of surplus water at least until such time as American Falls Reservoir was first filled and tested, and that whatever surplus water was available in 1927 be held for irrigation use as crop insurance for the million acres of agricultural land in the Snake River Valley. The assembled irrigation interests also voted to draft a resolution favoring construction of the first unit of the government power plant at American Falls.[575]

After meeting with F. A. Banks, the Bureau's construction engineer, IPC vice-president Putman stated that he would not favor the leasing plan, if it was opposed by the irrigators. In fact, he went further and proposed that the company would not use water contained in its

574 Stoutemyer to Chief Engineer, August 16, 1926. Folder: 301.1. Box 729, GAPR, 1919-1945, RG 115, NARA-Denver.

575 Shepherd to Mead, October 28, 1926; Resolution by Legal Representatives of the Water Users of the Snake River Valley, November 10, 1926. Folder: 301.1. Box 729, GAPR, 1919-1945 RG 115, NARA-Denver.

secondary storage capacity (i.e., 255,000 acre-feet) until after January 1, 1927, citing as his reason the company's interest in "seeing the irrigation districts have an ample supply of water for next year's crops."[576] The Bureau's district counsel Stoutemyer agreed that the surplus water should be saved for irrigation purposes until the Bureau was sure that enough water had been stored to fully protect the crops of the Snake River Valley in the upcoming irrigation season.[577] Banks counseled Walter, the chief engineer at the Denver Office, to conserve the winter water in storage at American Falls "as far as possible for irrigation purposes" and to hold any preparation of advertisements for the sale of stored water in abeyance until consideration was first given to leasing any surplus water to the irrigators. Walters forwarded Banks's recommendation to Commissioner Mead suggesting that plans to lease a portion of the storage at American Falls for power purposes be dropped from further consideration.[578]

From a water supply standpoint, the 1927 water year turned out to be one of the best on record. Normal stream flow fulfilled all irrigation requirements on the Snake River until the latter part of July. Only a small fraction of the total water stored in the upper basin reservoirs was needed to meet irrigation requirements and the irrigation season closed with large amounts of hold-over storage in all the reservoirs. Jackson Lake and American Falls reservoirs were practically empty at close of the 1926 season, but nevertheless each filled to capacity several days before there was any need to release stored water in 1927. During the year some 19,099 cfs of stored water and 6,354 cfs of normal flow was wasted past Milner Dam. Of these totals, only 1,635 cfs of the stored water was classified as ordinary or unavoidable waste and 207 cfs as loss due to maintaining Lake Milner at above normal stages to permit an unusually heavy draft by the North Side Canal Company. The balance of the waste was considered as "intentional waste" which could have easily been avoided had there been any need for larger carry-over storage.[579]

---

576 W. R. Putnam, vice-president & general manager IPC, to F. A. Banks, November 12, 1926. Folder: 301.1. Box 729, GAPR, 1919-1945, RG 115, NARA-Denver.

577 Stoutemyer to Commissioner, November 10, 1926. Folder: 301.1. Box 729, GAPR, 1919-1945, RG 115, NARA-Denver.

578 Banks to Chief Engineer, Denver, November 11, 1926 and November 23, 1926; Putnam to Banks, November 12, 1926; Walter to Commissioner Mead, November 27, 1926. Folder: 301.1. Box 729, GAPR, 1919-1945, RG 115, NARA-Denver.

The total hold-over of stored water at the end of the 1927 irrigation season was such that there was slightly more than one million acre-feet of unfilled capacity in all of the Snake River reservoirs (Table 7.1). As a consequence, under Article 24 of the June 15, 1923, contract as interpreted by the Bureau, no storable run-off was available to be "wasted" at the dam and used by IPC for power purposes.[580] Representatives of the Bureau met with Putnam and his chief engineer at the end of the irrigation season to discuss release of winter storage for power purposes in light of the high hold-over storage in the reservoirs. As the prospect for the reservoirs filling from flood run-off in the spring was high, Putnam desired that a large part of the normal flow of the river be allowed to pass through the reservoir during the winter for power use. IPC could productively use approximately 7,000 to 8,000 cfs to generate power at its American Falls plant, an amount of water close to equaling the normal winter discharge of Snake River at American Falls. Putnam wanted the water released immediately and in a uniform manner, free of charge, as waste water to avoid excessive flood flow passing over the dam in the spring that could not be used to generate power. Putnam suggested that IPC and the Bureau "enter into a stipulation setting out the understanding of the parties as to the terms of the contract, and covering the matter of release of storable run-off in such a way that failure to assert full rights under the contract would not establish a dangerous precedent." Company officials planned on checking the Bureau's calculations on hold-over storage and then suggest to the Bureau how the "wastage" could be used most advantageously by the company "so that an agreement can be reached if possible based upon mutually beneficial use."[581]

Table 7.1 Unfilled Capacity in the Upper Snake River Reservoirs, October 1, 1927

Reservoir	*Total Capacity (acre feet)*	*Holdover (acre feet)*	*Unfilled Capacity (acre feet)*
American Falls	1,700,000	1,390,000	310,000

579 Baldwin, "Water Distribution and Hydrometric Work, District No. 36, 1927," (typescript, 1927); Unsorted Water Distribution Reports; District 5: 1936, 1937, 1938, 1939; District 12A: 1915, 1919-1920, 1927; District 36: 1926, 1927, 1938, Accession no. 20071914, AR 20, ISHS.

580 Under Article 24 of the June 15, 1923 contract water would have been available for wasting in the winter of 1927 only if the unfilled capacity in the reservoirs was less than 1,000,000 acre feet.

581 E. B. Darlington, superintendent Minidoka Project, to Stoutemyer, September 29, 1927. Folder: 301.1. Box 729, GAPR, 1919-1945, RG 115, NARA-Denver.

Jackson Lake	847,000	520,000	327,000
Lake Wolcott	100,000	100,000	None
Henry Lake	75,000	47,000	28,000
Emma Matilda Lake	2,500	500	2,000
Two Ocean Lake	2,500	1,000	1,500
Portneuf- Marsh Valley	16,000	3,000	13,000
Blackfoot	450,000	126,300	323,700
TOTAL	1,005,100		

Under Article 24 of the contract of June 15, 1923, the U. S. Reclamation Service was to determine at the end of each irrigation season under a set formula spelled out in the contract how much storable run-off water was available for "wasting" during the winter months and to release this water at American Falls in as uniform manner as practicable, "or in such other manner as to secure its reasonable availability for beneficial use for power development." The time had arrived in October 1927, to determine for the first time what amount of "waste" water was available and for the engineer in charge of the reservoir to determine how to proceed in releasing the water in question. The Bureau took the point of view that in determining the storable run-off available for wasting, consideration be given to unfilled capacity in all existing government reservoirs in the Snake River drainage system, rather than to unfilled capacity at American Falls Reservoir alone.[582]

This reading of the contract was favorable to the Bureau and was advanced by the Bureau at a conference held with IPC officials in late September 1927, at American Falls. Contrary to the Bureau's expectations, neither Putnam, nor his chief engineer, Henry L. Senger, objected at the meeting. The Bureau's district counsel, Mr. Stoutemyer and its chief engineer, R. F. Walter were delighted that their legal theory had seemingly prevailed without argument. Walters later remarked in correspondence to the Minidoka superintendent:

> It is very gratifying to this office that Mr. Putnam of the Idaho Power Company has not objected to the interpretation of contract advanced by our forces at the conference, to the effect that all the storage reservoirs in the Snake River Basin would be considered in determining allowable winter waste and it is hoped that officials of the power company may be led to concur in our interpretation. Even if they

582 Stoutemyer to Darlington, October 1, 1927. Folder: 301.1. Box 729, GAPR, 1919-1945, RG 115, NARA-Denver.

> do not officially do so, acquiescence will unquestionably strengthen our position very materially.[583]

According to the Bureau's calculations, no water was available for winter waste under terms of the contract. Nevertheless, without a government power plant at American Falls, upwards of 2,000 cfs or more of continuous flow was required to be released for operation of the Minidoka power plant, all of which, in the absence of a government plant at the falls, would be passed through IPC's East Side American Falls plant first. Thus, operation of the Minidoka plant would give IPC released water at American Falls unavailable that year under Article 24, but significantly less than the 7,000 to 8,000 cfs it could beneficially use at its recently expanded American Falls plant.

Stoutemyer prepared a letter for the commissioner's signature explaining that the reason for releasing water to IPC in excess of what was required under the contract was "solely for the purpose of supplying the requirements of the Minidoka power plant and will not imply or be held to impose any obligation to continue such excess release at any future time during the present or any future season." If the company wanted use of additional water, Stoutemyer instructed the Minidoka project superintendent, IPC could call for it as part of its reserved 300,000 acre-feet or the company would need to pay an annual charge for its use.[584]

Adapting: The Gooding extension

Idaho Power was interested in obtaining use of any additional water that might be released at the dam and delivered to its American Falls power plant. In addition to utilizing any wastage or spill at American Falls, the company desired to obtain the right to any excess storage space in the reservoir not used by the irrigation interests.

In the years immediately following completion of American Falls Reservoir, the United States had about 520,000 acre-feet of surplus storage space in the reservoir earmarked for the Milner-Gooding and/or

583 Stoutemyer to Darlington, October 1, 1927 and Walter to Superintendent Darlington, October 5, 1927. Folder: 301.1. Box 729, GAPR, 1919-1945, RG 115, NARA-Denver.

584 Stoutemyer to Darlington, October 1, 1927 and Darlington to Walter, October 6, 1927. Folder: 301.1. Box 729, GAPR, 1919-1945, RG 115, NARA-Denver.

the North Side Pumping divisions of the Minidoka project that were not yet constructed. The North Side Pumping Division (Minidoka Extension) was not started until 1949 and its water source was primarily groundwater, as opposed to the surface water diversions from the Snake River below American Falls as originally intended. The promoters of the Gooding project, designed to irrigate some 83,164 acres in the vicinity of Shoshone and Gooding, moved forward with construction of the reservoir at American Falls. It was an all-gravity project with water being supplied through a 70.46-mile-long canal, heading at the same point and following the approximate alignment of a canal line surveyed by the Reclamation Service in 1918, and again in 1923, for the Mountain Home project.

About 46,664 acres on the proposed Gooding project had a partial water supply from decreed rights on Big Wood and Little Wood rivers, but the supply was insufficient for farming every year. The water of Big Wood River was stored in Magic Reservoir, having a capacity of 191,500 acre-feet, but the reservoir had been filled only six times in the previous fourteen years, and extreme water shortages occurred in 1919, 1920 and 1924. The Big Wood Canal Company requested an investigation by the Department of the Interior to ascertain the feasibility of a plan for obtaining a more reliable water supply from the Snake River and its probable cost. Under a cooperative agreement with the Bureau of Reclamation an investigation was carried out in 1925. The resulting plan called for construction of a canal heading on the Snake River at Milner Dam to supply supplemental water to the 46,664 acres, together with new lands having no water supply that could be reached by gravity flow from the canal; the total proposed project having a gross irrigated area of about 83,164 acres. The Bureau estimated that with a net duty of 3.2 acre-feet per acre, after accounting for transit losses, would require the diversion at Milner Dam of 401,733 acre-feet per annum. In July 1925, a new irrigation district, called American Falls Reservoir District No. 2 (AFRD No. 2), was organized under state law for the purpose of contracting with the federal government to obtain water from the Snake River.

The water supply for the proposed Gooding project was made possible only through the construction of the American Falls Reservoir. The water rights of the United States under its filings made on March 21, 1921, for the American Falls project included the right to divert

Snake River water to a portion of the Gravity Extension Division lands.[585] The requirements for initiated rights from Snake River, including the Minidoka Extension, would have exceeded the natural flow in occasional dry years like 1902, 1905 and 1924. In other cases, such as 1919, a small amount of natural flow would be available, while in a few years the Bureau estimated that fully one-half of the Gooding requirements would be available from natural flow. Some 520,000 acre-feet of storage was being held by the Bureau of Reclamation in American Falls Reservoir for the 115,000 acre Minidoka Extension. The Bureau assumed that because the Gooding natural flow right would be junior to the Minidoka Extension, the Gooding project in most years would derive virtually its entire annual diversion requirement of 401,733 acre-feet from storage.[586]

The majority of water users on the Minidoka project opposed the plan to reallocate a large part of the remaining 520,000 acre-feet retained by the government in the proposed 1,700,000 acre-foot reservoir for the Minidoka Extension in order to irrigate new lands near Gooding. Some of the lands to be irrigated by the Gooding project, such as those within the 125,000 acre Idaho Irrigation Company service area in the Big Wood River and Little Wood River valleys, near the towns of Richfield, Dietrich, Shoshone and Gooding, were privately owned. Some of the lands proposed for irrigation were of lesser quality

585 Not all of the lands within the proposed Gooding gravity extension were included in the list of lands to be irrigated under Permit No. 15134, filed by the United States for the American Falls Project on May 21, 1921. The list excluded about 46,419 acres. S. C. Harper, acting chief engineer, brought the matter to the attention of Commissioner Mead in a letter dated February 7, 1927 in which he urged the Bureau to make a new water filing for the Gooding project. These lands were later included in an existing water rights filing by means of "Application for Amendment to Appropriate the Public Waters of the State of Idaho, Permit No. 15134," filed by the United States with the Idaho Commissioner of Reclamation on July 27, 1928. Folder 031 Minidoka – Water Appropriations Thru 1929 (Folder 031), Box 742 Minidoka-Gooding .023 – 101.02 (Box 742), GAPR 1919-1945, RG 115, NARA-Denver. It was important to handle this matter by amendment to preserve the priority date. Otherwise, if a new filing was required, the diversion rights under Permit No. 15134 were so large that any additional filings on Snake River after the date of Permit No. 15134 would not have been of much value. Stoutemyer to Commissioner, February 15, 1928; and Stoutemyer to Commissioner, August 9, 1928. Folder 031, Box 742, GAPR 1919-1945, RG 115, NARA-Denver. The complicated and thorny issue of apportioning benefits from American Falls Reservoir to the different classes of water rights held within AFRD No. 2 still needed to be resolved. Darlington to Stoutemyer, October 1, 1927. Folder 032 Minidoka Gooding – Water Rights, Settlement of, Exchange, Transfer, etc. Thru 1929, Box 742, GAPR 1919-1945, RG 115, NARA-Denver.

586 Homer J. Gault, engineer USBR, *Department of the Interior, Bureau of Reclamation: Reports on the Gooding Irrigation Project, Idaho, Engineering Report* (Denver. CO, November 25, 1925), Folder: 301 Minidoka – Board & Engineering Reports on Construction Features, 1925-1929, Box 745, Minidoka-Gooding 223.02 – 531, GAPR 1919-1945, RG 115, NARA-Denver.

than the agricultural lands within the Minidoka Extension. Furthermore, because of high transit losses (40 to 50 percent) on the long canal, it would require twice the water per acre to irrigate on the proposed Gooding project.

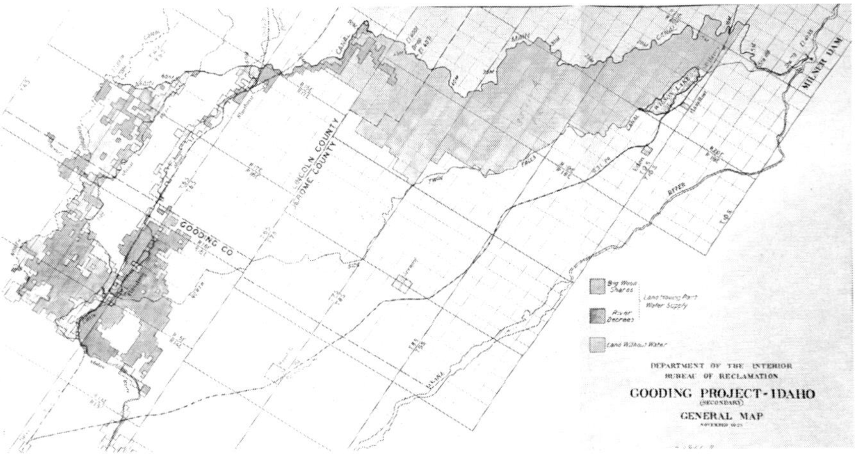

The Gooding Project as of November 1925.

Source: Folder: 301 Minidoka – Board & Engineering Reports on Construction Features, 1925-1929; Box 745, Entry 7, General Administrative and Project Files, 1919-1945,Minidoka-Gooding 223.02 – 531, Project Files, 1919-1929, RG 115, NARA-Rocky Mountain Region (Denver)

Given this situation, R. F. Walter, the Bureau's chief engineer in Denver, advised Commissioner Mead that if irrigators within AFRD No. 2 wanted to secure benefit from the storage in American Falls Reservoir, they should purchase water under the same conditions followed by other private canal companies contracting for water. If investigations demonstrated that a diversion from Snake River was feasible and the construction could be financed, they should contract with the Bureau and make the appropriate payments for any available storage space without encroaching on the 520,000 acre-feet being held for use in the Minidoka Extension. The Bureau still had not definitively settled on the size of American Falls Reservoir. If built to a larger size (the 1921 storage right filing was for 3 million acre-feet), it could furnish sufficient stored water to take care of the new district without encroaching on the Minidoka Extension.[587]

The Gooding Division, also known as the Gravity Extension Division, connected Milner Lake with various privately constructed canals. An irrigation district was created, AFRD No. 2, which embraced the privately-owned lands already irrigated, the partly developed land not irrigated and the public land susceptible of irrigation. The district secured 400,000 acre-feet of water out of the unsold capacity in American Falls Reservoir to acquire an assured water supply to 80,000 acres of land. The district entered into a contract to repay the entire cost of the development in accordance with the terms of the Reclamation Act within forty years.

In recommending approval of the Gravity Extension Unit of the Minidoka project to President Coolidge in July 1928, Interior Secretary Hubert Work described the water supply that would be made available to project lands as follows:

> The water supply will be obtained from Snake River, the natural flow of which in the irrigation season will be fully used, in low years, by rights initiated in advance of this project. The project lands will, therefore, have to depend on water stored in the American Falls reservoir. Four hundred thousand acre feet, or 5 acre feet of water for each acre of land, have been set aside for this purpose. The storable winter flow, together with the flood waters storable in most years and in part carried over from year to year, will cause the reservoir to fill, except in very dry years. Irrigation shortages will be infrequent and of minor extent.[588]

Construction on the Milner-Gooding Canal began in July 1928, but difficult, rocky conditions delayed completion of the canal until May 1932. The completed canal, extending from Milner Lake to the vicinity of the city of Gooding, had an overall length of seventy miles with a maximum carrying capacity of 2,700 second-feet.[589] Since it would be

587 Albert H. Lee, "Let Us Face the Facts," *Rupert News,* May 7, 1925; E. B. Darlington, Superintendent Minidoka Project, to Commissioner Elwood Mead, August 29, 1924, Darlington to R. F. Walter, Acting Chief Engineer, May 2, 1925 and Walter to Commissioner Mead, May 13, 1925. Folder 301, Box 745, GAPR 1919-1945, RG 115, NARA-Denver.

588 Work to Coolidge, July 2, 1928. Folder: 101.02 Minidoka-Gooding – Matters of Policy, Authority Thru 1929, Box 742, GAPR 191-1945, RG 115, NARA-Denver.

589 Bureau of Reclamation, *Annual Project History, The Minidoka Project,* 1928, 52; 1929, 41-50; 1931, 16, 41, 60; 19361932, 41. _ RG 115, NARA Denver, CO.

years before the Milner-Gooding Canal and the North Side Pumping project demanded their full allocation of stored water, on April 23, 1929, the United States entered into a temporary contract with IPC providing for a 10-year lease of the government's share of the carry-over stored water available in American Falls as an additional winter power release to the power company. In the discussion of the contract, the figure 500,000 acre-feet was mentioned as the amount probably available, but the contract was to furnish what was left of the government's stored water at the end of irrigation season. In return, the U.S., still without a government power plant at American Falls, received from IPC up to 7,500 kW of power needed for use on the Minidoka project whenever the government demanded during the irrigation season to meet peak loads. The cost to IPC was $31,000 per year, payable in advance. The right to this water was subject to being rescinded whenever the Bureau of Reclamation believed the water was needed for irrigation purposes and was subject to a protecting clause that the water be released to IPC only upon the condition that it be made available "without any danger of jeopardizing the filling of the reservoir."[590]

IPC made full use of its new water source the first year – more than 700,000 acre-feet. As the 1929 irrigation season drew near a close, the power company on September 16 requested release of its primary water right in an amount that when added to the total domestic and irrigation release at American Falls (6,600 cfs) would maintain a uniform flow of 7,700 cfs. Any excess above 45,000 acre-feet would be charged against hold-over storage under the lease of April 23, 1929. With the exception of a short cessation to accommodate repair work on the spillway at Milner Dam, the company requested that this flow rate be retained through October and into the next water year until all of the company's primary, secondary and hold-over storage water was released, excepting the last 50,000 acre-feet, which was to be reserved for emergency purposes.[591]

590 Agreement April 23, 1929 between United States and Idaho Power Company, June 23, 1929. According to Bureau chief engineer R. F. Walter, the contract did not specify that the power company was entitled to only 500,000 acre-feet of storage. Including hold-over storage in the unsold storage space, the quantity of water that could become available to the power company in any given year could exceed 500,000 acre-feet. Walter to Superintendent Darlington, January 31, 1930. Folder: 301.1 Minidoka – Correspondence re: Dams and Reservoirs, American Falls Dam, 1926-1929, Box 752 Minidoka Project, 301 – 301.1, PCF 1930-1945, RG 115, NARA-Denver.

591 Henry L. Senger, chief engineer IPC, to E. B. Darlington, superintendent Minidoka Project, September 17, 1929. Folder: 301.1. Box 729, GAPR, 1919-1945, RG 115, NARA-Denver.

By the end of calendar year 1929, dry conditions in the Snake River watershed brought forecasts of possible drought. In February, the water content in the snow pack in the Jackson Lake watershed measured 40 percent below normal. Expecting less available water supply than the previous year, water users in the basin re-evaluated their reservoir storage contracts and considered purchasing additional storage rights in American Falls Reservoir in 1930 if needed, in which case the Bureau needed to curtail its power releases to IPC under the 1929 lease. At the same time, the Bureau had completed its annual analysis of unfilled storage capacity in the Snake Basin reservoirs at the end of irrigation season and informed IPC that the unfilled capacity in its eight storage reservoirs was 1,718,850 acre-feet. With a calculated storable run-off of only 700,000 acre-feet available in the year beginning October 1, 1929, under the "wasting" formula of the 1923 contract, the Bureau informed the power company that no free water would be available for winter wasting at American Falls Dam that year.[592] Whereas in the two previous years, IPC had acquiesced to the Bureau's interpretation of the contract, IPC vice-president and manager, M. L. Hibbard now protested to Commissioner Mead:

> I must advise that our company cannot and does not agree with the construction placed by you upon the terms "existing storage capacity" or "unfilled storage capacity." As we understand your letter, these terms are so construed by you to include the unfilled capacity of any and all reservoirs (presumably confined to government control) in Snake River both above and below American Falls. We do not think that Article 24 in its application or reference to unfilled or existing storage capacity includes any other than the existing or unfilled storage capacity of American Falls reservoir.[593]

Apparently, this issue was not resolved.

In subsequent years, the Bureau issued its annual report on unfilled storage capacity in compliance with Article 24 of the 1923 contract showing the cumulative unfilled capacity of its various reservoirs and the power company responded with its protest against including any reservoirs other than the unfilled capacity of American Falls Reservoir

---

592 Mead to IPC, November 19, 1929. Folder: 301.1. Box 729, GAPR, 1919-1945, RG 115, NARA-Denver.

593 Hubbard to Mead, December 14, 1929. Folder: 301.1. Box 729, GAPR, 1919-1945, RG 115, NARA-Denver.

alone.[594] The issue never became a critical one for the company because as long as there was no government power plant at American Falls, IPC at no cost to itself continued to reap the benefit of the government's winter power releases at American Falls Dam used to operate its Minidoka plant.

American Falls in the low years

Implementation of the April 23, 1929, contract was followed by three successive low water years in the upper Snake River Basin. Even if no free "waste" water was available to IPC, it had access to a substantial amount of inexpensive purchased winter water through the 1929 contract. By the middle of January 1930, releases to the power company from American Falls Reservoir since October 1, 1929, under the June 15, 1923, and April 23, 1929, contracts, had totaled approximately 500,000 acre-feet, but there was still about 285,000 acre-feet under the lease contract to which the company was entitled. Under the 1929 contract, the Secretary of the Interior could withhold this water from the power company, if it was advisable to ensure an adequate irrigation supply for the ensuing season. If withheld, the company would be owed a rebate at the rate of $.06 per acre-foot. Superintendent Darlington believed that 650,000 to 750,000 acre-feet could safely be released to IPC without jeopardizing the irrigation supply for the upcoming irrigation season.[595] As it turned out IPC drew over 750,000 acre-feet from the reservoir with deliveries continuing long after it was evident that there might be a shortage of water and over the wintertime protests of the Twin Falls Canal Company.

The operator of the American Falls power plant later commented to R. E. Shepherd that "he was very much surprised at the quantity of water that kept coming down, at times even more than they could use."[596] While the storage reservoir was drawn down some 300,000

594 For example, see Mead to Idaho Power Company, December 23, 1931; Hubbard to Mead, December 29, 1931 and Mead to Idaho Power Company, January 6, 1933; Hubbard to Mead, January 13, 1933. Folder: 301.1, Box 752, PCF 1930–1945, RG 115, NARA-Denver.

595 Darlington to Walter, January 11, 1930; S. O. Harper, acting chief engineer, to Darlington, January 15, 1930; and Walter to Darlington, January 21, 1930. Folder: 301.1, Box 752, PCF 1930–1945, RG 115, NARA-Denver.

596 Shepherd to Sen. Thomas, June 21, 1930. Folder: 310.2 Minidoka – Correspondence Re. Distribution of Water, July 1930 thru 1932, Box 754 Minidoka-Project 310.2, PCF1930-1945, RG 115, NARA-Denver.

acre-feet, because the Bureau had so much unsold storage and some spaceholders had not reached their full demand, by borrowing from these entities there was sufficient water to meet all irrigation requirements. However, the water year ended with some 1,200,000 acre-feet having passed through American Falls Reservoir and spilt past Milner after irrigation season for power purposes. The total reservoir holdover in all of the Bureau's eight storage facilities going into 1930 amounted to less than 900,000 acre-feet.[597]

The Bureau drew intense criticism from Shepherd, the general manager of the North Side Canal Company, and other irrigators for alleged waste of water from Jackson Lake and American Falls Reservoir and for not shutting off water deliveries to the power company when it became apparent that the reservoir might not fill. As a result, water storage for irrigation was being jeopardized. [598] John W. Hart, chairman of the Committee of Nine, dispatched a telegram to Idaho Senator John Thomas, a member of the Senate Committee on Irrigation and Reclamation, stating that the farmers of the Snake River Valley were "up in arms" on account of the prospective water shortage and American Falls not filling. The reservoir had peaked with maximum storage of only 1,300,000 acre-feet in the spring of 1930. Hart blamed the Bureau for having sold an excessive amount of water to the power company and making "a careless waste of storage since last irrigation season." As a result farmers were faced with a 22 percent cut in their storage right for next irrigation season. Hart and Shepherd both suggested that the Bureau ought to do "the honorable and right thing" and make deliveries to the water users from the reservoir in the upcoming irrigation season "as if it had filled."[599]

Senator Thomas forwarded Hart's complaint about the reservoir's poor management to Commissioner Mead, who immediately sent off the following telegram to R. E. Walter:

---

597 Lynn Crandall, "New Storage on Snake River for Irrigation Use Above Milner, Idaho," Statement at Hearing by Corps of Engineers, Idaho Falls, Idaho, February 15, 1955 Folder: D-100-12G-9; WPB, 1894-1961; RG 57, NARA II

598 Shepherd to Senator John Thomas, June 16, 1930. Folder 301, Box 745, GAPR 1919-1945, RG 115, NARA-Denver.

599 Hart to Senator Thomas, June 19, 1930. Folder 301, Box 745, GAPR 1919-1945, RG 115, NARA-Denver.

> Senator Thomas presents complaint of country below American Falls that large amount of water was wasted last year from the reservoir after close of irrigation season. The result is shortage which the Bureau is asking to relieve by furnishing water free of charge belonging to the United States. If error of judgment was made believe unsold water should be furnished without charge but desire your advice.

Walter replied that the problem arose because of the complex water accounting system brought about by the exchanges of stored water between American Falls and Jackson Lake together with power and flood control releases.[600]

The Minidoka project superintendent's plan for operation of the reservoir during the 1929 season was to store water in such a manner as to guarantee a supply of 1,400,000 acre-feet by March 1, and reserve the remaining capacity of 300,000 acre-feet for flood regulation, taking precautions to capture the full capacity before the season for storage release began. The plan was approved by the Bureau's chief engineer on February 7, 1929. A miscalculation as to "the extent and duration of the flood flow" caused a large amount of water available for storage to be discharged from the reservoir, and it was not thereafter possible to capture sufficient storage to bring the reservoir to capacity. There was some speculation, confirmed by Darlington and Walter, that the regulation of the reservoir was implemented in part to protect Milner Dam from the full force of peak winter flood waters going over its weakened spillway. In essence, this action was taken to aid the irrigation interests below American Falls, specifically the Twin Falls canal companies, whose leaders were protesting the loudest.[601]

As a result of the Bureau's flood control regulation, when the spring rains were unusually light, the reservoir failed to rise on the tail of flood season. Initially, Walter, the Bureau's acting chief engineer, had instructed Darlington, the project superintendent, that the shortage should be prorated in the 1930 irrigation season among all the parties having contract rights in American Falls Reservoir.[602] The

600 Mead to Senator Thomas, June 24, 1930. Folder 301, Box 745, GAPR 1919-1945, RG 115, NARA-Denver.

601 Walter to Commissioner Mead, June 27, 1930. Folder 301, Box 745, GAPR 1919-1945, RG 115, NARA-Denver.

602 Since the error in judgment belonged to the United States, Walter ultimately agreed with the suggestion by Stoutemyer that all owners of storage space in American Falls and Jackson Lake be credited with full storage rights with the government making up any shortage with its unsold capacity in American Falls Reservoir. Such a decision, Walter cautioned, "should not be construed

complications in water accounting operations that resulted from reservoir exchanges was in no small part responsible for the claims and counter claims of improper regulation and operation of the reservoirs, as well as attempts to penalize the United States for such claimed errors. Mr. Shepherd, and others, went so far as to call for creation of an organization of representatives of all storage owners to sponsor the operation and management of the reservoirs.[603]

Commissioner Mead also wired Stoutemyer, the Bureau's district counsel, on June 24 requesting a full account of the water shortage controversies. Stoutemyer disagreed with Walter and Darlington's assessment as to the cause and remedy of the problems. Stoutemyer focused on a couple of the key legal questions involved. Two were concerned with the unwarranted discharge of water: first, for flood control purposes; second, to benefit the power company under the April 23, 1929, contract. The other issue concerned the North Side (Twin Falls) Canal Company's objection to returning the stored water that it had borrowed in 1929 out of the government share in American Falls Reservoir.

In Stoutemyer's opinion, use of the reservoir for flood control purposes was unwarranted, as was the sale of an excessive amount of water to the power company, and for the same reason. The Bureau "had no right to jeopardize the filling of the reservoir for such purposes" because American Falls Reservoir, maintained Stoutemyer, "is a storage reservoir for irrigation purposes." Irrigation users should always have the benefit of the doubt, argued Stoutemyer, and doubtful questions always should be resolved in favor of the irrigation users, rather than the power company. Irrigation spaceholders, in his opinion, had a right to insist that "water available for storage should be stored in the reservoir until it is filled." Stoutemyer quoted directly from the preamble of the April 23, 1929, contract with IPC to make his point:

> Whereas, holdover storage remaining in unsold storage space in the American Falls Reservoir at the end of each irrigation season may, to

as a recognition of the claims advanced, but merely a gesture of good will to silence the criticism of past operations, whether right or wrong." Walter to Commissioner Mead, June 27, 1930. Folder 301, Box 745, GAPR 1919-1945, RG 115, NARA-Denver.

603 Shepherd to Senator John Thomas, June 16, 1930; Walter to Commissioner Mead, June 27, 1930. Folder 301, Box 745, GAPR 1919-1945, RG 115, NARA-Denver.

> the extent and only to the extent hereinafter provided for, be used temporarily for the generation of hydro-electric power under conditions which will permit the United States to obtain a substantial return from storage which could not be used either for irrigation or for any other useful purpose.

Stoutemyer counseled Commissioner Mead that the Bureau had no right to jeopardize the filling of the reservoir, except at its own risk, and having failed to fill the reservoir, the deficit in filling of American Falls Reservoir should be charged to the government's share of the surplus storage water. While conceding that doubtful questions should be resolved to the benefit of the irrigators, he defended the government policy of selling its surplus stored water in ordinary years to the power company, which after all was the only entity then willing to buy it.

However, in doing so, the government needed to be careful to protect irrigation interests as its principal priority. It had done so in the April 23, 1929, contract by reserving the right to curtail or cease providing water to the power company in low water years, "whenever there is a demand for the rental of the water for irrigation purposes and to give first consideration to the rental for irrigation purposes of all water which is desired for that purpose."[604]

The final important issue addressed by Stoutemyer involved the contention of the North Side Canal Company that it should not be obligated to repay out of the stored water held in Jackson Lake Reservoir the 134,562 acre-feet of water which the company temporarily borrowed out of the government's share of the American Falls Reservoir. In 1929, in order to hold waters upstream as much as possible and to accumulate as much water as possible in Jackson Lake Reservoir (which of the two Bureau storage reservoirs was the less likely to fill), Minidoka project officials, as well as the Twin Falls companies, were permitted to use American Falls government storage and retain their storage in Jackson Lake, thus unduly depleting American Falls Reservoir.

The understanding when the company borrowed the water, stated Stoutemyer, was that it would be returned out of the water belonging to

---

604 Stoutemyer to Commissioner Mead, June 25, 1930. Folder 301, Box 745, GAPR 1919-1945, RG 115, NARA-Denver.

the company in Jackson Lake Reservoir. A total of 140,227 acre-feet of storage was sent down the Snake River to American Falls Reservoir, of which 76,050 acre-feet was the Jackson Lake right of the Minidoka project. The balance of 64,176 acre-feet was a partial repayment by the North Side Canal Company for its overdraft, leaving a balance of 76,051 acre-feet of water in Jackson Lake which belonged to the American Falls Reservoir. Mr. Shepherd, general manager of the canal company, claimed that under the storage contract adjustments and settlements had to be made before the end of the irrigation season, otherwise the borrower was exempt from the requirement of paying back the borrowed water. Under the contracts with the United States for stored water, the matter of handling the delivery of water from the reservoirs was left largely to the district watermaster. Thus, until the watermaster's annual report was received this type of large scale exchange and borrowing out of the government's share in American Falls Reservoir went unnoticed.

It greatly benefited the Twin Falls companies to hold over as much storage as possible in Jackson Lake and to first draw out all of their storage water from American Falls. As the holder of the upper rights in the reservoir at Jackson Lake, the loss if the lake failed to fill, fell upon the North Side Canal Company (three-fourths) and Twin Falls Canal Company (one-fourth). In addition, it was difficult to regulate the water supply desired by the Twin Falls projects during the irrigation season from Jackson Lake without experiencing frequent shortages or much wastage at Milner Dam.

To allow borrowing from the government at American Falls during the irrigation season and later making repayment from Jackson Lake was a valuable asset to these irrigation companies.

> ... if the repayment of the water borrowed from American Falls reservoir and held in Jackson Lake is postponed until the next spring or next summer, then, in the event that American Falls reservoir fills but there is insufficient flow to fill Jackson Lake without the retention of the borrowed water (which may frequently be the case), it would not be necessary to return the borrowed water and this would be a clear gain to the Twin Falls companies. This practice would involve no loss to the government if we could be sure that the borrowed water would be returned if and when needed to fill American Falls reservoir.

> The rule that a borrower who borrows and won't pay forfeits his credit and can't borrow again applies to the borrowing of water as well as the borrowing of money.

The practice of borrowing water not only affected the government's share of stored water, but also all of the other parties who had rights in American Falls Reservoir. If the borrowed water was not returned, the North Side Canal Company through its ownership of three-fourths of the top half of the Jackson Lake Storage would gain three-fourths of the 76,050 acre-feet. The same company as owner of about one-sixth of the stored water in American Falls would lose about one-sixth of 76,050 on its American Falls account. While the North Side Canal Company would experience a net gain of the difference, other parties who had an interest in American Falls storage, but none at Jackson Lake, would suffer a material loss if the water borrowed from American Falls and held in Jackson Lake was not returned to the reservoir where it belonged. If the North Side company refused to return its borrowed water, the government had a remedy in the American Falls contract, whose paragraph 34 provided for the deduction of the amount of wrongful diversion from the supply of stored water to which the contractor would otherwise be entitled. Under paragraph 18 of the Jackson Lake contract dated February 25, 1913, the United States had the right to substitute one acre-foot of stored water delivered for the company from Lake Wolcott at Minidoka Dam, as the equivalent of and in lieu of 1.5 acre-feet delivered at Jackson Lake. The substitution would involve a considerable loss to the company as the transmission loss from Jackson Lake was not that severe. Under the terms of its contracts, Stoutemyer concluded that the borrowed water should be returned by the North Side Canal Company, or in effect returned by charging the same amount against the Jackson Lake Reservoir and crediting it to parties who held rights in American Falls Reservoir.[605]

Walter recommended that contractors in American Falls storage be entitled in 1930 to a complete filing of their contract storage capacities. Given the attitude of contractors with regard to repaying their American Falls overdrafts with Jackson Lake water, Walter recommended that the Bureau curtail the privilege of exchange and that overdrafts on American Falls Reservoir no longer be permitted. He had little doubt

605 Stoutemyer to Commissioner Mead, June 25, 1930. Folder 301, Box 745, GAPR 1919-1945, RG 115, NARA-Denver.

that the advantage of the past method of operating American Falls and Jackson Lake would soon become readily apparent to the owners of the top storage rights at Jackson Lake and they would desire a reinstatement of this practice. Walter recommended that reinstatement be agreed to only upon a satisfactory agreement for the repayment of overdrafts on American Falls.[606]

Complex water accounting problems, water exchanges, and power and flood control releases all contributed to complaints registered by water users in the region below American Falls Dam in the 1929 and 1930 irrigation seasons.

Walter hoped that some new arrangement for operation of the reservoirs could be accomplished before the close of the 1930 irrigation season, in order to avoid any controversy regarding release of holdover storage in American Falls to the power company in the coming winter. The storage impounded in American Falls appeared that it would be materially in excess of all of the irrigators' space, other than that of the United States; therefore, the Bureau anticipated a market for the sale of a part of its storage capacity. If irrigators desired any of this water, Walter noted that the Bureau stood ready to supply it at the established rental rate of 50 cents per acre-foot. Otherwise, it would be released to the power company.[607]

Rather than criticizing the method of handling the reservoir in the winter of 1929-30, Stoutemyer believed that a fundamental change in government policy was at the root of the problem. The change in policy had put irrigators in a less favorable position than what was contemplated at the time the reservoir was built. The plan, as contemplated in all of the government condemnation suits for the Minidoka project, included construction of American Falls Dam, a power plant to supply the needs of pumping plants, and development of the Minidoka North Side Pumping Division. Under the plan, the same water would have been used to develop power – first at the government's American Falls plant and then at its Minidoka plant. The amount of winter waste or use for power development would have been reduced by one-half.

606 Walter to Commissioner Mead, July 2, 1930. Folder 301, Box 745, GAPR 1919-1945, RG 115, NARA-Denver.

607 Walter to Commissioner Mead, June 27, 1930. Folder 301, Box 745, GAPR 1919-1945, RG 115, NARA-Denver.

The departure from the original plan resulted from Congress substituting the Gooding Division ahead of the North Side Pumping Division, and thus temporarily postponing construction of American Falls power plant. Congress may have been justified in doing so from a settlement and needs perspective, but the change in plan resulted in less favorable operating conditions at American Falls than expected when the reservoir was built. The substitution of development projects left the Minidoka project with a growing need for power and without a means to supply it, except through the contract later made with IPC on April 23, 1929, which supplied the additional power requirements of the Minidoka project during the peak irrigation season demand.

The result was that until the government's carry-over storage contract with IPC was terminated, this water would not be available (without penalty) for irrigation purposes. Therefore, from Stoutemyer's perspective, the appropriate grounds for irrigators to request water from the government without rental charges rested not on a mistake in operating the dam, but on the grounds that the plan itself had changed in the time since irrigation spaceholders signed their contracts. Chief Engineer Walter agreed with Stoutemyer's assessment. Of course, in 1930 Stoutemyer and Walter viewed this matter merely as a temporary problem preceding construction of the government power plant at American Falls and the North Side Pumping Division. [608]

The Bureau offered an opportunity to storage contractors, who were concerned about the prospects for runoff into the reservoirs and the complete fill of their rights for the 1930 irrigation season, to reserve additional storage in the Bureau's unsold space by lease contract. In January 1930 the chief engineer suggested that 100,000 acre-feet be reserved and that water users who received water benefits from the reserve would pay for that privilege. Even though by late spring the Bureau and other water officials were predicting that American Falls Reservoir would not fill, no irrigation districts or canal companies accepted the offer. Protracted periods of cold weather retarded snow melt on the mountains and run-off came down so gradually that it was largely diverted for irrigation as normal flow. River discharges were too low to supply all irrigation demands and stored water had to be used to maintain the canal diversions. Early storage drafts alarmed project

608 Stoutemyer to Commissioner Mead, June 28, 1930; and Walter to Commissioner Mead, July 2, 1930. Folder 301, Box 745, GAPR 1919-1945, RG 115, NARA-Denver.

officials. The Bureau was cautious not to operate the dam in such a manner as to be subject to criticism for jeopardizing irrigation interests by the release of water for power purposes. The commissioner decided to permit filling of storage rights lost the previous season from unsold government storage. If too much water was called upon, the protective clause of the April 23, 1929, contract could be invoked and the government could reallocate enough stored water to fill out the rights of other irrigation storage owners. This meant that much less water would be leased for power purposes and that a considerable portion of the rental from IPC would be forfeited.[609]

Because of the lack of rainfall, the irrigation season began a month earlier in the spring of 1930 than usual. Conditions on the Snake River were such that the government had decided to use its unsold storage for the purpose of insuring an adequate irrigation water supply, but there was also concern over conserving as much water as possible for carry over storage as insurance against another dry year. Another dry year portended disaster with crop damages in the millions. By the middle of irrigation season, Stoutemyer projected that Jackson Lake could be near empty by the fall and American Falls Reservoir might carry over only one-half of what it had the previous year. If the following year was one of low water, the next irrigation season might start with no stored water for the Twin Falls companies in the top half of Jackson Lake Reservoir. To the extent that the reservoir did not fill, the companies would look to rent stored water from the government's surplus supply in American Falls Reservoir, perhaps as much as 400,000 acre-feet. In addition, the Gooding Canal was scheduled to be completed to Little Wood River by the spring of 1931, so there could be an additional 200,000 acre-foot demand on the American Falls water storage next season. [610]

While Stoutemyer advocated a cautious and conservative approach to managing the reservoirs, at the mid-point of irrigation season in 1930, R. F. Walter, the Bureau's chief engineer, expressed a great deal of confidence in the adequacy of the water supply. Exclusive of the Gooding Division, Walter noted that the total rights sold in American Falls Reservoir were about 700,000 acre-feet. If the Gooding Division took 200,000 acre-feet, the total requirement to satisfy storage rights in

609 Darlington to Commissioner Mead, June 28, 1930; Darlington to Commissioner (through Chief Engineer), June 28, 1930. Folder 301, Box 745, GAPR 1919-1945, RG 115, NARA-Denver.

610 Stoutemyer to Commissioner Mead, July 10, 1930. Folder 301, Box 745, GAPR 1919-1945, RG 115, NARA-Denver.

the reservoir would be roughly 900,000 acre-feet, an amount substantially below the minimum storable inflow experienced over the previous four seasons. Walter could see no valid reason for retaining any hold over storage belonging to the United States, nor any reason to withhold water from the power company under the April 23, 1929, contract. Why should the Bureau sacrifice the very substantial amount of money paid by the power company, noted Walter, merely to pile up a large surplus of water in excess of the amount that could be used for irrigation purposes? Even in a low water year like 1930, requests for additional rental storage water from the Bureau had been insignificant. If irrigators wanted to be assured of an adequate water supply in the upcoming irrigation season, Walter concluded, they could make arrangements to purchase additional stored water from the government at the agreed upon rate of 50 cents per acre-foot.[611]

In answer to Walter's letter to the commissioner, Stoutemyer countered that the irrigation season was only half over and irrigation companies usually waited to purchase water later in the irrigation season. Moreover, he argued, the only reason the water supply appeared adequate was due to the fact that Lake Jackson Reservoir began the year practically full as unused water was carried over from the previous year. Walter had failed to consider the inter-relations between the two reservoirs. Next year, Jackson Lake would begin near empty, or less than one-half full. Without this supply, crops would suffer significant shortages, particularly on the North Side Twin Falls project.

Stoutemyer criticized Walter for considering only the "financial standpoint." The public interest in saving crops in the Snake River Valley and "maintaining the prosperity of the whole community" should be the major consideration of the Bureau in managing the water supply of the basin. If 1931 turned out to be a year similar to 1930, Stoutemyer estimated that the need to supply farmers from American Falls storage would jump from 600,000 to 700,000 acre-feet more than what was needed in 1930. Stoutemyer asked the commissioner to set up a conference in the fall to be attended by representatives of the various companies and districts in the Snake River Valley to apprise them of the water supply situation and to urge them to secure some kind

611 Walter to Commissioner Mead, July 22, 1930. Folder 301, Box 745, GAPR 1919-1945, RG 115, NARA-Denver.

of contract or guarantee for rental water. Preferably, enough water to offset any loss the government might sustain by notifying IPC that the government would reserve the hold over storage.[612]

The meeting of storage owners at American Falls and Jackson Lake was held on October 16, 1930, in American Falls at an expanded meeting of the American Falls Reservoir Advisory Board. The meeting was attended by almost the full boards of AFRD and AFRD No. 2, by representatives from both the Minidoka and Burley districts, and by a large delegation from the upper valley. Once again, Stoutemyer apprised the storage contractors of the availability of 450,000 acre-feet of rental water at 50 cents per acre-foot. Apparently there was still little interest at that price, which was eight times the unit price paid by the power company.[613] Six months later, at the annual watermaster's meeting, water users, whether storage users or not, overwhelmingly voted to ask the Interior Department for the privilege of using the government's American Falls holdover storage at the rate of 6 cents per acre-foot.[614]

While the water year ending in October 1930 was generally characterized by low run-off, heavy August rains and storage at the two reservoirs made it possible to fill all demands on the Snake River and retain 247,000 acre-feet in Jackson Lake and 552,000 acre-feet in American Falls Reservoir at the close of irrigation season. Of the latter amount, IPC withdrew 392,000 acre-feet for power withdrawals during the winter months. The failure of American Falls to fill again in 1930, in view of the general belief that it should fill every year, led to more discussion of the government fulfilling rights in full from its share of available storage. Some critics laid blame on the failure to construct a government power plant at American Falls Dam to supply a portion of the winter power load. By running the same water through two government plants the amount of water released from storage could be cut in half and reduce the amount of water running to waste past Milner Dam during the winter months. Furthermore, if the government did not

612 Stoutemyer to Commissioner Mead, July 25, 1930. Folder 301, Box 745, GAPR 1919-1945, RG 115, NARA-Denver.

613 Stoutemyer to Commissioner Mead, October 21, 1930; and Minutes of the American Falls Reservoir Advisory Board Meeting, American Falls, Idaho, October 16, 1930. Folder 301, Box 745, GAPR 1919-1945, RG 115, NARA-Denver.

614 Darlington to Walter, March 11, 1931; and Minutes of Annual watermaster Election of Water District No. 36, American Falls, ID, March 2, 1931. Folder 301, Box 745, GAPR 1919-1945, RG 115, NARA-Denver.

need to lease power from IPC during the summer season, the hold-over storage at the end of the season could be saved and used only for irrigation purposes as planned. Failure to build American Falls power plant, contended attendees at the 1931 watermaster's meeting:

> … results in diminishing the supply available for storage in American Falls reservoir by several hundred thousand acre-feet annually below the amount that would be available under the plans proposed at the time the water users on Snake River decided to participate in the construction of the reservoir. The American Falls reservoir was the result of a great co-operative effort by the irrigation interests of Snake River valley, designed to secure them an adequate water supply in the dry years. While these irrigation rights are perfectly willing that the fullest possible use be made of the reservoir for power purposes they are insistent that such use be subordinate to the use for irrigation, and they will not view with equanimity any attempt to so operate the reservoir as to deprive them of the amount of water they would be entitled to receive if the plans and operations had proceeded according to the original representations.

It was unlikely that Congress would release funding for the power plant until economic conditions justified the building of the North Side Pumping Division of the Minidoka project. In the meantime, water users of Watermaster District No. 36 requested that the Bureau allot the irrigators a preferential right of purchase from the government's share of storage at 6 cents per acre-foot. A committee was appointed to confer with Bureau officials.[615]

As the winter of 1930-31 progressed, the snow pack in the Snake River watershed was at one of the lowest points in recorded history; concern was manifest that an insufficient amount of storage could be captured in the upper basin reservoirs to meet late season irrigation needs. The government's hold-over storage in American Falls Reservoir was about 380,000 acre-feet, with IPC having already used 130,000 acre-feet since October 1, 1930, under the 1929 lease contract. This left a balance of only some 250,000 acre-feet.

The executive committee of the American Falls Reservoir Advisory Board met with Bureau officials on January 5 to consider the water situation. They decided to reserve the remainder of the government's

615 Minutes of Annual watermaster Election of Water District No. 36, American Falls, ID, March 2, 1931. Folder 301, Box 745, GAPR 1919-1945, RG 115, NARA-Denver.

holdover storage until irrigation season as insurance against the reservoir not filling.

Idaho Power Company voluntarily discontinued delivery of water to its power plants, and within twenty-four hours storage of water in American Falls Reservoir commenced. Pursuant to the 1929 contract, at the request of the Snake River irrigators the Secretary of the Interior withheld 241,000 acre-feet in storage to insure an adequate supply for the upcoming irrigation season. The Interior Department paid the $14,460 refund due to the power company for relinquishing its leased rights to the stored water.[616]

The drought year prompted reconsideration of the 1929 contract. IPC wanted to leave the existing contract in place, but the irrigators, led by the executive committee of the American Falls Reservoir Advisory Board, wanted their own rental agreement to control hold-over water remaining at the end of each irrigating season. The unsold storage space, they argued, should be managed to ensure filling the reservoir for irrigation uses instead of power generation.[617] Bureau of Reclamation district counsel B. E. Stoutemyer wrote that M. L. Hibbard, general manager of IPC who represented the company at an April 8, 1931, meeting with the three interested parties, was "very reasonable and cooperative" and expressed his willingness to join in negotiating some arrangement with the irrigation and reservoir districts. After the meeting, Stoutemyer tried to reduce what he called the "hazy discussions" had at that meeting into a tangible three-party contract guided by the following general principles:

> I believe, in view of the fact that the American Falls Reservoir was constructed primarily for irrigation purposes and was made possible by the substantial cooperation of the irrigation companies and districts in

616 R. E. Shepherd, president and chairman Executive Committee, American Falls Reservoir Advisory Board to Members of the Advisory Board, January 8, 1931 and B. E. Stoutemyer, District Counsel USBR to Idaho Power Company, April 10, 1931. Shepherd's letter to members of the Advisory Committee left the impression with some water contractors that that they were to receive a proportionate share of IPC's holdover in return for their payment of a proportionate share of the rebate. However, the Bureau took the position that no contractor could secure a right to more than enough water to fill his share of the reservoir capacity. E. B. Darlington, Superintendent, to Stoutemyer, March 17, 1931; Stoutemyer to the Commissioner, March 19, 1931; Stoutemyer to IPC, April 10, 1931; and Stoutemyer to Superintendent Darlington, April 10, 1931. Folder: 301.1, Box 752, PCF, 1930-1945, RG 115, NARA-Denver.

617 Two such proposals were outlined by the American Falls Reservoir District. R. E. Shepherd to Elwood Mead, April 10, 1931. Folder: 301.1, Box 752, PCF, 1930-1945, RG 115, NARA-Denver.

> the valley, who paid to the government their share of the cost of the reservoir in advance, and that the value of the water for irrigation purposes when needed for such purposes is so many times greater than its value for power, that we should give favorable consideration to the proposals now being made by the irrigation interests of the valley, provided they are willing to make a reasonable annual payment each fall sufficient to offset the amount which the government will lose by holding the water for this purpose.[618]

Successive extreme low water years fed their anxiety. Upper basin irrigators faced the summer of 1931 with worries of an acute water shortage during the upcoming irrigation season. Little precipitation had fallen in the Snake River watershed in the winter and continued unfavorable spring weather portended a low natural flow in the river, meaning farmers would be more dependent on storage water. In early May, Jackson Lake Reservoir was only about one-half full (430,000 acre-feet) and once again the new reservoir at American Falls had not filled. Irrigation headgates on the Snake River opened early. American Falls Reservoir receded to below 1,500,000 acre-feet; it was still early in the month of May.

After attending a May 6 spaceholders meeting in Idaho Falls, Senator Thomas wrote to USBR Commissioner Elwood Mead asking that irrigators be allowed to lease the unsold space in American Falls Reservoir at a price not exceeding what the power company had paid for the same right ($31,000). Revenue derived by the power company from sale of surplus power realized from its rental of unsold storage was negligible, wrote Thomas.

In contrast, revenue from sales of storage water by the American Falls Reservoir District to irrigators would bring considerable economic return, assure a fuller reservoir because some irrigators would bank their storage, and any surplus irrigation water released at American Falls Dam would be used to generate power for project purposes at Minidoka. The power company was responsible in no small part for the poor water conditions at American Falls, argued Thomas, as the low reservoir level was "largely due to the increased amount of water drawn during the winter for power purposes." While he was critical of the Bureau's decision to sell surplus storage water to the power company, Thomas concluded his letter to Mead on a more optimistic note: "I

---

618 B. E. Stoutemyer to Commissioner USBR, April 14, 1931. Folder: 301.1, Box 752, PCF, 1930-1945, RG 115, NARA-Denver.

understand the Power Company will be willing to accept whatever arrangement will be made with the irrigation interests, and will continue their present contract arrangements for such water as may be available in future years after the irrigation interests have been cared for."[619]

The 1931 contract

With the completion of American Falls Reservoir, the Snake River Valley entered an era of rapid and substantial progress. A new period of confidence in the water supply led to farmers altering their cropping patterns to higher value crops, such as sugar beets, beans and potatoes, all of which needed late season water for irrigation. It was important to the agricultural economy of the region that confidence in the water supply be maintained. By 1931 the confidence of farmers in their water supply had been shaken; they were uncertain at planting time in the spring whether there would be an ample late water supply to mature their crops and produce the large returns per acre on their substantial investments. As the season advanced, it became apparent early that various canals in the upper valley, whose shareholders owned little or no storage, would be unable to mature their crops.

At a general meeting of water users and the Committee of Nine early in the irrigation season, participants unanimously adopted a resolution authorizing temporary storage of natural flow rights at Jackson Lake, provided such transfers would not deprive any canal of water that it would otherwise receive. The Committee of Nine appointed a Pool Committee whose task was to secure sufficient water to save the late harvest crops under upper valley canals with seriously inadequate supplies. These decisions greatly benefited certain canals whose normal flow rights were cut off or greatly reduced by midsummer, about 104,000 acre-feet of normal flow was stored at Jackson Lake through purchase, exchange, donation, and storage of normal flow.[620]

Another meeting of upper basin irrigators at Idaho Falls on May 8, 1931, was attended by representatives of most of the spaceholders in

619 Sen. John Thomas to Commissioner Elwood Mead, May 9, 1931. Folder: 301.1, Box 752, PCF, 1930-1945, RG 115, NARA-Denver.

620 Lynn Crandall, *Water Distribution and Hydrometric Work District No. 36, Snake River Idaho, 1931*,2-4, 24-29. Folder: 310.2 Special. Minidoka Project, Report on Distribution of Water, 1931 & 1932 (Folder 310.2), Box 758 Minidoka Project 310.2-320, PCF, 1930-1945, RG 115, NARA-Denver.

American Falls Reservoir, the executive committee of the American Falls Reservoir District, and Idaho Senator John W. Thomas of Gooding, who chaired the U.S. Senate Committee on Irrigation and Reclamation. R. E. Shepherd presented a plan providing for irrigation interests to lease carry-over storage remaining in the unsold storage space (about 465,000 acre-feet) in American Falls at the end of each irrigating season and to dispose of it as best served their interests in any given year. IPC and the Bureau of Reclamation both supported the plan. Most of the storage contractors expressed a willingness to participate at some level in the proposed plan.[621] After the meeting, Shepherd departed for Washington, DC to discuss the matter directly with Commissioner Elwood Mead.

Negotiations on a three-party agreement were consummated in the spring and the contract was signed by the United States, American Falls Reservoir District (acting as trustee for owners of storage rights in American Falls Reservoir) and Idaho Power Company on June 1, 1931. The contract, providing for a maximum five-year lease of the government's unsold storage space in the reservoir, so far as needed for irrigation, was signed by Idaho Power, so as to consent to the necessary modification of its existing contract. The United States had excess storage capacity in the dam reserved for the Gooding Division of the Minidoka project, then under construction, and for the North Side Pumping Division, not yet built. Under the June 1, 1931, contract, this water, destined in the future for the Minidoka project, was purchased by the irrigation concerns for $50,000 during the 1931 season, a year of almost unprecedented low run-off. The big AFRD assumed the liability for paying the $50,000 per year, but sub-lessees paid their proportionate share. If the contract was renewed, the district would make a like payment in 1932, and thereafter annually until 1935, in effect supplanting the April 23, 1929, contract between the United States and IPC. Each subscribing irrigation company or district was entitled to its *pro rata* share of the stored water supply to be furnished out of the government's unsold portion of the American Falls reservoir capacity. If water conditions improved on the Snake River and promised to provide sufficient water to irrigators, the American Falls Reservoir

621 Darlington to Commissioner Mead, May 8, 1931. Folder: 301.1, Box 752, PCF, 1930-1945, RG 115, NARA-Denver.

District had the opportunity to dispose of some, or all of their purchased water to IPC and recoup part of their rental payment.[622]

Lynn Crandall, watermaster of District No. 36 (in charge of the distribution of natural flow) and the special deputy commissioner of reclamation (in charge of the delivery of stored water on Snake River), in his annual report for 1931 confirmed that the water supply for that year was the lowest in 28 years, or for the entire period for which records were available. The run-off at Moran, the outlet of Jackson Lake Reservoir, was only 54 percent of normal. At Neeley, the gaging station below American Falls, it was 53 percent of the 28-year mean.[623] There would have been a complete crop failure in the greater part of the Upper Snake River Plain except for the water stored at American Falls and Jackson Lake Reservoirs. The alarmingly low levels of all reservoirs at the end of the 1931 irrigation season and the prospect of another inadequate season of run-off in 1932 virtually guaranteed a continuation of the June 1, 1931, lease agreement for the government's unsold American Falls storage. The Bureau calculated the approximate unfilled capacity of reservoirs in the upper basin including American Falls on October 31, 1931, at 3,119,640 acre-feet with all of the reservoirs combined having holdover storage of only about 88,000 acre-feet. Under the assumptions of Article 24 of the June 15, 1923, contract the Bureau assumed that the storable runoff available for filling the unfilled reservoir capacity was only 700,000 acre-feet. Thus, there was no winter waste water available for release from American Falls Reservoir to IPC for power generating purposes.[624]

In fact, a great deal of discussion was had over the winter of 1931-32 with respect to adoption of additional conservation measures to maximize the amount of storable water within the Snake River Basin and to make the best possible use of the limited water supply that was available.

---

622 Three-party Agreement (United States, American Falls Reservoir District and Idaho Power Company) for Lease of Reservoir Capacity in American Falls Reservoir, June 1, 1931. While the 1929 agreement remained intact, the power company waived the right to object to the giving of annual notice by the Secretary withholding from the power company the government's share of the carry-over storage water in American Falls and reserving it for irrigation uses, regardless of water supply conditions.

623 "Watermaster Makes Report," *Burley Bulletin*, January 28, 1932.; Minutes of the Annual watermaster Meeting, Idaho Falls, ID, March 7, 1932; and Annual Report of the Committee of Nine for 1931. Folder 301, Box 745, GAPR 1919-1945, RG 115, NARA-Denve

624 Lynn Crandall, *Water Distribution and Hydrometric Work District No. 36, Snake River, Idaho, 1931*. Folder 310.2, Box 758, PCF, 1930-1945, RG 115, NARA-Denver.

The extreme shortage of water gave rise to a careful study by various water users of methods of operation being followed on the Snake River, but upon careful review of the totality of factors involved in these cases, it was generally found possible to work out fairly equitable solutions to all water users. No water rights or water management controversies arose during the season serious enough to lead to any new litigation. The Committee of Nine, a voluntary organization of water users advising the watermaster, had rendered valuable assistance during the trying dry conditions of the 1931 irrigation season. Lynn Crandall, serving his first whole irrigation season as watermaster and special deputy to the commissioner, was widely praised for his service to the water users during what turned out to be the worst drought in the history of the state. State Commissioner of Reclamation R. W. Faris reported to Governor C. Ben Ross that Crandall, through "the exercise of rare judgment, tact and industry," had "surmounted the many perplexing and vexatious problems" with which he had to contend in handling and distribution of one-half of the usual water supply to more than one million acres of agricultural land, and did so "in a fair, efficient and constructive manner."[625] Crandall would become a key figure in the complex system of river operations in the upper basin, in charge of both normal flow and storage distribution on the Upper Snake River into the 1950s.

In December the Bureau, IPC and representatives of the irrigation and reservoir districts met to discuss run-off and conservation matters. The parties agreed to a temporary arrangement designed to restrict outflow from American Falls Reservoir to what was necessary to fulfill the Minidoka power requirement (1,600 cfs). At the request of the irrigation interests, IPC agreed to hold the remaining unused 165,000 acre-feet of its 300,000 acre-feet primary and secondary entitlements until findings of the January snow survey provided a more definite forecast of probable run-off in the watershed.[626] IPC reduced its output from 27,000 kW at its American Falls plant to 3,000 kW and, if necessary, the company offered to supply the Minidoka plant with

---

625 R. W. Faris to Governor C. Ben Ross, January 25, 1932, in Lynn Crandall, *Water Distribution and Hydrometric Work District No. 36, Snake River, Idaho, 1931*. Folder 310.2, Box 758, PCF, 1930-1945, RG 115, NARA-Denver.

626 E. B. Darlington, superintendent, to Commissioner USBR, December 11, 1931 and Darlington to Chief Engineer, December 16, 1931. Folder: 301.1, Box 752, PCF, 1930-1945, RG 115, NARA-Denver.

power made by steam.[627] Water users in District No. 36 also turned their attention to erecting a dam at Swan Valley, Idaho on the South Fork of Snake River to control flood waters in Madison, Jefferson and Bonneville counties, to store water to supply present irrigation and augment and supplement future agricultural use, and for development of electrical power.[628]

As in 1931, neither Jackson Lake, nor American Falls Reservoir filled in 1932. The irrigation season ended September 30, 1932, with approximately 690,000 acre-feet combined hold-over in the two reservoirs. Although the water storage situation improved somewhat over the next few seasons, supplemental agreements were made in 1932, 1933 and 1934 to effect a continuation of the original three-party agreement.[629] Between 1932 and 1935 no water was spilled to waste past Milner in excess of IPC's winter water released for its Shoshone plant and the reservoir hold-over on the upper Snake River never reached 700,000 acre-feet; it dropped to an all time low of just 49,200 acre-feet in 1934.[630]

Although discussions continued toward enlarging American Falls Reservoir to its ultimate capacity of 3,000,000 acre-feet, the Bureau of Reclamation initiated investigations to identify supplemental storage sites in the Upper Snake River Basin above Idaho Falls after 1932. A total of twenty-seven potential reservoir and power sites located between Ririe, Idaho, and Jackson, Wyoming were studied by the Bureau. A preliminary report on these sites was submitted by Bureau engineers E. B. Debler and J. R. Riter in June 1935 and field investigations to identify the most advantageous site continued as funding permitted through the remainder of the decade. Studies of the proposed dam sites included geologic examinations by the USGS,

627 *Minidoka County News,* December 3, 1931; Elwood Mead to Superintendent Darlington, December 5, 1931. Folder: 301.1, Box 752, PCF, 1930-1945, RG 115, NARA-Denver.

628 Annual Report of the Committee of Nine for 1931. Folder: 301.1, Box 752, PCF, 1930-1945, RG 115, NARA-Denver

629 Supplemental Agreement American Falls Reservoir District and Idaho Power Company Concerning Lease of Reserved Capacity in American Falls Reservoir, Minidoka Project Contract, October 31, 1932, American Falls Reservoir District; Supplemental Agreement, American Falls Reservoir District and Idaho Power Company Concerning Lease of Reserved Capacity in American Falls Reservoir, Minidoka Project Contract, December 1, 1933, American Falls Reservoir District; Supplemental Agreement American Falls Reservoir District and Idaho Power Company Concerning Lease of Reserved Capacity in American Falls Reservoir, Minidoka Project Contract, November 1, 1934, American Falls Reservoir District.

630 Crandall, "New Storage on Snake River for Irrigation Use above Milner, Idaho," February 15, 1955. Folder: D-100-12G-9, WPB, 1894-1961, RG 57, NARA II.

diamond drilling and test pit explorations, and rough cost estimates. Preliminary designs and cost estimates, studies of storable flows, irrigation yields, and power possibilities were prepared for several favored dam sites located in Swan Valley on the South Fork of the Snake River. Eventually the selection was narrowed to the Palisades (also known as Grand Valley or Calamity Point) Dam site and further studies of this location continued intermittently from about 1935 through 1939.[631]

Years of drought

The year of 1933 was the fifth consecutive year in which the run-off of Snake River had been below normal. Run-off at the Moran gaging station was 82 percent of normal and at Neeley about 66 percent of the recorded mean. Jackson Lake filled to capacity and American Falls Reservoir filled to 82 percent of capacity. At the close of irrigation season the two storage reservoirs contained 448,000 acre-feet in combined storage, which was 160,000 acre-feet less than in the fall of 1932. Shortages were experienced on a few of the tributary streams, but on the main Snake River canals that were able to rent storage rights had sufficient amounts of water to carry them through the irrigation season. Upper valley canals above Blackfoot diverted about 496,000 acre-feet of stored water, the largest amount of storage ever used in that section of the Snake River Plain. Regulation of water rights in the upper basin began on June 15 when natural flow rights with a priority later than 1921 were cut-off, and at the lowest point only 20 percent of the May 11, 1889, rights were being filled. [632]

Conditions the following year, 1934, would be worse.

In the early 1930s the Gooding project, owned by AFRD No. 2, came on-line during this extended period of drought. Each year, the

631 See E. B. Debler and J. R. Riter, "Report on the Upper Snake River Storage Investigations," Volume I: Snake River Above Idaho Falls, United States Department of the Interior, Bureau of Reclamation; and E. B Debler and J. R. Riter, "Report on Upper Snake River (South Fork) Storage Investigations," enclosed with August 18, 1939 letter to the Commissioner of Reclamation; Folder: 301._, Snake River, Report by Debler & Riter, June 1935 on Upper Snake River Investigations (Folder 301._), Box 1055, Snake River Basin, 301.-_301.1, PCF, 1930-1945, RG 115, NARA-Denver.

632 Annual Report of the Committee of Nine, Water District No. 36 (1933). Folder: 301.2 Minidoka – Correspondence Re. Distribution of Water, January 1, 1933 – December 31, 1937, Box 754, RG 115, NARA-Denver.

project demanded larger quantities of water as the canal system was extended across the project area to irrigate new lands. In June 1929 the United States District Court for the District of Idaho issued a decree adjudicating the rights to then-users of water from Snake River above Milner Dam in a decree known as the Woodville Decree. AFRD No. 2 and the Gooding project were not made a party to the action. However, the Secretary of the Interior was decreed water rights for 1.7 million acre-feet for storage in American Falls Reservoir and 8,000 cfs for direct diversion with a priority date of March 30, 1921, based upon the water right filings made by the United States in support of its American Falls Reservoir project.[633] On July 12, 1932 proof of completion of works was made under the permits, and the state commissioner of reclamation found that the United States had complied with the requirements of the permit to the extent of 1,700 cfs for direct diversion and 1,800,000 acre-feet for storage. The certificate of proof for 1,700 cfs was for use on about 80,000 acres on the Gooding Division of the Minidoka project through the recently constructed Milner Gooding Canal.[634]

The Milner Gooding Canal was partially completed in 1930 and during that year 3,000 acre-feet of natural flow was diverted from the river, as distinguished from storage water, into the main canal serving the Gooding District. In 1931 11,000 acre-feet of natural flow was diverted. In 1932 some 110,000 acre-feet of natural flow was released by the Bureau at American Falls Reservoir and delivered into the main canal. In 1933 no natural flow water was delivered to the Gooding District, whereupon AFRD No. 2 filed suit to require delivery of natural flow water to its main canal to the extent of its carrying capacity of 1,550 cfs. The basis for the claim was Section 4, Article 15 of the state constitution which protected the rights of irrigators who rented, bought, or received a distribution of water. The district argued the United States had dedicated to the use of AFRD No. 2 not only stored water under its contract with the United States, but also natural flow.

---

633 *Woodville Canal Company* v. *Clark & Edwards Canal & Irrigation Company, et al.*, U. S. District Court for the District of Idaho, Eastern Division, Decree 661 (June 25, 1929). Folder: 032 Minidoka – Gooding, Settlement of Water Rights, American Falls District No. 2 vs Crandall, et al., Box 765 Gooding-Minidoka 010-032, PCF, 1930-1945, RG 115, NARA-Denver RG 115, NARA-Denver.

634 R. W. Faris, Commissioner of Reclamation, Permit No. 15134 Certificate of Completion of Works, July 12, 1932 and Stoutemyer to Commissioner Mead, July 26, 1932. Folder: 031.2 Minidoka-Gooding, Water Appropriations, Proof of Completion, Box 765, PCF, 1930-1945, RG 115, NARA-Denver.

As soon as the Milner ooding Canal began drawing water controversy arose between AFRD No. 2 and the other twenty-seven spaceholders, represented by AFRD, who held an interest in the storage water in American Falls Reservoir. The problem concerned whether, during the part of irrigation season when American Falls Reservoir was in the process of filling, the reservoir or canal should have the preference right to use natural flow water available for filling the priority rights of March 30, 1921, decreed to the Secretary of the Interior. If AFRD No. 2 could draw upon natural flow, in the years when natural flow was available, it could restrict its use of water to natural flow at such times, and not charge it against its storage rights. The amount of unused storage water, under the hold-over rights provisions in its contract with the United States, would become an additional water supply upon which to draw the following year. The others holding similar storage contracts with the United States in American Falls Reservoir believed this deprived them of their contract rights.[635]

As a compromise, in 1932 government officials had decided that preference be allowed in alternate years first to one project, then to the other. The Milner Gooding Canal had the preference in 1932 and American Falls Reservoir in 1933. AFRD protested the delivery of natural flow water during the 1932 irrigation season in a hearing before R. W. Faris, Idaho Commissioner of Reclamation. Faris issued an order on May 20, 1933 in which he reiterated the findings in the Woodville Decree – all of the stream flow of Snake River available under the Bureau's natural flow and storage permits issued by the state Department of Reclamation were to be "stored in the American Falls Reservoir to the extent of the full capacity of the reservoir during the storage season." He determined that the Woodville Decree established that the two permits "represent and constitute one water right." Finally, he ordered that until American Falls filled each year, no water should be delivered to AFRD No. 2 as a direct flow right and any water used in the spring was to be charged as stored water unless there was a surplus over and above 1,800,000 acre-feet stored in American Falls, "in which event the surplus used shall not be charged as storage water."[636]

---

635 Opinion of Circuit Judge Haney, *American Falls Reservoir District No. 2* v. *Lynn Crandall, et al.*, U. S. Circuit Court of Appeals for the Ninth Circuit, San Francisco, Case No. 7906 (March 30, 1936). Folder: 032, Box 765, PCF, 1930-1945, RG 115, NARA-Denver.

The controversy resulting in the filing of a law suit in 1934 entitled *American Falls Reservoir District No. 2* v. *Lynn Crandall, et al.*, in which the plaintiff district sought a supplemental adjudication of its water right against Crandall, Faris and the Secretary of the Interior as administrative officers of water distribution on Snake River only, and not against other users. Oral arguments were heard on December 17, 1934 before Judge Cavanah in the U. S. District Court in Boise. At trial, the judge made clear that he would not reopen the Woodville Decree or make any changes to the rights decreed therein, but would only consider the question of whether or not the decree was being properly administered. As in the Woodville Decree, in his judicial opinion, Cavanah found that under the Secretary of the Interior's applications, permits, and proof of completion of works in its state water right filings the two permits (Permit No. 15134 and R-269) "must be considered indissolubly connected and cover the same project of storage water in the reservoir." Even if Permit No. 15134 represented a direct diversion right independent of the reservoir right, Cavanah opined, it could not grant AFRD No. 2 "any greater or other right than the other contract holders ... others have similar contracts with the United States to use the waters under these two permits which were entered into prior to the contract with the plaintiff." None of the contractors holding capacity in American Falls Reservoir had any deed or contract under which the government had granted to them any part of its water appropriation. During the trial, the attorney for AFRD No. 2 admitted that the district had no contract or any other instrument of conveyance granting it any right to any water other than 400,000 acre-feet of stored water from American Falls Reservoir.[637] Cavanah made certain other pronouncements with respect to state water law and the administration of water in American Falls Reservoir, including the following:

---

636 Crandall to Dana Templin, acting superintendent Minidoka Project, April 27, 1936. Folder: 301.2 Minidoka – Correspondence Re. Distribution of Water, January 1, 1933 – December 31, 1937, Box 754 Minidoka Project 310.2, PCF, 1930-1945, RG 115, NARA-Denver.

637 Opinion of Circuit Judge Haney, *American Falls Reservoir District No. 2* v. *Lynn Crandall, et al.*, U. S. Circuit Court of Appeals for the Ninth Circuit, San Francisco, Case No. 7906 (March 30, 1936); Stoutemyer to Commissioner Mead, October 27, 1934 Stoutemyer to Commissioner Mead, December 10, 1934. Folder: 032, Box 765; and Crandall to Dana Templin, acting superintendent Minidoka Project, April 27, 1936. Folder: 301.2, Box 754, PCF, 1930-1945, RG 115, NARA-Denver.

> Under the original decree and the contracts between the United States and the contract holders for storage water the right to direct use of the water made available under said permit rests with the United States. However, under the laws of Idaho, the Secretary of Interior has no discretion to allow plaintiff or other contract holders to use water in alternate years as such contract holders have the right to use it each year if they are complying with the terms of their contract.

In the end, the court merely dismissed the complaint without issuing any clear instructions or directions on how the water should be delivered – and whether the decision rested in the hands of the Secretary of the Interior or the state Commissioner of Reclamation or the watermaster of District No. 36. Both parties to the suit during the trial expressed a preference to divide the water each year, instead of alternating from one year to the next, as had become the practice. [638]

Following the ruling of the U. S. District Court, on March 14, 1934, Stoutemyer announced that the Bureau would continue to deliver normal flow and American Falls storage to the Milner Gooding Canal in alternate years according to the Secretary of the Interior's ruling assigning priority in the canal. Unless the court issued an order restraining the Bureau from doing so, it intended to divert normal flow into the canal in the 1934 irrigation season.[639] The issuance of conflicting orders and instructions from state and federal officials and the ambiguous decision of the court left watermaster Lynn Crandall in an embarrassing position and seeking resolution through a more definitive opinion from the court. In correspondence to E. B. Darlington in the spring of 1934, Crandall noted that he understood Stoutemyer and the Bureau's position, but as a practical matter, because of the severe drought, he suggested that there may not be any natural flow of 1921 priority to draw upon.

Since disagreements were already arising between the state and federal officials and the two American Falls reservoir districts on this issue, Crandall carefully laid out his position to the Minidoka project superintendent:

---

638 Stoudemyer to Commissioner Mead, November 7, 1935. Folder: 301.2, Box 754, PCF, 1930-1945, RG 115, NARA-Denver.

639 Darlington to Stoutemyer, March 22, 1934. Folder: 301.2, Box 754, PCF, 1930-1945, RG 115, NARA-Denver.

1. The Woodville Decree awarded two rights to the Interior Secretary in connection with the American Falls Development: one for natural flow and one for storage;
2. The Idaho Department of Reclamation had ordered that under the Woodville Decree no natural flow water was to be delivered to AFRD No. 2 until American Falls Reservoir was filled.

As watermaster and deputy commissioner of reclamation, Crandall felt obligated to follow the Woodville Decree and the order of the Idaho Department of Reclamation. Since the river was already under regulation in March 1934, it was Crandall's position that AFRD No. 2 not be permitted to divert any water until it ordered storage released, unless he was otherwise instructed by a court order.[640] On the other hand, in the administration of the American Falls Reservoir the superintendent of the Minidoka Project was bound to distribute water in accordance with the instructions of the Secretary of the Interior.

AFRD No. 2 appealed the U. S. District Court ruling to the Ninth Circuit Court of Appeals and a hearing was set for the October 1936 session of the court. March 30, 1936, Circuit Court Judge Haney's decision in *American Falls Reservoir District No. 2* v. *Lynn Crandall, et al.* dismissed the case without prejudice holding that the water appropriations involved in the case were the property of the United States, therefore the United States was an indispensable party to the action, but as it was not a party to the lawsuit, and could not be made to be a party, the court was without jurisdiction to adjudicate the water rights set up in the bill of complaint.[641] As for the merits of the plaintiff's case, the court held AFRD No. 2 was without an actionable cause because the delivery of extra decreed water rights by the Secretary of the Interior for a single season was a discretionary and gratuitous act, and did not constitute a dedication of a water right.[642]

640 Crandall to Darlington, March 21, 1934. Folder: 301.2, Box 754, PCF, 1930-1945, RG 115, NARA-Denver.

641 Stoudemyer to Acting Commissioner, October 12, 1936. Folder: 032, Box 765, PCF, 1930-1945, RG 115, NARA-Denver.

642 *American Falls Reservoir District No. 2* v. *Crandall, et al.*, U.S. Circuit Court of Appeals, Ninth District, Case No. 7906 (March 30, 1936); Stoutemyer to Crandall, April 15, 1936; Stoutemyer to Solicitor, Department of the Interior, October 2, 1936. Folder: 032, Box 765, PCF, 1930-1945, RG 115, NARA-Denver.

The board of directors of AFRD No. 2 decided to appeal the case to the U. S. Supreme Court, and in 1937 the case was sent to the U. S. Supreme Court on a writ of certiorari from the Court of Appeals.[643] It was, however, dismissed for lack of jurisdiction on account of the absence of an indispensable party – the Secretary of the Interior. The decision on distribution of water was therefore left in the hands of the Secretary of the Interior, and the reservoir was operated in conformance to these instructions until the *Eagle Decree* in 1968.

During the appeal process, expecting the appellant's complaint to be dismissed, Stoutemyer circulated among Bureau officials draft instructions to direct Crandall to deliver water under the water rights decreed to the Secretary of the Interior in a way that conformed as nearly as possible to the wishes of the parties involved in *American Falls Reservoir District No. 2* v. *Lynn Crandall* and in the conclusions of law filed by the district court. The instructions reflected the preferences of the parties for a division each year rather than the alternate-year plan. The proposed instructions read:

> That during the non-irrigation season of each year from October 1 of one year until May 1 of the next, all water available under the said priority rights of March 30, 1921 decreed to the Secretary of the Interior be stored in the American Falls Reservoir; that during the months of May and June, and so long thereafter during the irrigation season as there may be natural flow water available for filling the said priority of March 30, 1921, that the first 1700 second feet thereof be divided on a fifty-fifty basis and one-half thereof furnished to the Gooding Canal and one-half to the American Falls Reservoir and that all water available for filling the said priority rights decreed to the Secretary of the Interior during the said months of the irrigation season over and above the one-half of the first 1700 second-feet herein provided for the Gooding Canal be furnished to the American Falls Reservoir.[644]

The form of this agreement was satisfactory to AFRD and AFRD No. 2.[645] On March 31, 1936, Acting Commissioner of Reclamation John C. Page forwarded to Interior Secretary Harold L. Ickes, Mr.

643 Templin to Stoutemyer, March 4, 1937. Folder: 301.2, Box 754, PCF, 1930-1945, RG 115, NARA-Denver.

644 Stoutemyer to Commissioner Mead, November 7, 1935. Folder: 301.2, Box 754, PCF, 1930-1945, RG 115, NARA-Denver.

645 Stoutemyer to Commissioner, March 24, 1936. Folder: 301.2, Box 754, PCF, 1930-1945, RG 115, NARA-Denver.

Stoutemyer's proposed instructions on division of natural flow and storage rights under the 1921 priority of the United States for the Minidoka Project and recommended their approved.[646] On April 6, 1936, the secretary issued instructions that in years when American Falls Reservoir filled and spilled, the water diverted into the Milner-Gooding Canal during the time before the reservoir filled would be considered normal flow water. Therefore, it was not to be charged against AFRC No. 2 as part of its stored water supply. These instructions, he noted, were to remain in effect until further notice from the Secretary of the Interior, or unless the Circuit Court of Appeals ruled otherwise in *American Falls Reservoir District No. 2* v. *Lynn Crandall*. E. B. Darlington, superintendent of the Minidoka Project, informed Crandall, Idaho Commissioner of Reclamation R. W. Faris, and the secretary of AFRD No. 2 of the secretary's instructions concerning the division of the natural flow water.[647]

In view of Judge Cavanah's decision and Faris' order, Crandall felt he would be justified in refusing to be bound by the Secretary of the Interior's instructions on delivery of normal flow to the Gooding Project. Although the reservoir would easily fill in 1936, the other reservoir spaceholders were almost unanimous in their opposition to the secretary's proposal on principal, as it might set a dangerous precedent. They believed, along with the state Commissioner of Reclamation who was charged with general supervision of water distribution in the state, that the Milner Gooding Canal was not entitled to any natural flow water until and unless American Falls Reservoir filled.[648]

Nevertheless, the break in the long siege of drought reduced tensions on the river among competing water users, but would become a subject of contention again when the reservoir failed to fill and any natural flow was available after May 1.[649]

646 Page to Secretary of the Interior, March 31, 1936. Folder: 301.2, Box 754, PCF, 1930-1945, RG 115, NARA-Denver.

647 Notice to Lynn Crandall and R. W. Faris, from Darlington, April 6, 1936; John C. Page, acting commissioner, to Darlington, April 6, 1936; Stoutemyer to Darlington, April 16, 1936; and Dana Templin, acting superintendent Minidoka Project, to Stoutemyer, April 20, 1936. Folder: 301.2, Box 754, PCF, 1930-1945, RG 115, NARA-Denver.

648 Crandall to Stoutemyer, April 13, 1936. Folder: 032 Box 765, PCF, 1930-1945, RG 115, NARA-Denver; and Templin to Stoutemyer, April 28, 1936. Folder: 301.2, Box 754, PCF, 1930-1945, RG 115, NARA-Denver.

649 Crandall to Stoutemyer, March 3, 1937. Folder: 301.2, Box 754, PCF, 1930-1945, RG 115, NARA-Denver.

The year 1936 was the first since 1928 that the runoff of Snake River had been above normal at any of the measuring stations on the river. At the end of June when water users in the Snake River valley were making their maximum demand for irrigation water, 11,570 cfs was being diverted in the upper Snake River valley above Blackfoot, exclusive of the diversions on Henrys Fork and its tributaries. While some of the canals below American Falls were diverting stored water, most notably the Milner Gooding Canal, most still were operating at the end of June on natural flow.

As a result, there was still 1,670,870 acre-feet stored in American Falls Reservoir, only some 30,000 acre-feet below capacity. Both Jackson Lake and Lake Wolcott were full with no storage having been diverted from either of the two reservoirs. While the upper valley water storage and distribution system appeared to be nearly full, the amount of water wasting over Milner Dam measured a mere 18 cfs. Stoutemyer marveled at the degree of control exhibited over the river's water supply with several hundred canals diverting some 23,000 cfs and hardly anything wasting to the river past the last irrigation canals at Milner Dam.[650] In addition there was an ample supply of stored water in the American Falls and Jackson Lake reservoirs for all irrigation requirements in the upper Snake River valley to be served by water exchanges at American Falls with the Minidoka Project and the Twin Falls companies. [651]

The perilous drought years from 1929 to 1935 had proven to be a pivotal era in the management of water resources all along the Snake River in southern Idaho. It was an extremely dry period with six consecutive years of subnormal streamflow in the basin and with the lowest run-off ever recorded in 1934 on Snake River. Toward the end of the drought, the water shortage was magnified by unusually heavy river losses due to low groundwater levels and decreases in both return flow and waste into the river. At the Heise station above the upper valley canals the river run-off was 54 percent of normal in 1934 and at the Neeley station below American Falls it was 43 percent of normal. Because of the dry weather, irrigation was begun the latter part of March with a correspondingly heavy demand for water. Regulation of

650 Stoutemyer to Commissioner, July 1, 1936; Crandall to Stoutemyer, July 20, 1936. Folder: 301.2, Box 754, PCF, 1930-1945, RG 115, NARA-Denver.

651 Stoutemyer to Crandall, July 22, 1936; Stoutemyer to Commissioner, July 24, 1936. Folder: 301.2, Box 754, PCF, 1930-1945, RG 115, NARA-Denver.

the normal flow on Snake River by the watermaster began in early April, cutting rights to the June 1, 1889, priority. The river flow increased in early May so that rights before 1903 were served, but after that date river discharge decreased and by the beginning of July all of the 1889 rights were terminated.[652]

The peak of storage was reached at American Falls Reservoir in 1934 two months early on April 7 when some 1,143,683 acre-feet were impounded, about 75 percent of the peak storage in late June of the previous year. Jackson Lake peaked a week later at only 346,000 acre-feet. Some 223,000 acre-feet of storage belonging to the government was turned over to the water users during the summer in an attempt to save crops. Idaho Power Company contributed as well during the drought. In 1933, the company only drew on its primary storage after October 1 and used none of its secondary storage. Again in 1934, it conserved storage water for irrigation purposes releasing all of its 45,000 acre-feet of primary storage to various water users in the lower valley – North Side Canal received 15,000 acre-feet, Milner Low-lift Irrigation District 5,000 acre-feet, the Minidoka Irrigation District 10,000 acre-feet and the Burley Irrigation District 15,000 acre-feet. None of the power company's secondary storage was drawn until after October 1, 1934.[653]

One of the responses to the record drought was for the Bureau to search for and identify sites where storage water for irrigation use might be made available. Congress took notice of the water shortages on the Upper Snake River and allocated $4,000,000 of public works funds to hurry along feasibility studies of reservoir sites on the Upper Snake River. Lands irrigated from Snake River in the upper valley above American Falls had only Jackson Lake Reservoir as a source of stored water. Located on the headwaters of the stream, its dependable water supply was insufficient to meet the water needs of upper valley irrigators, especially in low runoff years like those of the early 1930s. People on Henrys Fork were desperately short of water in 1934 and in some cases farmers received no irrigation water at all. A great deal of

652 Annual Report of the Committee of Nine to the Water-Users of District No. 36, March 4, 1936. Folder: 301.2, Box 754, PCF, 1930-1945, RG 115, NARA-Denver.

653 USBR, Annual Project History: Report of Construction and Operation and Maintenance: Minidoka Project (1933), 55-56; and USBR, Annual Project History: Report of Construction and Operation and Maintenance: Minidoka Project (1934), 43-55. Box 302, PHFHR, 1902-32, RG 115, NARA-Denver.

trouble had been experienced in finding a reservoir site on the north fork that would hold water. In 1934, the Bureau of Reclamation began more detailed investigations of small reservoir sites on the north fork and tentatively approved reservoir sites of 110,000 acre-feet at Island Park (1935-38) and 10,000 acre-feet at Grassy Lake (1937-39). Considerable investigation, drilling of test holes, and excavation of tunnels and shafts were completed at major reservoir sites on the south fork, including the Palisades Dam site (1951-57) above Idaho Falls in Swan Valley. These reservoirs would provide additional storage water for upper basin water users. At the same time, a proposal lower down the river to enlarge the American Falls Reservoir to its ultimate capacity of 3,000,000 acre-feet also was under consideration by the Bureau.[654] The mammoth reservoir would afford carry-over storage space for flood water of high water years and holdover capacity to weather two or three consecutive years of low water. Inflows available for storing in a new upstream reservoir were subject to several senior water rights below American Falls:

- Power use by IPC at American Falls and below Milner Dam
- Power use by Minidoka Project at Minidoka Dam
- Irrigation requirements for canals diverting from Lake Walcott and Milner Lake
- Winter canal diversions required for stock and domestic water use
- Water required to fill Lake Walcott and American Falls Reservoir

To fill both the enlarged American Falls and proposed Palisades reservoirs, or either one of them, would require limiting domestic and stock water diversions and holding in storage for irrigation uses more of the natural flood flows that passed through American Falls during the winter months that were used for power production. Increased electric power production during irrigation season was a growing need vital to agriculture and the rural communities that supported it. Below

654 Debler and Riter, "Report on the Upper Snake River Storage Investigations," Volume I, Synopsis B-C. Folder 301, Box 1055 Snake River Basin, 301.-_301.1, PCF, 1930-1945; E. H. Neal, General Manager Aberdeen-Springfield Canal Company, to USBR Commissioner Elwood Mead, May 22, 1934. Folder: 301.1, Box 752, PCF, 1930-1945, RG 115; NARA-Denver; and M. A. Schnurr, acting commissioner, to Secretary of the Interior, July 6, 1934. Folder: Reclamation Bureau, Minidoka Project, Minidoka Irrigation District, February 2, 1931-May 13, 1936 (Folder: Reclamation Bureau, Minidoka Project), Box 1618, CCF, 1907-1936, RG 48, NARA II.

American Falls, expansion of irrigated agriculture onto higher lands back from the river brought ever-increasing requirements for electric power load to operate pumping plants on the Minidoka and Twin Falls projects. Furthermore, the mounting rural farm population on these newly irrigated lands, together with the growing towns and industries that supplied and served them, led to increasing electrical power demand for heating, lighting and a host of other domestic and commercial uses. Unlike local agriculture, power production was not totally dependent on the waters of the Upper Snake River. Electrical power could be produced at distant generating facilities on the Boise or Payette rivers or on the middle reach of the Snake River and be transmitted over high voltage lines to substations at Milner, Minidoka, American Falls, or other locations.[655]

3. The Woodville Decree awarded rights "for use upon the various projects which have heretofore or may hereafter become entitled to same by reason of contracts with the United States therefor." Crandall was only aware of one contract between the United States and AFRD No. 2 which operated the Gooding Canal. It was for 400,000 acre-feet in American Falls Reservoir. No contract conveyed a right to any portion of the 8,000 cfs of natural flow; therefore, Crandall believed that no canal was entitled to divert this water.

On October 1, 1934, the United States and Idaho Power Company entered into a water-saving contract; its main goal was to make more water available by reducing winter discharge past Milner Dam and storing the water conserved at American Falls for use by upper basin irrigators. It saved some 400 to 500 second-feet in the amount of water released from American Falls during the winter of 1934-35 as compared with the previous winter. The agreement sought to resolve several fundamental resource allocation issues that had beset upper basin water users over the previous decade. The Minidoka power plant, the

---

655 E. H. Neal, General Manager Aberdeen-Springfield Canal Company, to USBR Commissioner Elwood Mead, May 22, 1934. Folder: 301.1, Box 752, PCF, 1930-1945, RG 115, NARA-Denver; and Schnurr to Secretary, July 6, 1934. Folder: Reclamation Bureau, Minidoka Project, Box 1618, CCF, 1907-1936, RG 48, NARA II

government's only active electrical power plant on the Minidoka project, struggled to keep pace with growing power demand on the reclamation project. During the summer of 1932 the Minidoka station, operating at full capacity, fell short of supplying the full power demand of the Minidoka project. In 1933 the project purchased from IPC a peak demand of 969 kW for short durations of time in the summer months. Increased installations of pumping stations in the Burley Irrigation District foreshadowed a shortage during the summer months of May to September of 1,700 kW by 1934 and 2,500 kW soon thereafter. Increasing commercial power and lighting loads also contributed to peak demand. IPC estimated that these loads for the years 1935 to 1943 would increase 10 percent per year while irrigation pumping demand would increase 50 kW per year, thereby outstripping the capability of the Minidoka plant by more than 38,000,000 kilowatt hours (kWh). [656] One proposed solution to the power shortage was construction of a second government powerhouse for the Minidoka project at American Falls.

However, Idaho Power favored adoption of a plan of water conservation at American Falls which would dispense with the advisability of constructing any government power plant at American Falls – a strategy the company would employ on a number of occasions in the future.[657]

During the drought, elected officials, representatives of the irrigators and the Bureau of Reclamation, and IPC officials met on several occasions to discuss plans for the conservation and storage of water on the Upper Snake River. In November 1933 power company officials agreed to prepare and submit a proposal to the Bureau with the aim of effecting a substantial saving of water while at the same time meeting the increased power requirements and demands on the Minidoka project, at a lower cost than construction of a 7,500 kW government power plant at American Falls. The proposal required the company and

656 E. B. Darlington, superintendent Minidoka Project, to B. E. Stoutemyer, district counsel, November 18, 1933, and K.M. Robinson, Vice-President & General Manager, IPC to B.E. Stoutemyer, District Counsel, "Idaho Power Company's Proposal to United State Bureau of Reclamation for the Conservation of Water at American Falls for Irrigation Purposes," November 9, 1933, 13, enclosed with Stoutemyer to Chief Engineer, November 13, 1933. Folder: 320 Minidoka – Correspondence Re. Development of Power, Jan. 1, 1930 through 1933 (Folder: 320 Minidoka), Box 760 Minidoka Pro. 320, PCF, 1930-1945, RG 115, NARA-Denver.

657 R. E. Shepherd, president American Falls Reservoir District, to Sen. J. P. Pope, December 4, 1933. Folder: 320, Box 760, PCF, 1930-1945, RG 115, NARA-Denver.

the United States to agree, in effect, upon a "closely coordinated and synchronized operation of their respective plants" to conserve water for summer irrigation purposes. The company proposal was based on saving at least as much water as could be saved by the proposed government plant, but its proposal could take effect immediately. The offer made available for conservation purposes, subject to direction and control of the Bureau, the company's 255,000 acre-feet of secondary storage capacity at American Falls reservoir, less releases to meet minimum flows at Shoshone Falls between October 1 and June 10 as required by the June 15, 1923, contract. The company estimated that its proposed program of conservation would result in a minimum saving of 150,000 acre-feet in a dry year such as 1931, and up to 200,000 acre-feet in a normal water year.[658]

The Bureau's district counsel summed up the core elements of the IPC water-saving proposal as follows:

> I think the company's proposal may be stated from a water-saving standpoint as a proposed agreement under which the company will give up 50% of its reserved water right, or 150,000 acre-feet, in four years out of five provided the United States will refrain from building its own power plant at American Falls, will allow the company to use all of the water passing over the American Falls dam, and they will buy the power required for the Minidoka project from the company at a rate of 3.5 mills for power furnished during the irrigation season, with no charge for the power to be furnished during the non-irrigation season.[659]

The original plan for construction of a phased 30,000-kW government power plant at American Falls Reservoir never materialized. The Bureau intended to use the plant, in part, to relieve pressure on the Minidoka plant and conserve stored water to more fully guarantee filling of American Falls Reservoir from year to year. Over the previous decade, general population growth within the Minidoka project coupled with enlargement of canals and increasing demand for electric energy from larger and more irrigation pumps installed on the

658 Robinson to Stoutemyer, district counsel USBR, "Idaho Power Company's Proposal to United States Bureau of Reclamation for the Conservation of Water at American Falls for Irrigation Purposes," November 9, 1933. Folder: 320, Box 760, PCF, 1930-1945, RG 115, NARA-Denver.

659 Stoutemyer to Chief Engineer, November 13, 1933. Folder: 320 Minidoka, Box 760, PCF, 1930-1945, RG 115, NARA-Denver.

South Side Pumping Division of the Minidoka project (Burley Irrigation District) completely absorbed the power output of the government's Minidoka power plant. The irrigation district also built up a substantial and profitable business providing electric power for domestic, heating and other uses in the non-irrigation season to project towns and farms. Without importing power from outside sources or building a power plant at American Falls, the government could no longer meet its agricultural, commercial and domestic load within the Minidoka project.

Any long-term solution to water shortages, noted Crandall, had to address the larger need for supplemental irrigation water for existing and future acreage. Thus, power generation and water consumption in the non-irrigation season had to be mindful of this larger goal. To solve the problem, Crandall advised Commissioner Mead and others in the USBR, required "the elimination of discharge past Milner in years of deficient run-off" and making water so conserved available for storage:

> Inasmuch as all water passing Milner is waste as far as irrigation is concerned, a definite plan should be adopted to stop all such water or at least as much thereof as possible, whether such waste comes from the Idaho Power Company rights or from the Government power rights at Minidoka dam. Only by decreasing present discharge past Milner dam can the supply for American Falls reservoir be increased and to effect a permanent satisfactory solution of the problem this will require some negotiation with the Idaho Power Co. on account of its ownership of 300,000 acre-feet of water at American Falls.[660]

Because IPC already had equipment installed at American Falls, Crandall favored making some arrangement with the company to increase its power output while eliminating or reducing "to a very low amount" all waste past Milner. On the other hand, Stoutemyer favored prompt construction of the proposed government power plant at American Falls, as a conservation measure that would save 300,000 acre-feet of storable water and make available additional irrigation power below prevailing rates.[661]

---

660 Lynn Crandall to Commissioner Mead, February 21, 1934 Folder: 320 Minidoka, Box 760, PCF, 1930-1945, RG 115, NARA-Denver.

661 Stoutemyer to Minidoka Project Superintendent, February 28, 1934. Folder: 320 Minidoka, Box 760, PCF, 1930-1945, RG 115, NARA-Denver.

Others, like Acting Secretary of the Interior T. A. Walters, argued that the region was so dependent on farming and irrigation that nothing should compromise dedication of water supplies to these uses. Walters argued that the "public interest" demanded winter power uses of Snake River water be subordinated to uses for irrigation purposes:

> Under the Idaho Constitution use of water for irrigation purposes is considered a superior public use, to the use of water for power development, and the very serious water shortages which have occurred in large sections of the Snake River Valley during the last five years have emphasized the urgent importance of saving every acre foot of winter flow of Snake River which it is possible to save, and storing the water for use during the next irrigation season.
>
> Under the conditions prevailing in the Snake River Valley where farming is entirely dependent on irrigation and the towns are also dependent on the success of the irrigation farmers, it is believed that the furnishing of an adequate water supply for irrigation purposes should be the first consideration of those in charge of the irrigation works constructed by the United States and the power works incidental thereto, and that the public interest requires that the water supply needed for irrigation purposes should not be sacrificed for the purpose of providing increased power profits, and especially so in cases where ample power can be brought into the Valley from outside sources where the winter operation of power plants does not interfere with the storage of water needed for irrigation purposes.[662]

The October 1, 1934, contract did just that. It replaced Minidoka power with an outside source and saved the winter flow of Snake River so it might be stored for irrigation use. The agreement realized the goal of conserving waters of Upper Snake River above American Falls for irrigation purposes by coordinated use of underused power plants on the Boise-Payette project owned and operated by the federal government and the power transmission facilities of IPC in southern Idaho. Under the proposed contract, the power company agreed to deliver on the Minidoka project free of charge such power as was needed during the non-irrigation season up to 10,000 kilowatts. In exchange, the government gave the power company the winter power generated at the Black Canyon plant on Payette River and whatever benefit accrued to the company from the government agreeing not to construct its

662 T. A. Walters, acting secretary, "Finding of the Secretary of the Interior as to Net Profits from the Black Canyon and the Minidoka Power plants, through Sales of Power on the Minidoka Project and Towns Adjacent Thereto, During the Year 1935," March 12, 1936. Folder: Reclamation Bureau, Box 1618, CCF, 1907-1936, RG 48, NARA II.

proposed power plant at American Falls during the life of the contract. The company agreed to deliver free-of-charge such additional firm power, up to 4,000 kilowatts, as was needed on the Minidoka project during the irrigation season. The company also agreed to transmit free of charge during the irrigation season power generated at the government's Black Canyon and Boise Diversion plants to various Bureau pumping plants in other irrigation districts within reclamation projects in southwestern Idaho and Oregon. The company also surrendered, during the life of the contract, its 255,000 acre-foot secondary winter power right at American Falls by making release of its secondary storage subject to direction by the United States. Finally, the contract terminated the requirement for the release of water over Milner Dam between March 16 and June 10 to meet flow requirements at Shoshone Falls that supplied the company's downstream power plants.[663]

Originally a proponent of the U. S. building its own power plant at American Falls, B. E. Stoutemyer, district counsel for the USBR, became a strong advocate of the government executing the 1934 agreement with the power company because of the widespread benefits that would inure to Upper Snake River irrigators from its implementation:

> By securing the winter power requirements of the Minidoka project in this way and eliminating the company's reserved right of 255,000 acre-feet, we would be able during the non-irrigation season to close down the power plant at Minidoka dam (which has a decreed priority many years earlier than that of the American Falls Reservoir) and store all the winter flow of Snake River except the amount released to supply the winter stock water run by the Twin Falls companies.
>
> It is estimated that this contract will save each year from 400,000 to 600,000 acre-feet of the winter flow of Snake River, depending on how far the Twin Falls companies are willing to go in giving up the running of winter water for stock water and domestic purposes.[664]

663 Bureau of Reclamation, Snake River Valley Projects and Owyhee Project: Contract between the United States and Idaho Power Company for Conservation of Snake River Water and Furnishing Transmission Service, October 1, 1934 Folder: Reclamation Bureau, Box 1618, CCF, 1907-1936, RG 48, NARA II. To preserve the legal rights of IPC to the full use of 255,000 acre feet of secondary storage and the use of waters at Shoshone Falls power plant, the contract specified that water was to be released once every five years under the terms of the 1923 contract for these purposes.

664 The Twin Falls Canal Co. and the North Side Canal Co. diverted an average of about 600 second-feet in each canal (ca. 400,000 acre-feet per year) during the entire non-irrigation season

> Such a saving of winter flow would assure the filling of American Falls Reservoir practically every year. This contract would also provide the additional power needed on the Minidoka project, without the cost of building a power plant. The saving of water which can be secured under this contract would be somewhat larger than can be secured under the proposed plan to construct a power plant at American Falls, and without the cost of the power plant. The benefits of the proposed contract can be secured immediately so that the winter flow can be stored next winter, while the benefits from the proposed power plant would be postponed until a year later. If this contract had become effective two years ago, the American Falls Reservoir would have filled to full capacity this year and there would also have been at least 200,000 acre-feet more water in the Jackson Lake Reservoir.
>
> The reason why there would have been more water in Jackson Lake Reservoir is that if the projects in the lower valley [i.e., below American Falls] which have rights both in Jackson Lake and American Falls had a full supply from American Falls, they would leave a larger part of their Jackson Lake water as a carry-over supply in the Jackson Lake. This would be particularly true if it were known that the filling of American Falls Reservoir was assured every year. With the filling of American Falls Reservoir assured practically every year, the projects in the lower valley would have every motive for using their American Falls water first and holding as much water as possible in Jackson Lake, as hold-over, to aid in filling Jackson Lake the following year.
>
> As to the possible effect upon the proposed reservoirs in the Upper Snake River Valley, would say: The first essential in locating any feasible reservoir on the North or South Fork of Snake River (after finding a reservoir site that will hold water) is to find a water supply to put in the reservoir after it is built, especially in the low water years when the reservoir is needed.[665]

Thus, by managing available storage and power generation facilities as an integrated system, irrigators throughout the Minidoka project and the Upper Snake River Basin stood to gain by the water saving

---

for stock watering and domestic purposes, including the municipal water supply for the towns of Twin Falls and Buhl. Prior to the 1934 contract, there was little incentive to save this water because 600,000 acre-feet were used during the non-irrigating season developing power at the Minidoka plant and of this amount, 400,000 was diverted into the Twin Falls north and south side canals for stock water, domestic and municipal uses. Therefore, no additional loss of water supply occurred. Under the 1934 contract, the American Falls and Minidoka power plants could be shut down in the non-irrigating season. Therefore, to save this water supply the Twin Falls farmers and the cities of Twin Falls and Buhl turned to drilling wells and installing pumps to provide a substitute water supply. Elwood Mead, commissioner USBR, to Secretary of the Interior, January 25, 1935; Folder: Reclamation Bureau, Minidoka Project, American Falls Reservoir, Contracts, Dec. 9, 1930 to Sept. 16, 1935, Box 1614, 8-3, Milk River – Minidoka A, CCF, 1907-1936, RG 48, NARA II.

665 B. E. Stoutemyer to W. H. Philbrick, President, Aberdeen-Springfield Canal Co., June 6, 1934. Folder: 301.1, Box 752 Minidoka Project, 301 – 301.1, PCF, 1930-1945, RG 115, NARA-Denver.

agreement. Furthermore, the proposed contract with IPC, by saving a large amount of winter flow of Snake River, made it more feasible to guarantee a water supply for the proposed new reservoirs on the forks of Snake River above eastern Snake River plain.

The contract of October 1, 1934, was tentatively approved by the Secretary of the Interior, upon the condition that appropriate contracts would be secured to reimburse the United States for the use and disuse of its power plants in carrying out the proposed plan, the amount required for this purpose being $50,000. The IPC board of directors passed a resolution approving the contract on May 9, 1935, and the Interior Secretary also executed the contract.[666] Beginning in 1935, the secretary ceased operating the Minidoka power plant during the non-irrigating season to conserve water for irrigation purposes, but continued to supply commercial customers in the Burley and Minidoka irrigation districts with a firm electrical power supply by contracting with Idaho Power to supply the commercial demand in the districts from its own plant and by transmitting power generated at the Bureau's Black Canyon power plant on the Payette River.[667]

Practically shutting down the Minidoka project power plant during the non-irrigation season was opposed by the primary group of water power users on the Minidoka project – the irrigation pumpers within the Burley Irrigation District. The proposed contract with IPC provided that the company would furnish the Minidoka plant with enough electricity to supply the needs of the project and equal to the present capacity of the plant. But it also arranged for the distribution of this power to the customers of the Minidoka Irrigation District for their own profit, and left the Burley Irrigation District, whose water users had the right to 95.6 percent of the profits from operation of the Minidoka plant, without additional winter power and without the $45,000 annual profits

666 Bureau of Reclamation, Snake River Valley Projects and Owyhee Project: Contract between the United States and Idaho Power Company for Conservation of Snake River Water and Furnishing Transmission Service, October 1, 1934; Folder: Reclamation Bureau, Box 1614, RG 48, NARA II. The government entered into a contract with Minidoka Irrigation District by which it guaranteed a $50,000 return on power furnished to its customers by virtue of the IPC water saving contract.

667 The proceeds from the sale of commercial power brought in from outside the Minidoka Project as part of the Secretary of the Interior's plan of conservation became the subject of a lawsuit filed by the Burley Irrigation District. In defending the Secretary's actions, the court noted that the reclamation act was primarily an irrigation program, and that development of commercial power was an "incidental phase of reclamation" that could not be allowed to "impair the efficiency of the irrigation project." *Burley Irrigation District v. Ickes*, 116 F. 2d 529 (September 30, 1940), 530-31.

from its power market. The Burley district strongly opposed adoption of the water-saving contract and threatened litigation which, if successful, jeopardized the whole winter water conservation plan of the Bureau and with it the future water supply for expansion of irrigated agriculture in the upper Snake River Basin.[668]

Burley Irrigation District vs. Harold Ickes

The Minidoka plant was built primarily for developing 10,000 kW of power to pump irrigation water. The landowners who benefited were charged with the cost of the power plant, as well as the reclamation project's dam and canal systems. They were also charged with the cost of the plant, in proportion to their anticipated use of power for irrigation purposes. Over the first decade of its operation, the Minidoka power plant developed into an important commercial unit in addition to its purpose of supplying electrical power for irrigation pumping. The profits from the brisk sale of electrical energy to settlers and municipalities throughout the project area became considerable. In 1924 Congress passed legislation authorizing the Secretary of the Interior to use the accumulated net profits derived from operation of project power plants to offset the construction charges of the project to water users and to reduce their annual project operation and maintenance charges; thereby, aiding irrigators to meet their repayment obligations. Secretary of the Interior Hubert Work rendered a decision in 1927 apportioning the profits of the Minidoka plant in the same ratio as the cost of the power plant had been charged – 95.6 percent to water users in the Burley Irrigation District and 4.4 percent to the Minidoka Irrigation District. The Court of Appeals for the District of Columbia in *Wilbur v. Burley Irrigation District* affirmed the secretary's decision in 1932. [669]

Profits aside, the issue of using storable water in the upper basin for the generation of power for commercial purposes to the possible detriment of irrigation remained controversial. The Reclamation Act, as

---

668 Although the United States had title to the plant and water right, the Burley Irrigation District had contracted to pay for 95.6 percent of the construction cost of the Minidoka power plant, and had a right by court decree to 95.6 percent of the profits from operation of the plant. *Wilbur v. Burley Irrigation District* (1932), 58 Fed. (2nd), 871.

669 Act of December 5, 1924, Ch. 4, Sec. 4(1), 43 Stat. 703, 43 USCA 501; and *Wilbur v. Burley Irrigation District* (1932), 58 Fed. (2nd), 871.

amended, specifically provided that no water was to be furnished for purposes other than irrigation "if delivery of such water shall be detrimental to the water service for such irrigation project." Where there was a serious shortage of water, secondary or surplus uses could not be allowed to jeopardize primary ones.[670]

The plant at Minidoka Dam had a decreed water right to 2,700 cfs of natural flow for power, light, manufacturing and pumping purposes, but this use in the non-irrigation season consumed a vast volume of water to generate power for commercial sale that could be stored and saved for irrigation use. Water passing through the Minidoka dam exhausted its usefulness except for domestic, stockwatering or municipal uses downstream and was lost forever to fulfill its primary purposes – irrigation and irrigation pumping. Citing the recent shortage of water, on October 1, 1934, Secretary of the Interior Harold Ickes had entered into the water conservation contract with IPC designed to save up to 250,000 acre-feet for irrigation, preserve the commercial power business on the Minidoka project and ensure a dependable source of power. The secretary also entered into a contract with Minidoka Irrigation District by which it guaranteed to reimburse the United States $50,000 for use of the government's Black Canyon plant to furnish electrical power to the district's consumers. The contract had the effect of curtailing power production at Minidoka Dam during the months of October through March. Burley Irrigation District sued, arguing that it "owned" 95.6 percent of the Minidoka power plant and deserved its proportionate share of the profits allocated by Secretary Ickes to the Black Canyon plant.

The US Circuit Court of Appeals ruled in September 1940 in the secretary's favor. The judge found Burley's contentions without merit. In rendering the court's opinion, Judge Wiley Blount Rutledge, Jr. reviewed the legislative history of the Reclamation Act and the Secretary of the Interior's power to execute the proposed plan of conservation for the Minidoka project. "The primary object of the statute, and of the project," Rutledge wrote, "is the reclamation of arid lands through irrigation." Power was an essential use, but commercial sale of power was not.

---

670 Reclamation Act, February 25, 1920, Ch. 86, 41 Stat.451.

> ...the development and sale of such power is authorized only as an incidental phase of reclamation, not as a primary or independent end in itself. The statute and its amendments are *reclamation* [emphasis in original] acts, not commercial power development acts. The legislation is entirely inconsistent with the development and sale of commercial power whenever that may interfere with the development of irrigation or "impair the efficiency of the irrigation project." To assure that the secondary conservation of power will not impair the primary conservation of water for irrigation, the authority to dispose of surplus power is vested exclusively in the Secretary and is circumscribed by limitations upon the manner, terms and purposes of the disposal wholly inconsistent with permanent appropriation of power to non-irrigating uses and with subordinating irrigation to commercial sale.[671]

The litigation initiated in 1935 and ending with the circuit court decision in *Burley Irrigation District v. Ickes* left no doubt about the broad powers of the Secretary of the Interior, acting on behalf of the United States as holder of title and water rights to the Minidoka power plant, to manage, limit, dispose of or acquire a water supply for power generation in connection with the execution of a plan to conserve water for irrigation and reclamation of arid lands. However, use of project power plants for commercial purposes was an incidental use that could not harm "or impair the efficiency" of the primary purpose of the reclamation project.

Managing the 1934 contract

During each year in the early 1930s, water shortages prevented American Falls Reservoir from filling, and limited storage supplies contributed to late season crop losses. These drought years taught irrigators that conservation of winter flow was necessary to provide sufficient water to supply lands currently being irrigated throughout the basin above Milner Dam. Furthermore, any future storage development would be integrated into overall management of water resources on the upper river system in Water District 36, and therefore, had to consider the broader interests of the entire Snake River Basin above Twin Falls, rather than local or sectional interests. Water saved by impounding winter flow formerly used to operate American Falls and Minidoka power plants was to be dedicated by the Bureau for irrigation and other incidental purposes. Most of it was to be stored in American Falls

671 *Burley Irrigation District v. Ickes*, 116 Fed. 2d, 530-531.

Reservoir for the benefit of irrigation contractors who had purchased rights to store water in the reservoir. The smaller balance of the saved water was to be dedicated to supplement the supply of stored water in a proposed new Bureau reservoir on the North Fork of the Snake River, if and when it was built.

In the interim, before another reservoir was built upstream of Idaho Falls, irrigators who were short of water in the upper part of the basin increasingly looked to obtain storage rights in American Falls Reservoir to exchange for rights in Jackson Lake. In the interest of finding more exchange water for irrigators in the upper basin above American Falls, Idaho Senator J. P. Pope asked in May 1938 for "the complete disposition and ownership of all waters stored in the American Falls Reservoir." Commissioner John C. Page of the Bureau of Reclamation responded:

> Of the 1,700,000 acre-feet of reservoir capacity at American Falls, 433,593 acre-feet is retained by the United States for the Minidoka North Side Pumping division…, 477,421 acre-feet is held by the American Falls Reservoir District No. 1, 400,000 acre-feet by the American Falls Reservoir District No. 2, and the remaining capacity [389,000 acre feet] was held by 26 other canal companies and irrigation districts ranging from 70 to 50,000 acre-feet each.[672]

The original American Falls rights purchased or held under option by upper valley canal companies and irrigation districts amounted to 265,853 acre-feet and the contracts between the government and these parties provided for exchange of this amount with the bottom water in Jackson Lake belonging to the Minidoka project (325,810 acre-feet).[673] By the end of 1936 on account of the additional water secured by upper valley interests under lease contracts, the total amount of water to which upper valley companies were entitled from American Falls was more than was available for exchange out of the bottom rights at Jackson

672 John C. Page, Commissioner of Reclamation to Hon. J. P. Pope, United States Senate, June 8, 1938. Folder: Reclamation Bureau, Minidoka, American Falls Reservoir, Idaho State Water Rights, General, Part 4, April 5, 1938 to Dec. 4, 1947, Box 3612 8-3 Minidoka-American Falls Res. Contracts Gen., 8-3 Minidoka-Homestead Entries, CCF 1937-1953, RG 48, NARA II. Pope's letter of May 4 is referenced by Page in his June 8 response.

673 Lynn Crandall, district engineer, to B. E. Stoutemyer, December 19, 1936. Folder: 032.1 Minidoka – Correspondence re. Exchange of Idaho State Water Rights Between Reservoirs (Folder: 032.1 Minidoka), Box 719 Minidoka Project, 032. – 070, PCF, 1930-1945, RG 115, NARA-Denver.

Lake in years when American Falls filled to capacity. As a result, exchanges for part of the water had to come out of the Twin Falls companies' top rights at Jackson Lake. This type of exchange had always been accomplished by informal agreements made between the consenting parties on a year-to-year basis.[674]

Management of water resources on the Upper Snake River involved a complicated mosaic of water entitlements, especially during years of subnormal runoff when the storage reservoirs did not fill. Permanent contracts for storage space in American Falls Reservoir did not provide for exchange of rental water, for the reason that rental contracts were not contemplated at the time permanent contracts were made. The number of acre-feet included in a lease contract was for the *pro rata* share of the rental supply of stored water to which the party may become entitled which was the *pro rata* share of the amount of stored water actually available out of the total rental capacity in American Falls Reservoir. In the case of those parties diverting above American Falls, their right was further subject to the amount of stored water available for exchange from Jackson Lake Reservoir, the available amount to be prorated if there was insufficient supply to make a complete exchange.[675]

For a number of years after completion of American Falls Reservoir, the Twin Falls companies by informal annual agreements had exchanged their water rights in Jackson Lake for rights held by upper valley companies in American Falls on an equal acre-foot to acre-foot basis. Upper valley companies wanted formal written contracts to give them greater assurance in the exchange privilege which was the only way they could derive any benefit from their rental water in American Falls Reservoir. On the other hand, the Twin Falls companies had less to gain from a formal written exchange contract, although acre-foot for acre-foot exchanges were beneficial to the lower valley companies because water stored in American Falls was less subject to seepage, evaporation, and adverse diversions.

In addition to their regular contracted storage space, in 1936 upper valley companies owned approximately 123,000 acre-feet or almost

674 B. E. Stoutemyer to R. E. Shephard, manager of the North Side Canal Company, December 12, 1936. Folder: 032.1 Minidoka, Box 719, PCF, 1930-1945, RG 115, NARA-Denver.

675 Stoutemyer to Crandall, December 22, 1936. Folder: 032.1 Minidoka, Box 719, PCF, 1930-1945, RG 115, NARA-Denver.

one-third of the government's more than 400,000 acre-feet of unsold storage capacity in American Falls Reservoir. Providing inducements for upper valley companies to participate in renting the surplus water in American Falls Reservoir was important to the government in that without their participation it would be more difficult to sell the full amount of unused storage at a favorable price.

In 1936, by means of these informal water exchanges, the upper valley projects were able to secure about 100,000 acre-feet more water than they could have secured without the exchange. This extra late season storage water helped mature crops on a number of irrigated tracts and districts in the upper valley. Thus, until additional storage was secured in the upper valley, this already strained informal water exchange program, as unsatisfactory as it was during dry seasons, was destined to become an increasingly important means by which a precious late season water supply was secured by upper Snake River valley irrigators when natural flow rights had been exhausted.[676]

By cooperation between the various water users, the Committee of Nine, the Idaho Department of Reclamation, the Bureau of Reclamation, the District No. 36 watermaster, Idaho Power Company, and the Governor's Emergency Drought Committee, it was possible to work out various plans for development, conservation, and use of available water supplies that somewhat ameliorated the effects of the prolonged drought. The year 1934, however, served to emphasize the need for additional dry year water supplies on Snake River. While the USBR investigated potential reservoir sites on the North Fork and South Fork of Snake River to bring more permanent relief to irrigators by adding to the available water supply, R. W. Faris, Idaho Commissioner of Reclamation, told the Idaho State Planning Board in the summer of 1934 that a comprehensive solution must reduce all types of non-irrigation use and include a strong water conservation element in the plan. American Falls Reservoir had filled only twice since 1926, he noted, therefore there would be little water available to store in new federal reservoirs unless a plan could be developed "to release or transfer some of the rights on the lower river at and below American Falls." He had several specific recommendations on how to achieve these savings:

676 B. E. Stoutemyer to Commissioner USBR, December 29, 1936. Folder: 032.1 Minidoka, Box 719, PCF, 1930-1945, RG 115, NARA-Denver.

> The rights that might be looked at to supply this deficiency are those of the Idaho Power Company, with a primary storage right of 45,000 acre feet of water in the American Falls reservoir, and a secondary right of 255,000 acre feet to be drawn upon under certain limitations and restrictions, the right of the Minidoka project to divert something like 2,000 second feet, continuously, for the operation of its power plant at the Minidoka dam, and that of the canals diverting water at the Milner dam for domestic purposes during the non-irrigation season. During the non-irrigation season, from October 15, 1930, to April 15, 1931, the several canals heading at the Milner dam diverted 519,827 acre feet of water from Milner Lake for domestic and stock watering purposes, and during that time 644,108 acre feet passed over Milner dam to the Pacific Ocean, making a total of 1,163,915 acre feet, of which amount 80,815 acre feet was gain between Minidoka dam and Milner which could not have been recaptured for storage, leaving a net loss of 1,083,100 acre feet.
>
> The American Falls Reservoir lacked about 150,000 acre feet of filling in 1931, so if the gates could have been closed from October 15, 1930, to April 15, 1931, the reservoir would have filled and there would have been 933,100 acre feet of water left for storage on the upper river, or in the enlarged American Falls reservoir, which would not only have afforded complete relief to that area but would have supplied the anticipated deficiency to the Michaud Flats and left 583,100 acre feet for carryover storage. [677]

Commissioner Faris believed that with cooperation among all water users, the gates of American Falls reservoir could remain essentially closed during the entire non-irrigation season and nearly the entire flow of the river above the reservoir left available for storage. Among the specific solutions offered by Commissioner Faris were:

- Build a government powerhouse at American Falls to be operated in tandem with the Minidoka plant, thereby saving 200,000 acre-feet of water during the non-irrigation season;
- Construction of carry-over storage reservoirs in the upper basin;
- Greatly reduce the non-irrigation season draft by the Twin Falls South Side Canal;
- Supply towns with water by drilling community or individual wells;

677 R. W. Faris, State Commissioner of Reclamation, "Supplementary Water for Irrigation in Idaho, With Particular Reference to Boise and Snake River Valleys," paper delivered to the Idaho State Planning Board, June 1934; Folder: 302.12 Idaho Surveys and Investigations, 1930-1934 (Folder 302.12); Box 560, GAPR, 1919-1945, RG 115, NARA-Denver.

- Use the inflow to the river between American Falls and Milner Dam to supply domestic water and meet the demands of the power plant at Shoshone Falls.

Some of the savings that Commissioner Faris advocated were partially achieved in the 1934 agreement. The need to conserve water for irrigation during the series of dry years that gripped the valley in the early 1930s required limiting water supplies for power generation at American Falls Reservoir. Serving the water requirements of the Minidoka power plant and the flow requirements at IPC plants below Milner Dam also compelled adoption of methods for further conservation. Under the 1934 water-saving contract the United States agreed not to construct a power generating facility at American Falls and, in return, IPC released control of its 255,000 acre-feet of storage in American Falls to the United States. Electrical energy was transmitted over IPC's power network to the Minidoka project from outside the region. This allowed limiting outflows from the reservoir primarily to flows that were incidental to the operation of the reservoir system for existing and proposed irrigation development and for satisfying domestic water rights. More conservation measures were implemented in the ensuing decades when Palisades Dam was built.

The year 1935 was the seventh, and final, year in the consecutive string of subnormal runoff seasons on the Snake River. Runoff at the upper basin measuring stations at Moran and Heise exceeded 80 percent of normal, but further down the river the percentages declined markedly – 40 percent at Blackfoot, 55 percent at Neeley, only 52 percent on Henry's Fork near Rexburg, and a scant 4 percent at Milner. According to watermaster Lynn Crandall, the summer months brought only one-half of the moisture received in the previous dry year of 1934. In addition to the shortage of precipitation, what water fell was absorbed in partially restoring soil moisture and groundwater basins depleted by years of drought. In 1936 for the first time since 1928, run-off measured above normal on the main Snake River. Both Jackson Lake Reservoir and American Falls Reservoir filled and 620,000 acre-feet of surplus flood waters spilled past Milner Dam below all of the irrigation diversions.

Crandall pointed out in his annual watermaster's report that one-quarter of this waste could have been retained in the small storage reservoirs under construction by the Bureau at Island Park and Grassy

Lake on the North Fork of the Snake River if they had been completed. They were both scheduled for completion the following year. "The occurrence of a year of fairly good run-off in 1936," Crandall warned the water users of District No. 36, "should not be permitted to result in any decrease in efforts toward further water conservation. Dry years to be expected in the future will require all the water that can be made available from any source at reasonable cost if the established irrigated agriculture is to be adequately maintained in the Snake River Valley."[678]

The failed American Falls power plant

Since completion of American Falls Dam in 1926, the federal government had anticipated building its own power plant some 200 feet downstream from the dam. The Bureau of Reclamation had gone so far as to install four 15-foot penstocks in the west side of the 1926 American Falls Dam to deliver water to the government's power plant once it was built. Power from this facility was intended for the 115,000 acre North Side Pumping Division of the Minidoka project. These lands located near Rupert, Idaho and situated at higher elevation north of the Snake River were to be supplied with water lifted out of the river by large electrical pumps. Until the pumping division was completed, additional hydroelectric power developed by the government at American Falls could provide reserve power for the Minidoka project; provide power to increase the capacity of the south side pumping plant; permit saving of water required in the winter at Minidoka Dam; provide power for construction purposes; and finally to dispose of surplus power by sale to Utah Power & Light or IPC.[679]

As part of the authorization for the American Falls Reservoir, the Bureau had asked Congress for $1,250,000 to initiate construction of the American Falls power plant with a view of using a large portion of the power developed to pump water onto the North Side Pumping

678 Lynn Crandall, "Water Distribution and Hydrometric Work, District No. 36, 1935," (typescript, 1935); Unsorted Water Distribution Reports; District 5: 1936, 1937, 1938, 1939; District 12A: 1915, 1919-1920, 1927; District 36, 1926, 1927. 1938, Accession no. 20071914; and Crandall, "Water Distribution and Hydrometric Work, District No. 36, 1936," (typescript, 1936); Unsorted Water Distribution Reports; District 36: 1926 (2), 1933, 1935, 1936, 1956, Accession no. 20071916, AR 20, ISHS.

679 Elwood Mead, US Reclamation Commissioner to R. E. Shepherd, president of the American Falls Reservoir District, January 7, 1927. Folder 320 MP Corresp. Re Development of Power, Jan. 1, 1927-Dec. 31, 1929 (Folder 320 MP), Box 734 Minidoka 320, GAPR, 1919-1945, RG 115, NARA-Denver.

Division of the Minidoka project. In December 1926, Congress authorized an initial $700,000 as partial payment toward construction of a 30,000 kW American Falls power plant, but the remainder of the congressional funding was diverted to development of the Gooding-Milner and Black Canyon irrigation projects. What was left, proved too little to initiate construction, but was reserved for later use. Interior Secretary Hubert Work later announced that the North Side Pumping Unit would be put off for six years; thus, negating the primary need for the power facility. Rumors persisted that the American Falls power plant had been undermined by IPC.[680]

In January 1928, at the urging of his constituents, Rep. Addison T. Smith looked into this matter with the interior secretary's office. He found that Idaho Power had raised issues that had stalled action on the matter within the Interior Department. According to the *Salt Lake Tribune,* Smith replied by wire:

> On receipt of your wire I took up matter of construction of power plant with Secretary of Interior and find that legal questions have been raised by Idaho Power Company which are now under immediate consideration. Outcome of which will determine action to be taken by the Secretary. Am greatly disappointed over turn of affairs.[681]

The *Tribune* noted that there were a couple of problems with the government's American Falls powerhouse proposal.

First, the government planned to proceed with construction of a 30,000 kW power plant when it only had a market for 2,500 kW. Second, building the plant would be a violation of the winter power limitations placed in the 1923 contract between the government and Idaho Power that were made when the company sold its interests to the United States.[682] Under the terms of the 1923 contract, the United

---

680 Asher B. Wilson, President and Newell S. Wright, Secretary, Twin Falls Chamber of Commerce , Protest, November 17, 1927; E. J. Feldsted, Pocatello Chamber of Commerce, November 17, 1927; Herbert Work, to Hon. Addison T. Smith, House of Representatives; enclosed with Addison T. Smith, House of Representatives to Hon. Hubert Work, Secretary of the Interior, November 25, 1927; Protest, Adopted at American Falls on Sept. 28, 1927; and "Minidoka Reclamation Association Gives Statement of Facts Regarding Power Development at American Falls and North Side Extension; enclosed with Smith to Work, November 26, 1927. Folder 320 MP), Box 734 Minidoka 320, GAPR, 1919-1945, RG 115, NARA-Denver.

681 Quoted in "Idaho Power Blocks Work on plant At Falls," *Minidoka County News*, March 15, 1928; USBR-Boise.

682 Clipping from *Salt Lake Tribune*, March 11, 1928. USBR-Boise.

States could not exceed a half-hour peak of 10,000 kW to supply the power demands and uses on the Minidoka project and its proposed extensions during the period of November to March, inclusive. Any federal power plant built at American Falls would have to be coordinated with the Minidoka plant to produce 10,000 kW without wasting irrigation water past Milner Dam.[683] The Interior Department had planned to build only the first 7,500 kW unit, but needed a firm market for the surplus 5,000 kW to justify the plant. The power company was in a strong position to bar construction of the plant, concluded the *Tribune:*

> Instead of offering to buy this power, Idaho Power Company reminded the government that it has a contractual relation with that company which it cannot ignore, and which, if observed, would prevent the government from selling commercial power from its American Falls plant if built.
>
> As a counter-proposal, the power company offered without cost to furnish 2500 kW for project use in exchange for water now going over the American Falls spillway.
>
> If this offer be accepted, the government would escape the investment of $1,250,000.00 and the needs of the Minidoka project would be fully met at no expense in cost to water and power users.
>
> At the time the department made its recommendations to congress, for an appropriation for a power plant appropriation, it completely overlooked its contract with the Idaho Power Company of five years ago. Having been reminded of its obligation under that agreement, department officials admittedly are in a quandary.[684]

Representatives of seven upper valley chambers of commerce and the Burley and Minidoka Irrigation Districts sent a resolution to Secretary Work protesting against any contract that might cause postponement or delay in advancement of irrigation projects on the Snake River. Specifically, they protested against the department entering into "any contract with the Idaho Power Company that will defer or retard the building of the Government Power plant at American Falls immediately as provided for...and authorized by Acts of Congress."[685] The Bureau decided to develop irrigated lands within the Minidoka Gravity Extension tract before initiating the North Side

683 Contract between the United States and the Idaho Power Company Relative to Power Rights at American Falls, Idaho, June 15, 1923, Section 23.

684 Clipping from *Salt Lake Tribune*, March 11, 1928; USBR-Boise.

Pumping Unit. Realizing that limitations imposed by the 1923 contract made government power development unprofitable until that power could be used for irrigation pumping, there was no need to hasten construction on the American Falls power plant. Instead, the Bureau arranged for a contract with IPC for use of the government's surplus unsold water in the reservoir, pending the time this water was needed for irrigation purposes.[686]

In making its appropriation request to Congress for fiscal year 1929, the Interior Department had secured an additional $500,000 earmarked for the construction of a power generating facility at American Falls. However, the April 23, 1929, contract with Idaho Power for use of the government's surplus unsold water in American Falls Reservoir provided for IPC to furnish up to 7,500 kW of power to the Minidoka project. Of this, 3,000 kW was for emergency or standby use, the remaining 4,500 kW being firm power available during irrigation season. In return for receiving this power during the period of peak demand, the U. S. provided surplus power in the off-season at the exchange ratio of 2 kW of off-season power to the power company, in return for 1 kW received by the government during peak irrigation demand.[687] Thus, at no direct expense to the project, the U. S. received the power it needed to operate its Minidoka project until the North Side Pumping Unit was authorized for construction. In the meantime, the 1929 agreement with Idaho Power had rendered construction of the government powerhouse at American Falls unnecessary.

The 1929 contract was temporary; it was to last a maximum of 10 years, but could be terminated by the government with one year advanced notice after October 1, 1932, and by the power company after October 1, 1934.[688] Therefore, in theory, the contract would not prohibit

685 Asher B. Wilson, President and Newell S. Wright, Secretary, Twin Falls Chamber of Commerce , Protest, November 17, 1927; E. J. Feldsted, Pocatello Chamber of Commerce, November 17, 1927; Herbert Work, to Hon. Addison T. Smith, House of Representatives; enclosed with Addison T. Smith, House of Representatives to Hon. Hubert Work, Secretary of the Interior, November 25, 1927; Protest, Adopted at American Falls on Sept. 28, 1927; and "Minidoka Reclamation Association Gives Statement of Facts Regarding Power Development at American Falls and North Side Extension; enclosed with Smith to Work, November 26, 1927 Folder 320 MP), Box 734 Minidoka 320, GAPR, 1919-1945, RG 115, NARA-Denver.

686 Elwood Mead, Commissioner to Hon. William E. Borah, December 5, 1928. Folder: 233.03 Lease of Power Contract, Idaho Power Co. thru 1929 (Folder 233.03), Box 709 Minidoka 223.038-223.031, GAPR, 1919-1945, RG 115, NARA-Denver.

687 Agreement, April 23, 1929, between United States and Idaho Power Company.

688
Agreement, April 23, 1929, between United States and Idaho Power Company.

building the power plant, if Congress decided to fund the North Side Pumping Unit. Although the decision to make the agreement with IPC was criticized by many irrigators and civic organizations in the upper basin who wanted public power, in a communication to Idaho Senator William E. Borah, Commissioner Mead explained why the Interior Department favored the temporary agreement:

> The financial benefit of leasing the power privileges are so great that they can be ignored only on political rather than economic grounds I can comprehend the anxiety and apprehension of the people interested in the North Side area, but I am convinced that waste of money on this plant will not advance the construction of their project, but will certainly retard it. The postponement of this project would remove it from the program of development. It is still there and the water for its irrigation is being reserved in the reservoir. The Government has refused offers for the purchase of this stored water because it was reserved for that particular area. That fact alone is of far more importance as a guarantee of early development than the existence of a partly used pumping plant would be.[689]

In 1930 the appropriation act for the department re-allocated the $1.2 million set aside for the American Falls power plant to the Minidoka project's Gravity Extension Division (American Falls Reservoir District No. 2), which started construction in 1933. Four years later, as part of the 1934 water-savings contract negotiated with Idaho Power, the federal government agreed not to build a power plant at American Falls in exchange for the power company releasing control of its 255,000 acre-feet of secondary storage to the United States. This agreement remained in force until the federal government negotiated the new contract with the power company on December 15, 1950.[690]

These developments notwithstanding, the Bureau of Reclamation did not completely abandon its plans for a federal power plant at or near American Falls. In June 1936, the Bureau's chief engineer conducted a preliminary study of the potential benefits of both an American Falls

689 Mead to Borah, December 5, 1928. Folder 233.03, Box 709 Minidoka 223.038-233.01, GAPR, 1919-1945, RG 115; NARA-Denver. [D1-1-09]

690 American Falls Powerplant, Minidoka Project, Idaho, Definite plant Report, Volume 1, General Plan, July 1956, 2-3. A marked-up copy of this report appears, without a cover or abstract, in the Frank Church Papers at Boise State University. See Report regarding proposed power plant to provide irrigation pumping power for the North Side Pumping Division of the Minidoka Project, undated. Folder 8, Box 108, Series 1.1, MSS 56 Frank Church Collection, Boise State University, Special Collections (BSU Spec. Coll).

power plant and one located a short distance downstream at Eagle Rock.[691] Thirteen years later, the Interior Department completed a financial analysis of an American Falls power plant that concluded construction and operation of the proposed American Falls power plant by the United States was "feasible from the engineering and financial points of view."[692]

In the meantime, as local interest in groundwater pumping increased in the wake of the discovery of the East Snake River Aquifer and federal attention to power developments at American Falls and Palisades returned, Idaho Power Company officials worried about the potential economic impact of these project on IPC's ability to generate power both at its American Falls plant and its hydroelectric power plants below Milner. Following publication of a news article in the summer of 1948 quoting long-time Upper Snake River watermaster Lynn Crandall as endorsing "construction of a government power plant, perhaps at American Falls, or possibly using Palisades power" to meet groundwater pumping power needs, B.C. Russell, IPC's general superintendent, took note and wrote to company president T.E. Roach to express his concern.

Russell's memorandum, dated June 4, 1948, and circulated to company counsel, A.C. Inman, warned that neither prospect boded well for the company. Under present conditions, the power company had the right to use whatever water passed through American Falls dam and became available to it at its East Side power plant. Construction of a government power facility at American Falls would alter this situation to the detriment of the power company. The company's 2,500 cfs primary *summer* power right under the 1923 contract supplied enough water to produce only about 7,500 kW at its 27,000 kW-capacity East Side plant. Before the company could exercise its additional (tertiary) summer power rights under the contract, the United States would have

691 American Falls Powerplant, Minidoka Project, Idaho, Definite plant Report, Volume 1, General Plan, July 1956, 3.

692 "Financial Study of Proposed Completion of American Falls Power plant" (U. S. Department of the Interior, Bureau of Reclamation, Region 1, Boise, Idaho), enclosed in a letter from J. A. Krug, Secretary of the Interior to Hon. Joseph C. O'Mahoney, Chairman, Committee on Interior and Insular Affairs, August 10, 1949, House Committee on Public Lands and Senate Committee on Interior and Insular Affairs, *The Palisades Dam and Reservoir Project: Hearings before a Subcommittee on Irrigation and Reclamation of the Committee on Public Lands, House of Representatives, and a Special Subcommittee of the Committee on Interior and Insular Affairs, United States Senate*, 81st Cong., 1st sess., 8, 9, 10, 11, 12, 19, and 19 August 1949 (GPO, 1949), 99.

the opportunity to use its 6,000 cfs secondary summer power right to generate 30,000 kW at its proposed American Falls power plant. If the government built a hydroelectric power plant at American Falls, Russell observed, the company had the right to cancel the 1934 contract in which it had relinquished its 255,000 acre-feet secondary *winter* storage right to the United States. Nevertheless, if the contract was canceled and the government built its power plant at American Falls, IPC would lose its current right to use for power generation all of the winter water that passed through American Falls dam; the company would be largely limited to what could be drawn from its storage space under the terms of the 1923 contract. Russell concluded that government development of a 30,000-kW plant at American Falls would have the effect of reducing Idaho Power Company's annual power production "to considerably less than half what we now develop at this point." Thus, the outcome was considered by Russell to be "a very important matter to Idaho Power Company." Construction of Palisades dam, even without a government power plant at American Falls, gave Russell little solace. In the long run, once Palisades dam became available to store additional winter water for irrigation, Russell observed, "there is certain to be a great deal of pressure put on the Company to give up its Secondary storage rights for the benefit of irrigation."[693]

On the heels of the Bureau's favorable feasibility study on American Falls power plant, in August 1949 congressional consideration for construction of the American Falls power plant re-emerged along with deliberations on authorization of the North Side Pumping Division during congressional hearings on the Palisades Dam and Reservoir re-authorization legislation. As it had during the 1934 water-savings contract negotiations, Idaho Power came out strong against a federal power plant at American Falls and any accompanying high-voltage transmission system. Ralph Gale, assistant to Idaho Power's president, argued that his company was already meeting all electric power needs of local groundwater irrigators, including those of the Minidoka North Side Pumping Division. Since satisfying those groundwater pumping demands was the chief reason for renewed interest in a power plant at

693 B.C. Russell, General Superintendent, to Mr. T.E. Roach, cc: Mr. A.C. Inman, regarding Power Requirement for Pumping in the Vicinity of Rupert, June 4, 1948, B1_IPC0013881, 2010-06-01 USBOR Response to IPCo. Subpoena Duces Tecum, AGO-06-10-2010-1 data disk, Idaho Attorney General's Office.

American Falls, Gale argued authorization of a federal power plant would be "an unnecessary expenditure of Federal funds."[694]

While a bill to authorize Palisades Dam and the American Falls power plant failed to pass that year, it passed as part of the measure that re-authorized the Palisades Dam project in the fall of 1950. During the early 1940s, a seventh generating unit was added to the Minidoka power plant to furnish power to the Milner Low-Lift Irrigation District. Growing demands on the project absorbed this additional power by the end of the decade. By this time, the summer power requirements of the United States on the Minidoka project had outstripped the 4,000 kW allowed under the 1934 contract by some 2500 kW. IPC expected to be paid handsomely for the excess at rates equitable for peaking power, delivered for a short term of service under low load factors.[695]

On December 15, 1950 the United States and IPC entered into a contract that provided for the interconnection of federal powerhouses constructed by the Bureau on the Mindoka and Boise projects (at Black Canyon, Diversion, Minidoka and Anderson Ranch dams) and Corps of Engineers (Lucky Peak Dam) with the power company's system "to permit the fullest practical utilization of the Federal powerplants consistent with the conservation of water for irrigation use and to make the benefits of such interconnection available to power customers and water users." The new contract terminated several existing power agreements, including the 1934 water savings contract, between the parties with the aim of exchanging and purchasing power and energy between the parties "on terms other than are provided in the existing contracts between the parties."[696] Among the customers to be served by IPC under the 1950 contract were the North Side Pumping Division, not to exceed a maximum demand of 2,000 kW and the maximum energy of 6 million kWh per year. The additional needs of the Minidoka power plant were also to be serviced under the agreement. The term of the

694 House Committee on Public Lands and Senate Committee on Interior and Insular Affairs, *The Palisades Dam and Reservoir Project*, 198.

695 Acting Commissioner to Secretary of the Interior, July 11, 1950. Folder: Division of Power 655 Bureau of Reclamation – Minidoka, Box 259 Records of the Division of Water & Power, Classified Files, 1941-1951, Project File, USBR, Contract Files, Kendrick –Minidoka (Records of the Division of Water & Power), RG 48, NARA II.

696 Article 31 (b) of the contract noted that the following contracts were terminated by the agreement: April 23, 1929; April 10, 1930; October 1, 1934, as amended; August 13, 1942, as amended; February 7, 1950; and July 28, 1950. USDI, Bureau of Reclamation, Boise and Minidoka Projects, Contract with Idaho Power Company for the Sale of Electric Power and Energy and for Service to Loads of United States or its Customers, December 15, 1950.

contract was to run to October 31, 1963, subject to the right of either party to terminate it on October 31, 1956, or thereafter with three years advance written notice.[697]

In the ensuing years, Idaho Power continued to argue that a government power plant at American Falls was unnecessary and that it could accomplish the same task of supplying the reclamation project with electricity at more affordable rates. Following congressional approval of the Palisades Dam project in 1950, senior Interior Department officials urged the Bureau to move on construction of a new government power plant at American Falls, as provided for under that act. Assistant interior secretary William Warne on November 5, 1951, stressed the need to build the American Falls power plant as expeditiously as possible in a memorandum he sent to the Bureau's commissioner.[698] Construction of the American Falls power plant, however, continued to languish into the 1950s. As of the fall of 1951, Congress failed to provide any money for construction of the plant.[699] In part, as Assistant interior secretary Fred G. Aandahl later explained to Senator Herman Welker in a June 1953 letter, the act itself was to blame. In May 1953, Welker had written to the Interior Department, forwarding a letter that he had "received...from the Cassia County unit of the Hells Canyon Association inquiring about the status of the American Falls Power plant of the Minidoka project." Funding for construction of the power plant had been eliminated prior to congressional approval of the Bureau's budget in April; Welker and his constituents wanted to know why. According to Aandahl:

> An estimate of funds for the American Falls Power plant for the F. Y. 1954 was deleted from the revised budget due to the need to reexamine the feasibility of this proposed installation because of its increased cost. The act authorizing construction of the power plant of

697 Ibid.

698 Memorandum to Commissioner, Bureau of Reclamation, from Assistant Secretary Warne, regarding American Falls power plant, November 5, 1951; and Fred G. Aandahl, Assistant Secretary of the Interior, to Hon. Herman Welker, United States Senate, June 9, 1953. Folder: 5000, Palisades Project (Folder: 5000), Box 11 Regional Solicitor, Portland, Oregon, Special Projects Files, 1951-1970 (Special Projects Files, 1951-1970), RG 48, National Archives at Seattle, WA (NARA-Seattle)

699 Robert R. Rose, Jr., Acting Secretary of the Interior, to Mr. G. W. Bunting, Manager, Central Farmers Fertilizer Company, November 27, 1951. Folder: Reclamation Bureau Projects, Palisades Dam, File no. 8-3, Part 2, Box 3619 8-3 Paint Rock – Withdrawals & Restorations, 8-3 Payette-Boise – Claims, CCF, 1937-1953, RG 48, NARA II.

> the American Falls Reservoir (Act of September 30, 1950, 64 Stat. 1083) limits appropriations for the plant to $6,600,000. Current estimates of the Bureau of Reclamation show that the completed plant would cost substantially in excess of this amount.

In addition to the high cost of construction, the US-IPC contract of December 15, 1950 lessened the urgency for additional government power at American Falls. Aandahl assured the senator that the long-term need for the plant "as a source of power for pumping on the North Side Division of the Minidoka project is appreciated," but that "satisfactory arrangements can be made with the local utility for the provision of the necessary power for pumping, at least on an interim basis, pending a final decision" by the federal government on the construction of the power plant.[700]

The "local utility" to which reference was made was IPC. During senate appropriation hearings on the Interior Department's 1954 budget in May 1953, Idaho Power changed the emphasis of its objections. Rather than directly opposing a federal American Falls power plant and a federal transmission line, the power company stressed that the added financial burden on the federal government was unnecessary because the co-operative private-public power and electrical transmission exchange system operating under the 1950 contract made the investment unnecessary altogether. Ralph Gale, assistant to IPC's president, once again representing the power company, noted that IPC was currently working with the federal government to transmit federal power not only from the future Palisades facility, but also from the Black Canyon and Anderson Ranch plants of the Boise project to the pumping division as well as local groundwater irrigators, and that the company had agreed to supply such power through fiscal year 1955.[701] To build a federal transmission system, Gale asserted, would be "a waste of taxpayers' money, because our system at the present time adequately serves everyone who has requested or asked for power in that area."[702]

---

700 Aandahl to Welker, June 9, 1953. Folder: 5000, Box 11 Special Projects Files, 1951-1970, RG 48, NARA-Seattle

701 Senate Committee on Appropriations, *Interior Department Appropriations for 1954: Hearings before a Subcommittee of the Committee on Appropriations, United States Senate, Part 2*, 83rd Cong., 1st sess., 11 May-3 June 1953, 1760-1761.

702 Ibid., 1766.

The Upper Snake River Water Users' Protective Union seconded Idaho Power's stated position on the American Falls power plant. Writing to both the Idaho congressional delegation and Idaho's governor Len Jordan in February 1951, the union's president Ival V. Goslin urged that money earmarked for the American Falls power plant be re-allocated to construction of Palisades Dam.[703] Goslin contended the proposed federal power plant was "an unnecessary project which would produce practically no net increase in power because it would be duplicating already existing power facilities." Moreover, the power plant "would actually be detrimental to the economy of our state because of loss of taxes from the power company."[704]

Congress agreed with both IPC and the Upper Snake River Water Users' Protective Union, and while the Bureau of Reclamation continued to submit annual budgets that provided for a federal power plant and transmission system in the 1950s, no such facility or system was authorized. The arrangement that Idaho Power had worked out with the Bureau through the December 15, 1950, contract was apparently adequate enough in the eyes of Congress to make funding either an American Falls power plant or a federal transmission system from such a plant unnecessary.[705]

Some local interests favorable toward public power did persist in agitating for a federal power plant into the late 1950s. In July 1955 a group of 16 municipal and cooperative power agencies forwarded a resolution to Senator Frank Church declaring their support for low-cost, federal power and a government-owned and operated American Falls

---

703 Ival V. Goslin to Governor Len B. Jordan. February 10, 1951. Folder: Department of Reclamation X (Folder: Department of Reclamation X), Box 6, AR 2/23, Governors' Papers: Len B. Jordan (AR 2/23), ISHS.

704 Goslin to Hon. Henry C. Dworshak, February 10, 1951; see also Goslin to Hon. Hamer Budge, February 10, 1951. Folder: Department of Reclamation X, Box 6, AR 2/23, ISHS.

705 This was true at least through fiscal year 1956, according to C. Frank Keyser of the Economic Division of the Library of Congress' Legislative Reference Service. Keyser noted in a report forwarded to Rep. Gracie Pfost (D, ID), that for the Bureau's 1956 appropriations, funding for transmission lines to the Minidoka North Side Pumping Division was purposely "eliminated" and subsequently construction was to occur "only if satisfactory arrangements cannot be made to provide power for the division deep wells and pumping plant." C. Frank Keyser, "Private Power Companies' Proposals in Minidoka-Palisades-Michaud Flats Area," April 27, 1956, enclosed in a letter from Ernest S. Griffith, Director, Legislative Reference Service, the Library of Congress to Honorable Gracie Pfost, April 26, 1956. Folder: 1772, Legislative, Committee, Interior and Insular Affairs: Michaud Flats Project, 1954-1956; Box 47, MG 44, Gracie Pfost Papers, 1950-1962, University of Idaho Special Collections. See also Edwin C. Schlender, Manager, Raft River Electric Cooperative, Inc., to Hon. Frank Church, U. S. Senator, January 28, 1957. Folder 8, Box 108, Series 1.1, MSS 56, BSU Spec. Coll.

power plant.[706] Two years later, in January 1957 Edwin C. Schelender, manager of the Raft River Electric Cooperative, wrote to Church ostensibly to inform him of the long history of failed opportunities to build the American Falls power plant and to recommend that the senator contact the Bureau for the full story of why the power plant had not yet been built.[707] Four months later in April, John L. Brog, manager of the Rural Electric Company, wrote Senator Henry Dworshak and sent copies to Church and Representatives Gracie Pfost and Hamer Budge to solicit support for a federal power plant.[708]

With the completion of Palisades Dam and its powerhouse to supply power to the Minidoka project and other preference power users in the region, the issue of a federal power plant at American Falls went dormant for the next fifteen years. By the time the 1950 contract was set to expire in 1963, the need to replace American Falls dam had become more important than construction of a government powerhouse at the site.

The issue briefly re-emerged once again in the 1970s when the Bureau contemplated replacing the deteriorating American Falls Dam, but the construction of a joint American Falls Reservoir District-IPC dam and powerhouse complex at American Falls sealed the fate of any proposal for a government owned and operated power plant at American Falls.[709]

706 Resolution of sixteen municipal and co-operative agencies, July 14, 1955. Folder 8, Box 108, Series 1.1, MSS 56, BSU Spec. Coll.

707 Edwin C. Schlender, Manager, Raft River Electric Cooperative, Inc., to Hon. Frank Church, U. S. Senator, January 28, 1957. Folder 8, Box 108, Series 1.1, MSS 56, BSU Spec. Coll.

708 John L. Brog, Manager, Rural Electric Company, April 25, 1957. Folder 8, Box 108, Series 1.1, MSS 56, BSU Spec. Coll.

709 American Falls Dam Replacement, Upper Snake River Basin, Idaho-Wyoming: Executive Summary Report, December 1972, 11. Box 442 8NS-115-95-083, FY 96, Project Reports, 1910-1995, Minidoka, RG 115, NARA-Denver. See also "Chapter 13: Related Power Potential," excerpt from internal 1972 Bureau report, enclosed in a letter from Norman Moore, Acting Regional Director to Mr. Philip H. Holm, Attorney at Law, November 26, 1974. USBR-Boise.

8

Development on the Middle Snake

While much of the focus of reclamation efforts in the Snake River Basin in the early twentieth century wa s in the upper basin area above Milner Dam, private interests and the federal government also took a good look at irrigation in the area below Milner Dam, in the "Middle Snake" region. They did so despite the fact that – as recognized by USGS hydraulic engineer W. G. Hoyt in the 1920s – the Snake River and its canyon was better suited to hydroelectric development than to irrigated agriculture. Downstream from Milner the Snake River lies so far below the level of the plain that direct diversion onto large areas above the rim of the Snake River Canyon was not practical, except in a few select areas.

When private irrigators and federal reclamation officials first sought to expand irrigated agriculture on a large-scale to the reach of the river below Milner, they turned to more accessible tributary streams with the most reliable water supplies.

Only a dozen or so pumping plants were established on the Snake River below Milner prior to 1935 that lifted water directly from the river to an elevation varying from about 40 to nearly 180 feet in order to irrigate bench lands above. Some were not successful and only a few met their full potential. The largest and most successful of these pump irrigation systems was operated by the Gem Irrigation District; it covered a gross area of 60,000 acres on the south side of the river near the Oregon-Idaho border. Organized in 1909, the major portion of the development was built in 1913 and included approximately 28,000 irrigated acres by 1928. The district lifted water through two pumps: The upper pump lifted water into three canals, a distance of 70, 103 and 177 feet, with a combined capacity of 270 second-feet; at the second, lower pump 33 second-feet were lifted 110 feet into one main canal. At the outset, electric power for pumping was purchased from IPC, until 1924 when the irrigation district contracted with the Bureau of

Reclamation to obtain power from its Black Canyon hydroelectric plant on Payette River.[710]

In the Boise and Payette valleys irrigation in the nineteenth and early-twentieth centuries was achieved by diverting from the major tributary streams, the Boise River and Payette River which flowed in a generally southwesterly direction into the Snake River. These streams commanded a large body of irrigable land that might be thought of as a wide area of the great Snake River Plain; they also joined the Snake River below all of Idaho Power's existing hydroelectric plants.

The Reclamation Service tapped the Boise River in the Middle Snake River region for the Boise Division of its so-called "Boise-Payette project," comprised of something over 240,000 acres of land, in the early 1900s. The first land withdrawals for the project were made in 1903. Contracts for construction of dams and the main canals in the Boise Division were let in 1906 and water deliveries began in 1909. Between 1900 and 1930 the irrigated acreage in Boise Valley grew from 84,000 acres to about 356,000 acres.[711]

The initial federal Boise Valley reclamation project was confined to a compact body of land extending south from the Boise River westward and southwestward across the Snake River Plain to the Snake River. As early as 1908, however, private canal companies and the Reclamation Service looked with eager anticipation at the chance to introduce irrigation to the stretch of the Snake River plateau between Twin Falls and Boise north of the Snake River – the 400,000-acre sagebrush desert of the Mountain Home area. An ambitious $250 million comprehensive Snake Basin Plan offered by the Bureau of Reclamation in the late 1940s proposed to supply the Mountain Home region with water by tapping central Idaho's mountain water through a trans-mountain diversion from the Payette River. Nearly one-half of the financing for the tunnels, aqueducts, canals and headgates of the Mountain Home Reclamation project would be raised through proceeds from the hydroelectric power installations at the proposed federal high dam at Hells Canyon.[712]

---

710 Hoyt, *Water Utilization* (1935), 131-132.

711 "Annual Project History and Operation and Maintenance Report of the Boise Project, Idaho for 1930." Folder: Department of the Interior, Bureau of Reclamation, Boise Project, Annual Project History, 1930-1932, Volume 8, Box 256 ANN-115-90-039, Engineering & Research Center, Project Histories, Boise 1926-1932, RG 115, NARA-Denver.

The State of Idaho withheld its endorsement of the Bureau's plan, which was opposed by leading business and agricultural organizations, the Idaho Chamber of Commerce, Idaho Power Company, and the Upper Snake River Basin watermaster. Although all of the various reclamation blueprints advanced for a Mountain Home irrigation project failed to materialize, the ideal of maximizing available water supply for development of irrigable land in the Snake River Basin remained a core value among those entities opposing the Bureau's Snake River Plan. If a high dam was to be constructed at Hells Canyon, noted Idaho Governor Charles A. Robins (1947-51), the water right granted by the state to store water would need to be subordinated to future upstream diversions for irrigation by carrying "a permanent legal servitude in favor of Snake Basin irrigated farming."[713]

Robins and his allies sought to retain local control over decisions about vital land and water resources in Idaho. The supremacy of future agricultural expansion over hydroelectric power development, whether by public or private power interests, remained central to the debate over the best use of the waters of Snake River in the latter half of the twentieth century.

The Hells Canyon controversy gave legal and political leverage to Idaho Power Company's alternative hydroelectric power proposals for development of the Middle Snake River which recognized the Idaho constitutional irrigation preference and the state's control over the river. Nevertheless, proposals for irrigation expansion in the Mountain Home region, as elsewhere in the Snake River Basin, still clashed directly with the more modest, lower cost run-of-the-river dam strategy of Idaho Power for its Middle Snake River power development. The legal, political and economic battle among the various contestants for the limited water resources of the Snake River reached a climax in the decades following the World War II as the Hells Canyon controversy came to national attention.[714]

---

712 Memorandum from The Regional Director, Region I to The Commissioner, Bureau of Reclamation, regarding Supplemental Report on the Payette Unit, Mountain Home Project, Idaho, May 18, 1949, enclosed with R. J. Newell, Regional Director, Region I, "Payette Unit, Mountain Home Project, Idaho: A Supplement to the Columbia River Basin Report," Project Planning Report No. 1-5.5-7.S., May 1949. File Folder 100-2, Project No. 1971, Formal, Part 2, Box 253, FERC; and Stacy, *Legacy*, 139-142.

713 C. A. Robins to Julius A. Krug, August 6, 1947 and September 27, 1947. Part IV, Box 2, Series 1, Governor C. A. Robins Papers (AR 2/22), ISHS.

Reclamation in the Boise and Payette Valleys

Irrigation had been introduced to the Boise Valley as early as 1864, but construction of the Oregon Short Line Railroad in the early 1880s made agriculture an important industry in the region. The major irrigation canals and irrigation institutions in the Boise Valley date from the railroad construction era. By 1900 the irrigating canals taking water from Boise River were numerous, but most ditches served the bottom lands extending a mile or two from the river. Only a few of the largest canals diverted water onto bench lands.

As in the Upper Snake River Basin at Twin Falls, irrigation engineer A. D. Foote developed the first proposal for an irrigation project on the Boise River. The engineer envisioned irrigating between 300,000 and 500,000 acres lying south of the river. Working as chief engineer to the Idaho Mining and Irrigation Company in 1882, Foote drew up plans for a diversion dam and canal system similar to those later developed by the U. S. Reclamation Service in the early twentieth century.

John H. Burns came to Boise City from New York in the fall of 1882 with the idea of establishing a commercial irrigation enterprise in Boise Valley. Burns filed a water claim on November 13, 1882 for 3,000 cfs and in the spring of 1883 conducted some preliminary surveys. In the fall of that year, A. D. Foote arrived from New York to take charge of the irrigation project. Acting as company engineer, Foote filed for another 1,500 cfs in October 1883. Soon thereafter, Foote surveyed a line for an elaborate main canal running at the highest practical elevation and seventy-five miles in length, along with an extensive web of lateral ditches crisscrossing the countryside and extending some 5,000 miles in length. To serve such an extensive area would require a large reservoir to store the flood waters of Boise River.

As Foote's surveyors conducted their work and the company published its promotional material, the project attracted eastern capital and settlers, even though the prospects for water being delivered through the New York Canal were five years off in the future. A national financial panic in early 1884 led to the failure of a key Baltimore investment firm which had taken a large stake in the project, and thereafter the private project was never fully financed. For several

714 Karl Boyd Brooks, *Public Power, Private Dams: The Hell's Canyon High Dam Controversy* (Seattle: University of Washington Press, 2006), 49-59, 140-158.

years Foote pressed on with intermittent assessment work in an effort to keep the company's appropriative water claim alive under state water law. Finally, a mechanic's lien against the Idaho Mining and Irrigation Company forced liquidation of the property at a sheriff's sale in February 1894. For the next five years two groups of builders, the Ada County Farmers Irrigation Company and the Farmers' Canal Company, competed for the right to take over the project. After litigation was carried all the way to the Idaho State Supreme Court, Charles Fifer in 1898 unified both canal companies into a new combine called the New York Canal Company. Water was turned into the canal at a temporary rubble diversion dam built of hay, straw and loose rock in June 1900 and into a canal system that had a capacity of only 300 cfs, a stunted version of A. D. Foote's original grandiose plan.[715]

Only a small portion of the project received water over the next five years. Farmers used only about 60 second-feet in 1901, but each year the amount used grew until by 1905 some 240 second-feet was put to beneficial use irrigating about 10,000 acres of land. In 1906 irrigated acreage increased to approximately 28,000 acres, but the irrigation system had reached its effective supply capacity without storage and permanent diversion works.

That issue was remedied by placing the project under the U. S. Reclamation Service. D. W. Ross, who was Idaho State Engineer from 1899-1903, and then became district engineer for the Reclamation Service, had been associated with A. D. Foote as an assistant engineer. It was Ross who initiated the federal irrigation project and following plans originally laid out by Foote brought irrigation water to the remaining non-irrigated areas of the Boise-Payette project. In the *Second Annual Report of the Reclamation Service,* Ross reported on general conditions in the Boise and Payette valleys and concluded that the vast area to be irrigated and expense of developing a suitable reservoir site made it impossible to develop the region through "private enterprise." He recommended that construction of the irrigation project be undertaken by the government and that the Boise-Payette project was

715 "History Payette-Boise Project, Part I: Early Investigation, Reports and Estimates,"15-18. USDI, BOR, Boise Project, Annual Project Histories, Through 1912 & Early History (Annual Project Histories), Box 6, Vol. 1, Box 6, Engineering & Research Center, Project Histories, 1911-1991 (ERC PH, 1911-1991), RG 115, NARA-Denver. "The Beginning of the New York Canal," Idaho State Historical Society Reference Series, No. 190 (March 1972), 1-5.

"worthy of first place in the consideration of feasible irrigation projects in this State."[716]

The Boise Division

The early plans for the Boise Division Boise-Payette project as adopted by the Reclamation Service had been published in its first annual report and stipulated a project of 243,000 acres, comprised of 164,000 acres of new irrigated land and 79,000 acres of existing irrigated land. The major engineering works associated with the proposed project were: a 250,000 acre-foot storage reservoir formed behind Arrowrock Dam (1915) on the headwaters of Boise River; a 170,000 acre-foot Deer Flat Reservoir (1910) south of Caldwell to serve 52,000 acres on the lower end of the project; a rubble concrete diversion dam on Boise River about eight miles above the town of Boise at the head of the Main South Side Canal; and a power plant located at the Boise diversion dam which was constructed in 1912 to supply power for construction of Arrowrock Dam. After the dam was completed, the Reclamation Service leased the plant to Idaho Power Company to be operated and maintained at the expense of the power company. Incorporated into the Payette-Boise project were formerly irrigated lands lying south of the river, including 18,000 acres irrigated by the old New York Canal (Main South Side Canal), 25,000 acres served through the Ridenbaugh Canal in the Nampa-Meridian Irrigation District, and about 36,000 acres in the Pioneer Irrigation District. [717]

Since a large part of the lands under the Payette-Boise project were in private ownership, the Reclamation Service had to determine if the landowners favored a government irrigation project. Public educational meetings were held with the general public, local governments, irrigation organizations, landowners and business leaders within the proposed project area through the winter of 1903-04. Based on the strength of their favorable response, the director of the Geological Survey recommended to the secretary of the Interior that more detailed

716 "The Beginning of the New York Canal," Idaho State Historical Society Reference Series, No. 190 (March 1972), 4-5; "History Payette-Boise Project, Part I, 20-22.

717 Gilbert H. Hogue, assistant engineer, "History of the Payette-Boise Project, Idaho," March 1916. Annual Project Histories, 1912-1916, Vol. 1, Box 6, ERC PH, 1911-1991, RG 115, NARA-Denver; Department of the Interior, United State Geological Survey, *Second Annual Report of the Reclamation Service* (GPO, 1903), 297-325; and "Payette-Boise Project," *Eighth Biennial Report of the State Engineer to the Governor of Idaho, 1909-1910* (1910), 190-191.

surveys and examinations be made.[718] Government action was conditioned upon prospective water users standing united and pledging their co-operation and financial support in reimbursing the government for construction costs estimated at more than $3 million. Local interests formed the Payette-Boise Water Users Association on August 10, 1904. Older, existing irrigation districts, like the Riverside Irrigation District, voted to co-operate with the federal government, if some credit was given to them for their irrigation works already completed.[719]

After a decade of growth, the amount of irrigable land, number of farms and the rural population within the Boise project area stabilized. The Reclamation Service's 1920 project history listed 2,652 farms, all but 632 farmed by their owners. By 1920 about one-half of the irrigated area (140,000 acres) within the Boise project was irrigated without any water right other than the water provided by the government. These irrigated acres received water from canals constructed by the government. In addition, supplemental stored water was sold by the federal government to eight canal companies and irrigation districts, who had constructed their own irrigation ditches and canals prior to the government reclamation project; these lands totaled some 133,000 acres.[720]

As early as 1900, 1901 and 1902, during low water season irrigators at the west end of Boise Valley were experiencing dwindling water supplies and consequent crop losses. Settlers were being forced to abandon formerly fertile and productive agricultural land. These irrigators blamed junior upstream diverters and turned to the courts to solve their problem. The Farmer's Cooperative Ditch Company brought suit against the Riverside Irrigation District and all other ditch and canal companies using water from the Boise River for depriving the plaintiff of the lawful use of its senior 240 second-foot appropriative water right and the resultant injury to the crops grown under Farmer's Cooperative Ditch.[721] After hearing thousands of pages of sworn

---

718 *Third Annual Report of the Reclamation Service, 1903-1904* (GPO, 1905), 238.

719 "History Payette-Boise Project, Part I: Land and Water Problems," 26-32. By 1926 the Payette-Boise Water Users Association was a defunct institution, having been replaced by irrigation districts organized on the reclamation project.

720 "Annual Project History and Operation and Maintenance Report of the Boise Project, Idaho for 1920," USDI, BOR, Boise Project, Annual Project Histories, 1919-1921, Volume 5, Box 257, ERC PH, 1919-1925, RG 115, NARA-Denver.

721 *Farmers Cooperative Ditch Company v. Riverside Irrigation District Ltd., et al.*, 16 Idaho 525 (1909).

testimony from local irrigators and other technical information submitted in writing to the court, on January 18, 1906, District Court Judge George H. Stewart issued his opinion, known as the "Stewart Decree." The decree adjudicated all of the natural flow appropriative rights on the river with priority dates from June 1, 1864 through April 1, 1904.

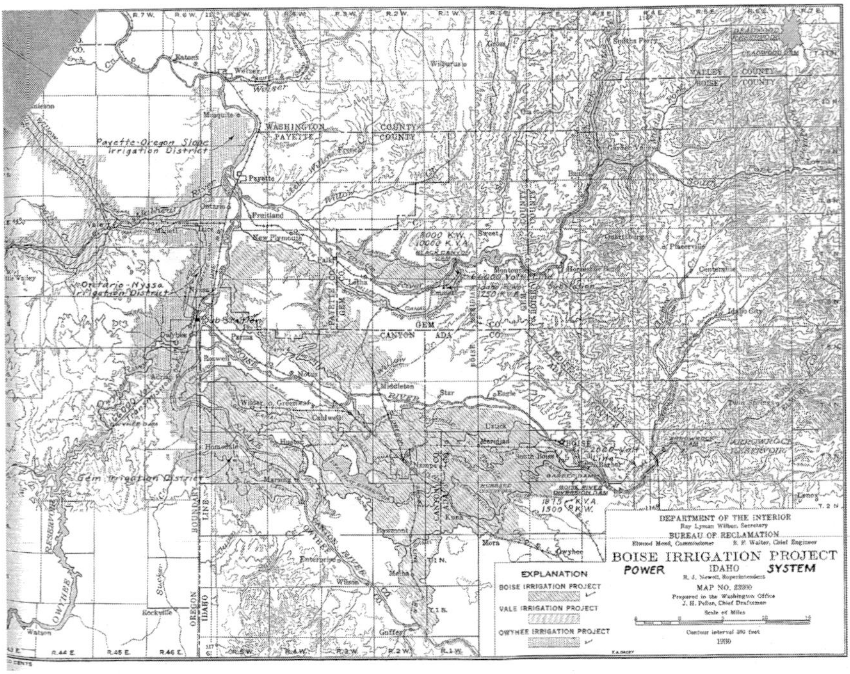

Map . The Boise Project as of 1934

Source: Boise Project, Idaho, Bureau of Reclamation, U.S. Department of the Interior, "Boise Project History, 1934." Folder: Department of the Interior, Bureau of Reclamation, Boise Project, Annual Project History, 1933-1936, Volume 9, Box 255, 8NN-115-90-039, FY 91, Engineering & Research Center, Project Histories, Boise, 1933-1939, RG 115, NARA-Rocky Mountain Region (Denver).

The case was appealed to the Idaho Supreme Court where it was affirmed but returned to the district court to determine the quantity of water required on bench and bottom lands. No final decree on this matter was issued.[722] However, in 1919 Judge E. L. Bryan issued an innovative order instituting a sliding-scale of cuts to be used in the distribution of water among irrigators on the Boise River in times of shortage, as follows:

> The various rights, as adjudicated in the so-called "Stewart Decree," shall receive 100 percent, until the natural flow of the waters of the Boise River shall decrease, until all the rights in said decree cannot receive 100 percent, at which time the various rights as adjudicated in the so-called "Stewart Decree" shall be cut to 75 percent of the amount of water decreed by the "Stewart Decree" as the natural flow of the Boise River decreases, beginning with the latest right and proceeding to the earliest rights in the order fixed by the said "Stewart Decree" …

If the natural flow fell below the amount necessary to supply 75 percent, then the amount would be reduced to 60 percent and the same formula of reductions applied. Once the oldest appropriative rights were cut below this 60 percent level, if the river's natural flow continued to decrease, then junior rights were shut-off completely in order of priority.[723]

The judgment in *Farmers Cooperative Ditch Company v. Riverside Irrigation District* provided only a partial answer to the problems of the Boise Valley irrigators, where a growing number of water users sought a limited quantity of natural flow water in the stream. Water shortage proved a more vexing problem than its equable distribution. The large-scale federal reclamation project and construction of storage reservoirs to conserve flood flows for use during irrigation season further complicated administration of water rights on the river. In the late 1920s, a second adjudication was commenced to eliminate wasteful water use and to enter a new decree fixing the rights held under appropriations and permits on the Boise River initiated since the Stewart Decree.

In the case *Pioneer Irrigation District v. American Ditch Association, et al. v. United States of America, Intervenor* the plaintiff, Pioneer Irrigation District, filed a complaint against fourteen canal,

---

722 Paul L. Murphy, "Early Irrigation in the Boise Valley," *Pacific Northwest Quarterly* 44 (October 1935), 177-184.

723 William F. Welch, "Stewart and Bryan Decrees – Sliding Scale Cuts – Boise River," *Address before the Association of Western State Engineers* (September 10, 1947), 11-12.

ditch and water companies to determine the priorities and duty of water for water rights accruing since the Stewart Decree.

The United States, which was not a party to the earlier decree, because of its subsequent purchase of the New York Canal Company interests and its construction of a reclamation project on Boise River, intervened in the suit. Judge Bryan handed down the opinion, known as the "Bryan Decree" in Idaho state District Court in February 1929. All of the water rights adjudicated in the Bryan Decree were junior in priority to water rights decreed in the Stewart Decree. The United States, the New York Irrigation District, and Boise-Kuna Irrigation District appealed the Bryan decision to the Idaho Supreme Court which heard arguments and remanded the case back to the district court in 1931 with respect to certain rights decreed to seven irrigation districts and ditch companies and the United States. The Bryan Decree was superseded by an order issued by Judge Sutton on January 30, 1932 establishing priority dates and quantities of water to be distributed to the parties in the suit, subject to the sliding scale established by Judge Bryan.[724] The United States received certain natural flow rights with 1903 and 1909 priorities (63-0301, 63-0373 and 63-302) and a storage right for 8,000 second-feet in Arrowrock Reservoir with a priority of January 13, 1911 (63-303).[725]

In 1937, the Bureau of Reclamation completed a project to raise Arrowrock Dam five feet in elevation. The end of several years of drought and construction of sugar beet factories resulted in an increase in sugar beet crops, truck farming, and orchards in the Arrowrock Division, thereby increasing late summer water demand. By the end of 1938, the Bureau was seriously investigating construction of additional storage capacity on the Boise River. The following year a topographical survey was conducted at the Anderson Ranch site on the South Fork of Boise River. Ultimately, the Bureau selected this site, but construction progress was slow and water was not stored at the

---

724 In the District Court of the Seventh Judicial District of the State of Idaho, In and For Canyon County, *Pioneer Irrigation District v. American Ditch Association, et al. v. United States of America, Intervener*, February 16, 1929, Decree File No. 459, IDWR Backfile; *Pioneer Irrigation District v. American Ditch Association, et al. v. United States of America, Intervenor*, 50 Idaho 732 (1931)

725 In the District Court of the Seventh Judicial District of the State of Idaho, In and For Canyon County, *Pioneer Irrigation District v. American Ditch Association, et al. v. United States of America, Intervener*, "Order for Distribution of Waters of Boise River for 1932 (Sutton Decree), January 30, 1932, IDWR Backfile, 63-0301.

reservoir site until late 1945. The dam, built to an active storage capacity of 423,200 acre-feet, was not completed until the end of 1950. Like the Boise diversion dam and Black Canyon Dam, Anderson Ranch Dam had a power plant with a capacity of 27,000 kW.[726]

The Payette Division

The lands of the Black Canyon Irrigation District in the Payette division were located in a compact body of 80,000 acres situated between the Boise and Payette rivers and extended westward thirty-five miles to the Snake River. In 1921 the 6,600-acre Notus district became the first irrigated unit of the Payette division. It was irrigated by return and seepage water from drains in the Arrowrock division and siphoned across the Boise River at Caldwell. The Reclamation Service proposed to irrigate the remainder of the Payette division by diverting water from Payette River at Black Canyon Dam and connecting to a thirty-mile long main canal to a distribution point commanding 26,500 acres. The power plant at Black Canyon Dam was constructed primarily to provide electricity to lift pumps to boost water into a highline canal above the gravity line to reach an additional 20,500 irrigable acres.[727] The power plant served many functions in addition to generating electricity for transmission to the pumping plants of various irrigation districts within the Payette division. It provided power for construction activities and other government uses on both the Boise and Owyhee projects, and delivered electrical power to the Idaho Power Company for use on the Bureau's Minidoka irrigation project. The Boise River diversion dam power plant was used during peak irrigation season to supplement the output of Black Canyon when demand for irrigation pumping required additional power.[728]

The Black Canyon Dam – the first major facility of the Payette division – was authorized by Congress in 1922. Located six miles

726 "Boise Project History, 1920-1945," Idaho State Historical Society Reference Series, Number 193 (n.d.), 1-5.

727 "Annual Project History of the Boise Project, Idaho for 1927,"27. USDI, BOR, Boise Project, Annual Project Histories, 1926-1929, Volume 7, Box 256, Engineering & Research Center, Project Histories, 1926-1932, RG 115; NARA-Denver. [D4-6-135].

728 "Annual Project History of the Boise Project, Idaho for 1937, 1938, and1939," USDI, BOR, Boise Project, Annual Project Histories, 1933-1939, Volume 10, Box 255, and "Annual Project History of the Boise Project, Idaho for 1941 and 1942," USDI, BOR, Boise Project, Annual Project Histories, 1940-1942, Volume 11, Box 254, ERC PH, 1940-1944, RG 115, NARA-Denver.

above Emmett, Idaho, the site had been surveyed in 1915 and resulted in a number of settlers flocking to the Emmett area in anticipation of the dam being built. The dam was completed in 1924 with the immediate purpose of supplying the canals of the Emmett Irrigation District. The canals for distributing the stored water to some 48,000 acres of land opened for settlement under the federal reclamation project in the Black Canyon Irrigation District were hardly begun for another decade.[729] In the meantime, Deadwood Dam was built on a tributary of the South Fork of Payette River between 1929 and 1931 as a storage facility for the Black Canyon Canals, in spite of the fact that no construction had begun on the irrigation delivery system. The dam had a capacity of 161,900 acre-feet.[730] The Black Canyon Canal, finally constructed over the years 1936 to 1940, delivered some 1,300 cfs of water to tracts located west of the dam for a distance of twenty-nine miles. Four main laterals were added to this irrigation network by 1948.

The Bureau of Reclamation added Cascade Reservoir, located on the North Fork of the Payette River, to the Payette division in 1948. The largest storage reservoir in the system, the Cascade Dam, held some 653,000 acre-feet of active storage.

Plans for the Mountain Home Desert

Reclamation plans for development of the Mountain Home desert, an area of about 400,000 acres of land lying north of the Snake River between Glenns Ferry and Boise, date back to the early 1900s.

In 1908, the Twin Falls North Side Land & Water Company – also responsible for the Twin Falls North Side project – put together a plan to irrigate this land by means of a canal diverting water at Milner and running from "Gooding to a point west of Little Canyon Creek near Glenns Ferry." This particular route, however, was found unsafe and

729 Under a February 5, 1925 contract agreement between Gem Irrigation District, Idaho Power Company and the United States, power from the Black Rock Canyon plant was delivered to the lines of IPC for transmission to Gem Irrigation District pumping plant during irrigation season (April 15 to October 15). During the remainder of the year, IPC had an option to take power from the plant for a payment of $600 monthly rental. "Operation and Maintenance Black Canyon Dam, 1927." "Annual Project History of the Boise Project, Idaho for 1927,"121-122. USDI, BOR, Boise Project, Annual Project Histories, 1926-1929, Volume 7, Box 256, ERC PH, 1926-1932, RG 115, NARA-Denver.

730 "Annual Project History of the Boise Project, Idaho for 1930," 12-31. USDI, BOR, Boise Project, Annual Project Histories, 1930-1932, Volume 8, Box 256, ERC PH, 1926-1932, RG 115, NARA-Denver.

construction was deemed prohibitively expensive. The plan was thus abandoned. Roughly two years later, another proposal surfaced: This one would have diverted the waters of the Salmon and Boise rivers to service the tract. This plan likewise did not come to fruition.[731] Considering the water-supply difficulties that the North Side project faced in the 1910s – difficulties that ultimately led to the expansion of the Jackson Lake Reservoir – it is unsurprising that a proposal to expand the irrigated acreage of the project was abandoned.[732]

The Reclamation Service took up the idea of a Mountain Home project in the 1910s. During studies leading up to construction of American Falls Dam, the federal government contemplated developing the Mountain Home area in conjunction with the dam.[733] Between 1914 and 1920, on behalf of the Reclamation Service, consulting engineer A. J. Wiley made several investigations of the Mountain Home project. In 1914, he proposed pumping water from the Snake River near Glenns Ferry to serve the Mountain Home area. Costs, however, made such a venture prohibitively expensive.[734]

In a 1918 report, "Irrigation Possibilities in Idaho," Wiley recommended a Mountain Home project and offered an alternative idea: Studying the possibility of diverting water at Milner Dam, in conjunction with the creation of a reservoir site near American Falls, Wiley identified both a 400,000-acre "Big Bruneau" project on the south side, or a 500,000-to-600,000 acre Mountain Home irrigation

731 Department of the Interior, Bureau of Reclamation, Report on Investigation of Mountain Home Secondary Project, Idaho, Made in Cooperation with the Boise Chamber of Commerce, Under Contract Dated May 12, 1921, 4. Box 92 8NS-115-93-001 (Box 92), FY 93, Engineering & Research Center, Studies, Reports, & Projects, 1899-1978 (FY 93, ERC, SRP, 1899-1978), RG 115, NARA-Denver. See also Substantiating Materials, 18, enclosed with *Letter from the Secretary of the Interior, Transmitting a Report on the Hillcrest Unit, Mountain Home Division, Snake River Project, Idaho, Pursuant to Section 9(a) of the Reclamation Project Act of 1939 (53 Stat. 1187)*, House Document No. 41, 85th Cong., 1st sess. (GPO, 1957).

732 Arrington, "Irrigation in the Snake River Valley," 3-11; Lovin, "Free Enterprise," 2-14; and Hoyt, *Water Utilization*, 40-41, and 125-129.

733 See A. J Wiley, Consulting Engineer, Report: Irrigation Possibilities in Idaho, April 12 1918, 8-33; and Department of the Interior, Bureau of Reclamation, Report on Investigation of Mountain Home Secondary Project, Idaho, Made in Cooperation with the Boise Chamber of Commerce, Under Contract Dated May 12, 1921, 7. Box 92, FY 93, ERC, SRP, 1899-1978, RG 115, NARA-Denver.

734 "Mountain Home Project," 2, enclosed with R. B. Williams, Acting Commissioner, to Hon. D. Worth Clark, United States Senate, February 24, 1939; Folder 302.12 – Idaho, Surveys & Investigations, Mt. Home Project, 1930 thru September 1941 (Folder 302.12); Box 564, Entry 7 (Box 564); 302.12 Idaho; General Files, 1930-1945, GAPR, 1919-1945, RG 115, NARA-Denver.

project on the north side of the Snake River as viable projects.[735] The engineer favored the Mountain Home project for its "smoothness of surface" – similar, he observed, to the successful, private Twin Falls South Side project with comparable soil characteristics and consisting of a compact and continuous area of irrigable land extending over one-half million acres. The Mountain Home area also had direct access to the Oregon Short Line Railroad, and Wiley suggested that the North Side project would require a "less expensive distribution system" for water, and afford "great saving in the cost of preparing the farms for irrigation."[736] In this and in subsequent investigations for the federal government, the consulting engineer suggested construction of holdover reservoirs on Boise River, as well as diversions from Boise River to the Mountain Home tract. These plans, however, were rejected as impractical.[737]

In March 1918, the Reclamation Service initiated a more detailed feasibility study. The investigation was undertaken "intermittently," as funding allowed, by regional engineer F. A. Banks. In August 1918, his initial field studies were completed; two months later, Banks released a report and plan for reclamation and irrigation of the Mountain Home plateau. His particular variation on the plan called for diversion of Snake River water at Milner Dam and construction of a 215-mile long canal to service the Mountain Home area. On March 22, 1919, the interior secretary withdrew a large tract from entry so as to safeguard its use for a future federal reclamation project, and the service began drawing up engineering plans for the necessary irrigation facilities.[738]

In May 1921, the Reclamation Service secured an additional $20,000 from the Boise Chamber of Commerce to advance preliminary surveys and engineering studies for the Mountain Home irrigation project. Banks, along with E. B. Debler, was assigned the task. The two engineers investigated the best possible route for a canal from Black Canyon, north of Gooding, on the east end of the project to Little Canyon Creek, north of Glenn's Ferry, on the west. They hoped to

735 Wiley, Irrigation Possibilities, April 12, 1918, 8 and 20. Box 92 FY 93, ERC, SRP, 1899-1978, RG 115, NARA-Denver.

736 Ibid., 32-33. Box 92 FY 93, ERC, SRP, 1899-1978, RG 115, NARA-Denver.

737 "Mountain Home Project," 1. Folder 302.12, Box 564, GAPR 1919-1945, RG 115, NARA-Denver.

738 Report on Investigation of Mountain Home Secondary Project, 5-6. Box 92 FY 93, ERC, SRP, 1899-1978, RG 115, NARA-Denver; and Substantiating Materials, 18, House Document No. 41 (GPO, 1957).

reserve sufficient storage space in the proposed American Falls Reservoir to accommodate the needs of the Mountain Home project. Most of their extensive survey work and hydrological studies were conducted between the spring of 1921 and summer of 1923.

These studies confirmed Wiley's earlier favorable assessment of a Mountain Home project: the soil was "deep and fertile," well-suited "to the dairying industry, and ... the production of wool and livestock." Apples, prunes, potatoes, beans and beets were recommended as crops. "Where properly farmed and irrigated," Debler and Banks concluded, "this land is well up to the average of the Boise and Minidoka projects and there is no reason to doubt but that the yield would equal the average of the these two well and favorably known projects."

The 1921 Interior Department plan for Mountain Home melded aspects of the earlier canal and Salmon and Boise diversion proposals of the Twin Falls North Side Land & Water Company with the federal government's own plans for American Falls Dam. In its broadest outlines, the irrigation plan for Mountain Home was to:

> consist of the American Falls Reservoir, constructed in cooperation with other irrigation interests, the Milner Dam that would have to be used jointly with the two Twin Falls Projects, a headgate just above the Milner Dam, a main canal about 210 miles long extending from the Milner Dam on Snake River to the vicinity of the United States Reclamation Service Dam on Boise River, several extensive wasteways connecting the main canal with Snake River and a distribution system to cover the irrigable area. [739]

Seven-and-a-half acre-feet of water per irrigated acre would need to be stored in the American Falls Reservoir, at a cost of $175 to $185 per irrigated acre.[740] Although 420,000 acres within the Mountain Home project were classified as irrigable, it was not feasible from a water supply perspective to irrigate the entire area from American Falls Reservoir. The report originally projected that some 325,000 acres could be irrigated, but after due consideration for canal transmission losses, evaporation, and a three acre-foot duty of water, by 1923 Banks

739 Report on Investigation of Mountain Home Secondary Project, 6, 9-11, quote on 11. Box 92 FY 93, ERC, SRP, 1899-1978, RG 115, NARA-Denver.

740 "Mountain Home Project," 1, enclosed with R. B. Williams, Acting Commissioner, to Hon. D. Worth Clark, United States Senate, February 24, 1939. Folder 302.12, Box 564, GAPR, 1919-1945, RG 115, NARA-Denver.

and Debler had reduced the feasible estimated project to between 230,000 and 265,000 acres. The 1921 report had assumed a 3,000,000 acre-foot American Falls Reservoir with 2,480,000 acre-feet of storage devoted to initial development of new land. Even with the reduced acreage proposed in 1923, seventy to eighty percent of the new land to be served by the reservoir would be located within the proposed Mountain Home project area under the Banks and Debler plan. The Bureau estimated the total cost of the Mountain Home project as a little over $60 million.[741] At the time, the cost was considered infeasible.

The same year that Debler and Banks's report on American Falls Reservoir and the Mountain Home project was released, J. B. Hays, of the engineering firm of Hays, Martin, and Cameron, conducted a reconnaissance survey for an alternative proposal using a different water source. Hays's Mountain Home survey:

> proposed the construction of a high dam below the Stanley Basin on the Salmon River, a tunnel from the Salmon River to the Boise River Basin, with storage on the Boise River below Big Smoky Creek, and diversion to the Mountain Home lands. In connection with these investigations, very few additional surveys were made and no geologic investigation was made to determine the feasibility of the plan.[742]

American Falls Dam was constructed in 1923 with a storage capacity of only 1.7 million acre-feet and without any consideration for supplying the Mountain Home irrigation project. Instead, the focus of new upper basin irrigation development centered on completion of the 115,000 acre North Side Pumping Unit and the 80,000 acre Gooding division of the Minidoka project and in providing supplemental irrigation water to hundreds of thousands of acres in the Twin Falls, Minidoka and Ashton-Rexburg areas.[743] With these demands on surface water supplies, irrigation of the Mountain Home area from the upper Snake River became increasingly unrealistic.

741 Report on Investigation of Mountain Home Secondary Project, 1-16; and memorandum from Engineers E. B. Debler and F. A. Banks to Chief Engineer, Subject: Report on Mountain Home Secondary Project Investigations, Sept. 13, 1923 enclosed with Department Report on Investigation of Mountain Home Secondary Project. Box 92 FY 93, ERC, SRP, 1899-1978, RG 115, NARA-Denver.

742 Mountain Home Project," 1. Folder 302.12, Box 564 302.12 Idaho, GAPR, 1919-1945, RG 115, NARA-Denver; Substantiating Materials, 20, House Document No. 41 (GPO, 1957).

743 Ibid., 2. Folder 302.12, Box 564 302.12 Idaho, GAPR, 1919-1945, RG 115, NARA-Denver.

Throughout the 1930s and 1940s, the Bureau conducted more preliminary reports and investigations on Mountain Home irrigation development, some in cooperation with the state Water Conservation Board and Idaho Commissioner of Reclamation. Most of these reports focused on water sources other than the Snake River, which had failed to meet upper basin irrigation demand alone during the series of drought years that visited the Snake River Basin in the early 1930s. In 1937, Bureau engineers L. C. Walker and J. R. Riter, investigating the Payette watershed, endorsed the diversion of water from the Boise River to service some 350,000 acres of land between the Boise and Snake rivers. Water diverted from the Payette, Salmon, and Snake rivers would supplement and replace Boise River water under this plan. A year later, in a separate report, Riter proposed using runoff from the Salmon and Payette rivers along with diversions from the Boise River to realize the reclamation of Mountain Home.[744] A 1939 Bureau memorandum, observing that "[r]ecent water supply studies of the Upper Snake River indicate periods as long as five consecutive years when no surface water will be available in the Snake River at Milner Dam," concluded "[i]rrigation of the Mountain Home area from the Snake River, as proposed in the 1923 report, is, therefore, impracticable."

Nevertheless, given the results of the Payette watershed investigations, the memorandum went on to note that:

> ... a study is being made of the possibility of diverting the surplus water from the Payette River to the Mountain Home area. Consideration is also being given to the possibility of supplementing this surplus water by importing some water from the Salmon River in the Stanley Basin to the Boise or Payette Rivers for rediversion to the Mountain Home area.[745]

A 1941 memorandum sent from the Bureau's chief engineer S. O. Harper to the commissioner, regarding the project, confirmed that the Bureau had settled on diverting water from the Payette, Boise, and Salmon rivers to service the Mountain Home roject. However, despite identifying "certain features' of the proposed project," Harper indicated

744 Substantiating Materials, 19, House Document No. 41 (GPO, 1957).

745 "Mountain Home Project," 2. Folder 302.12, Box 564 302.12 Idaho, GAPR, 1919-1945, RG 115, NARA-Denver.

that "investigations of the project have not yet progressed to the point where the usual detailed project report can be submitted."[746]

Over the next five years, Bureau plans for Mountain Home began to solidify. In October 1942, a preliminary report on the project was issued. The report called for diverting water from the Payette and Salmon rivers into the Boise River, "then rediverting that water either directly or by exchange for an equivalent amount of Boise River water to the Mountain Home area." From 1944 through 1946 at least three more plans were introduced by the Bureau that called for inter-basin water transfers using the Payette, Boise and Salmon river watersheds to supply 350,000 to 400,000 acres of Mountain Home lands with irrigation water.[747]

As the Bureau formulated its plans, local support for the development of the Mountain Home area grew.

In January 1940, irrigation interests in communities from Boise to Weiser joined with Idaho Power Company to form the Southwestern Idaho Water Conservation project, Inc. (SWIWCP). It was concerned principally with development of the Boise and Payette watersheds, but also took a strong interest in reclamation of the Mountain Home desert.[748] Through the 1940s and early 1950s, the organization contributed money and technical resources to state and Bureau investigations and lobbied members of Congress to advance the project. In January 1940 IPC tasked engineer T. A. Purton to provide the Bureau with "available engineering data" on "various sites."[749] During the 1941 fiscal year, SWIWCP contributed $10,000 to state and Bureau studies of the Mountain Home project.[750] In late 1950, members of the group

---

746 Memorandum from Chief Engineer to Commissioner, Subject: Mountain Home Project, Idaho, July 21, 1941. Folder 302.12, Box 564 302.12 Idaho, GAPR, 1919-1945, RG 115, NARA-Denver.

747 Substantiating Materials, 19, House Document No. 41 (GPO, 1957); "Mountain Home Project to Add 3,000 Irrigated Farms," *Twin Falls Times News*, December 7, 1945.

748 "Group Will Promote Irrigation," *Idaho Statesman*, January 23, 1940. Folder: 302.12 Idaho, Surveys & Investigations, Jan. 1940 Thru Nov. 1940 (Folder: 302.12); Box 561 302.12 Idaho (Box 561), GAPR, 1919-1945, RG 115, NARA-Denver.

749 C. J. Strike to R. F. Walter, Chief Engineer, January 17, 1940, enclosed with memorandum from Chief Engineer, to Commissioner, Subject: Boise-Payette-Weiser Investigations, Idaho, February 12, 1940. Folder: 302.12; Box 561; 302.12 Idaho; GAPR, 1919-1945, RG 115, NARA-Denver.

750 S. O. Harper, Acting Chief Engineer, Bureau of Reclamation, to Mr. James Spofford, State Commissioner of Reclamation, July 6, 1940. Folder: 302.12; Box 561; 302.12 Idaho; GAPR, 1919-1945, RG 115, NARA-Denver.

traveled to Washington, DC to lobby for the project in conjunction with the establishment of a federal high dam at Hells Canyon.[751]

Over this same time, from the 1920s to the 1950s, Idaho Power continued development of the hydroelectric potential of the Snake River Basin – particularly below Milner Dam.

By the mid-1930s, it had not only an expanded power plant at American Falls, but also a proposed plant at Twin Falls, and plants at Shoshone Falls, Upper Salmon Falls, Lower Salmon Falls, Swan Falls, and Oxbow; by the early 1950s, it had refurbished Lower Salmon Falls and added new facilities at Bliss and C. J. Strike.

As most of these facilities were located in the stretch of the Snake River between Milner and Weiser, they troubled proponents of federal reclamation and irrigation in Idaho who still held hopes for the Mountain Home project. In a February 12, 1937, letter Will Simons, chair of the Idaho State Planning Board, informed John Page, commissioner of reclamation that he was "apprehensive of the effect" of Idaho Power's Lower Salmon Falls facility on water rights and water use not only on the proposed Bruneau and Boise-Mountain Home projects, but also "on the whole reclamation situation in the Snake River Basin." A more comprehensive approach to water development and planning was required, Simons wrote:

> I believe quick action, in effecting a study of this situation is most necessary to provide for the proper planning coordination between the proposed Bruneau project, the Lower Salmon Falls project, and the rest of the Snake River Basin. If this is not done it is quite obvious that confusion in planning and development will occur, resulting in the retarding of irrigation development in the basin or its proceeding under a haphazard and disconnected program, all of which will be most detrimental to the welfare of the State.[752]

---

751 Statement by Executive Committee of Southwestern Idaho Water Conservation Project, Inc. in Re Snake River Basin Project, For Release July 9, 1951, 2, enclosed with J. L. Driscoll, Chairman, Executive Committee, to the Honorable Len Jordan, Governor of the State of Idaho, May 1, 1951. Folder: Southwestern Idaho Water, Water Conservation Project, Box 7, AR 2/23 Governors' Papers – Len Jordan; *Idaho Statesman*, December 19, 1950. Folder: Washington Trip, December 18th to 22nd, 1950; Box 19980207; AR 53, Department of Water Administration, ISHS; and Stacy, *Legacy*, 139-141.

752 Will Simons, Chairman, to the Honorable John C. Page, Commissioner, United States Bureau of Reclamation, February 12, 1937. Folder 302.12 Idaho, Surveys & Investigations, 1935 Thru 1937 (Folder 302.12), Box 509 302.12 Idaho (Box 728), General Files, 1930-1945, GAPR, 1919-1945, RG 115, NARA-Denver.

Page's response, coming a month later, sought to reassure Simons: "I will call this situation to the attention of the Chief Engineer, who is well informed as to the whole program on the Snake River, and it is hoped that a plan can be worked out which will be most satisfactory to all interests."[753]

No subsequent response, from either Page or the chief engineer, however, has since come to light.

For its part, Idaho Power sought to assuage any concerns that might have existed over its development of the middle Snake and joined with irrigation interests to form SWIWCP. In offering his company's assistance to the Bureau in January 1940, IPC president C.J. Strike was careful to write:

> Idaho Power Company has no ulterior motive and is only interested in the coordinating of the reclamation program with the development of additional power requirements … Idaho Power Company has a number of sites on which additional power could be developed, however such developments are being held in abeyance in that we feel that the market for additional power should be reserved, so far as possible, for the multiple use of water in connection with the irrigation development.[754]

The drought of the 1930s and the realization that the water supply of the Snake River above Milner, including the water stored at American Falls Reservoir, was dedicated fully to irrigation uses in the upper basin, forced promoters of the Mountain Home project to search elsewhere for water. They looked northwest to the headwaters of the Boise, Payette, and Salmon rivers and finally to a storage reservoir at Hells Canyon.

Following the World War II, concern over the impact of Idaho Power on development of the Mountain Home project re-emerged – particularly in the Hells Canyon reach. Reclamation of the Mountain Home desert was integral to the Bureau of Reclamation's proposal for the construction of a federal high dam in the Hells Canyon reach.

---

753 Page to Simons, Chairman, March 12, 1937. Folder 302.12, Box 509 302.12 Idaho, GAPR, 1919-1945, RG 115, NARA-Denver.

754 C. J. Strike to R. F. Walter, Chief Engineer, January 17, 1940, enclosed with memorandum from Chief Engineer, to Commissioner, Subject: Boise-Payette-Weiser Investigations, Idaho, February 12, 1940. Folder: 302.12; Box 561; 302.12 Idaho; GAPR, 1919-1945, RG 115, NARA-Denver.

Under the Bureau's final plan, known as the Snake Basin Plan and released in May 1949, such a dam would subsidize the construction and operation of the Bureau's reclamation projects and generate the electricity needed to pump water from the Payette and Boise watersheds through mountain tunnels to irrigate 400,000 acres of sagebrush plain between the east-west mountain range along the northern margin of the Snake River plain and the Snake River. Thus, rather than diverting from the Snake River itself, the Mountain Home project planned to draw off central Idaho mountain water and transferred it via mountain tunnels into the southwest Idaho desert. The Bureau estimated that this former Mountain Home wasteland would support 2,500 new irrigated farms, many of them occupied by veterans of World War II.[755]

Idaho Power Company unveiled its own plan for smaller run-of-the-river Hells Canyon dams in 1947 to counter the Bureau's high federal dam proposal. Idaho Power Company, in league with Idahoans who opposed the growing federal involvement in the power business of the Pacific Northwest, as reflected in the Snake Basin plan and Hells Canyon's federalization, ultimately thwarted the Bureau's postwar plans for its Mountain Home development.[756]

Idaho Power's postwar middle Snake designs

Since the 1920s, knowledgeable observers had noted that the Middle Snake was best suited to hydroelectric power development – not irrigation. The 1920 joint state-federal Board of Engineers' report on American Falls had stressed that geology, hydrology, and topography dictated that all of the waters of the Snake River above Milner be devoted to irrigation; below Milner Dam, as the river receded into a deep canyon, irrigation development was technologically and economically impractical.[757] USGS engineer W.G. Hoyt made a similar

755 Memorandum from The Regional Director, Region I to The Commissioner, Bureau of Reclamation, regarding Supplemental Report on the Payette Unit, Mountain Home Project, Idaho, May 18, 1949, enclosed with R. J. Newell, Regional Director, Region I, "Payette Unit, Mountain Home Project, Idaho: A Supplement to the Columbia River Basin Report," Project Planning Report No. 1-5.5-7.S., May 1949. File Folder 100-2, Project No. 1971, Formal, Part 2, May 1, 1951 to June 30, 1952; Box 253; Accession No. 138-88-006; Idaho Power Co. – Snake River, FERC; and Stacy, *Legacy*, 139-142.

756 Substantiating Materials, 11, 18, and 20, House Document No. 41 (GPO, 1957).

757 Board of Engineers to Honorable D. W. Davis, Governor of the State of Idaho, April 10, 1920, 5-6. Box 728 Minidoka 301.1 (Box 728), General Files, 1919-1929, General Administrative & Project Correspondence, 1919-1945 (GAPC, 1919-1945), RG 115, NARA-Denver.

point in a separate report to the Federal Power Commission in 1921. Fourteen years later, in *Water Utilization in the Snake River Basin*, he again emphasized that almost all of the waters of the Snake River above Milner could be diverted and used for irrigation without significantly hindering the generation of hydroelectric power below Milner.[758]

Idaho Power was no less aware of the hydroelectric potential of the Middle Snake – which was precisely why they sought to develop new facilities in this reach of the river, beginning with Twin Falls and Upper Salmon Falls in the early 1920s. IPC's predecessor companies had already built four power facilities along the middle Snake at the start of the century: Swan Falls (1900), Shoshone Falls (1907), Lower Salmon Falls (1909), and Oxbow (1914).[759] After the power company acquired these facilities in 1916, they upgraded nearly all of them – obtaining additional water rights, installing new generators, or a combination of the two – so as to make greater use of the river.[760]

Twin Falls and Upper Salmon Falls were important elements of the company's plan to capitalize on the hydroelectric potential of the middle Snake, seemingly without coming into conflict with upstream irrigators. Nevertheless, Idaho Power's designs for the two power sites when presented to the Federal Power Commission ran headlong into local and federal desire to maximize irrigation above Milner Dam. Upper Snake irrigators and the Reclamation Service (the predecessor to the Bureau of Reclamation) had two reasons to object to the Twin Falls facility. First, they wished to preserve as much water above Milner Dam for irrigation as possible – effectively turning the river dry at the dam during the summer months. Possessing a water right prior in time

---

758 Hoyt, "Report Relative to Application of Idaho Power Company for a Preliminary Permit to Develop Upper Salmon Falls Site," July 1921, 5. Folder: D-100, 12-G-44, WPB, 1894-1961, RG 57, NARA II; and Hoyt, *Water Utilization in the Snake River Basin*, 35-36, and 231-244.

759 The company also acquired the Thousand Springs facility, but this power plant made use of water from Idaho's "Lost Rivers" and not the Snake River itself. See Hoyt, "Report Relative to Water Power Resources of Snake," August 1922, 57-63. Folder: D-100, 12-G-51, WPB, 1894-1961, RG 57, NARA II; E. A. Wilcox, "Electrical Development in Southern Idaho," Twin Falls News, January 9, 1913, and "Harnessing A Thousand Springs So the People May be Served," *Twin Falls Daily News*, May 24, 1920.

760 Hoyt, *Water Utilization*, 201; Stacy, *Legacy*, 28-29 and 44; Idaho Power Company, *History*, 12-13, and 64; State of Idaho, License and Certificate of Water Right, Water License No. 14881, December 10, 1925, Idaho State Water Right 02-2036 Back File; "Exhibit Q," Application for License (Act of June 10, 1920, 41 Stat. 1063), Lower Salmon Falls Power Project (Idaho). Idaho Power Company, November 22, 1923. File Folder 100-2, Project No. 457, Formal, Part 1, December 19, 1923 to December 1946, Idaho Power Company (Lower Salmon Hydro-Electric plant), Snake River, Box 2, Accession No. 138-62A-0938, FERC]

to many irrigators upstream, IPC could prompt the watermaster to pass water at Milner Dam to service its Twin Falls plant that otherwise would be used for irrigation purposes. Second, local water users and the service/Bureau wished to acquire Twin Falls as a public power site, to compensate for any future reduction in winter releases from the then-proposed American Falls Reservoir that would curtail the production of cheap federal power at either Minidoka or a proposed government facility at American Falls.[761] There was less concern for Upper Salmon Falls as a potential public power facility – but once again, the service and local irrigators saw IPC's proposed plant as a threat to the federal dam at American Falls, which was intended to "completely control and use Snake River above Milner dam."[762] Ultimately, after more than a decade-and-a-half of debating with the Bureau and irrigators before the Federal Power Commission, IPC managed to secure a federal license for its Twin Falls plant – a license limited by a stipulation, Article 14.

Article 14 restricted the water the company could use at the facility to its primary and secondary rights under the June 15, 1923, contract, as well as the return flows below Milner Dam.[763] Upper Salmon Falls, by contrast, was not licensed by the FPC; Idaho Power decided to drop its proposal to build such a facility on public lands in lieu of developing a different site on private land (outside of the FPC's purview) in the 1930s.[764]

---

761 George Hoffman, Report on Application Before Federal Power Commission by Idaho Power Company, Serial No. 18, for Preliminary Permit covering Twin Falls site, on Snake River, 5-6, enclosed in a memorandum from Project Manager, Minidoka Project to Chief Engineer, Denver, Colorado, Subject: Inclosing report on the application before the Federal Power Commission by the Idaho Power Company, Serial No. 18, for Preliminary Permit covering Twin Falls Site, on Snake River, March 7, 1921, enclosed in a memorandum from Chief Engineer to Director, Washington, D. C., Subject: Report on the application before the Federal Power Commission by the Idaho Power Company for preliminary permit covering Twin Falls site on Snake River, March 12, 1921. Folder: 412.12 Idaho, Federal Power Applications (Application #18-Twin Falls Site on the Snake River), Idaho Power Company, 1930 thru (Folder 412.12), Box 809 412.06_412.12 - Colorado – Idaho, GCF, 1930-1945, RG 115, NARA-Denver; and Minutes of Public Hearing, October 7, 1921, 13 and 24. Formal File 100-2, Part 1, Federal Power Commission No. 18, July 7, 1920 to December 31, 1927, FERC.

762 A. P. Davis, Director, to O. C. Merrill, Executive Secretary, Federal Power Commission, March 26, 1921. Folder 412.12, Box 809, GCF, 1930-45, RG 115, NARA-Denver.

763 Federal Power Commission, License on Government Lands, Project No. 18, Idaho, Idaho Power Company, June 11, 1934. File Folder 100-2, Project No. 18, Formal Pt. 2, January 1, 1928 thru December 31, 1948 (File Folder 100-2, Formal, Pt. 2), Box 13, Accession No. 138-88-006, FERC.

764 Application for Permit to Appropriate the Public Waters of the State of Idaho, Application No. 24072, Permit No. 17069, April 2, 1929. Idaho State Water Right 02-2057 Back File.

Idaho Power Company's clash with federal reclamation officials over their proposed Twin Falls and Upper Salmon Falls power plants in the 1920s foreshadowed the conflicts over the Lower Salmon, Bliss, and C.J. Strike facilities and especially, the Hells Canyon complex (Oxbow, Brownlee, and Hells Canyon dams) in the 1940s and the 1950s. Whereas the earlier disagreements among the power company, the federal government, and local water users had focused the potential impact of Idaho Power's downstream facilities on the Upper Snake, these later ones concerned irrigation development along the Middle Snake proper.

By the 1940s, IPC was by far the largest distributor and producer of hydroelectric power in Idaho. It was interconnected by transmission lines to the power plants of the Utah Light & Power Company, which supplied a large part of the eastern Idaho and northern Utah markets, and to the three power enterprises of the Bureau of Reclamation at Black Canyon (Payette River), Boise River Diversion (Boise River) and American Falls (Snake River). The power company also had twelve plants spaced along the Snake River and its tributaries – ranging from American Falls to Oxbow below Weiser. With a total installed capacity of 103,270 kW in these plants, Idaho Power produced about seventy-five percent of all the hydroelectric power used in the state.

Table 8.1 lists the company's plants as of 1941.[765]

Table 8.1. Idaho Power Company Hydroelectric plants, 1941[766]

Name	Location	Installed Capacity (kilowatts)	Static Head (feet)
Snake River			
American Falls	American Falls	27,500	50
Twin Falls	Kimberly	13,500	149
Shoshone	Twin Falls	10,880	214
Upper Salmon	Hagerman	18,000	47
Lower Salmon	Hagerman	6,080	36
Swan Falls	Kuna	9,060	23
Salmon System	Salmon	1,100	
Malad River			

765 E. V. Berg, Commissioner of Reclamation, *12th Biennial Report of the Department of Reclamation, State of Idaho, 1941-1942,* 31-34. The installed capacity of IPC's generating system was increasing rapidly. In 1935, it was reported as 71,900, kW and by 1941 the total has risen to 103,270 kW – an increase of more than forty percent in only six years. Hoyt, *Water Utilization,* 192.

766 E. V. Berg, Commissioner of Reclamation, *12th Biennial Report of the Department of Reclamation, State of Idaho, 1941-1942*, 32.

Malad	Bliss	4,400	125
Clear Lakes Springs-			
Clear Lake	Buhl	2,500	78
Thousand Springs			
Thousand Springs	Wendell	8,000	182
Boise River			
Barber	Boise	1,050	20
Payette River			
Horseshoe Bend	Horseshoe Bend	1,200	36
Total			**103,270**

Despite a robust system, growing postwar power demands spurred Idaho Power to expand the capacity of existing plants and construct new ones. In 1946, the company calculated that over the past two years, its system load had grown by nearly 20 percent, approximately 32,000 kW, and over the past ten years, the load had increased by nearly 113 percent. In due course, the company further estimated that its "[a]verage load growth" would soon reach 15,000 kW.[767] IPC's system had to increase, and as no other power sites were available to the company in the Upper Snake, the company turned its attentions once again to the Middle Snake. The company sought to develop all of the remaining power sites on the river between Milner and Lewiston.

At the same time, eschewing the historic division of the Snake River at Milner, with the river above Milner Dam devoted to irrigation and the river below the dam given over to power, the Bureau sought to develop irrigation projects on the Middle Snake. In the early decades of the twentieth century, the agency had established projects along tributaries to the Snake River – the Boise and Payette rivers, in particular – and had initiated studies of the Mountain Home desert area.

By the mid-1940s, federal proposals for the area's development involving the Middle Snake began to take shape. These proposals would ultimately involve construction of a high dam at Hells Canyon that precluded Idaho Power's own designs on the site.

The stage was set for a decade-long series of clashes between Idaho Power Company and the federal government over the future development of the Middle Snake – clashes that would reverberate into the Upper Snake River in the 1960s, 1970s and beyond.

---

767 Before the Federal Power Commission, Application for Amendment of License (Project No. 457 – Idaho), August 1, 1946, 2. File Folder 100-2, Project No. 457, Formal, Part 1, Box 2, FERC.

The Authors

Stephen R. Wee (M.A., History, University of California, Davis, 1985) is the principal author of Working River. Mr. Wee is a retired founder of JRP Historical Consulting, LLC, one of the oldest historical consulting firms in the United States, located in Davis, California. In his four-decade-long career, Mr. Wee undertook historical investigations of alleged riparian / littoral and appropriative water rights, groundwater, and Native American and federal reserve water rights throughout the American West, and provided expert witness testimony in administrative hearings and judicial proceedings in Idaho, California, Oregon, and New Mexico.

Scott A. Miltenberger (Ph.D., United States History, University of California, Davis, 2006) assisted Mr. Wee in readying this book for publication in 2022. Now a principal with JRP, Dr. Miltenberger's first project out of graduate school was aiding in the research and preparation of the expert historian report that is the basis of Working River. Dr. Miltenberger specializes in historical research and analysis of environmental and natural resources issues – including all manner of Western water rights – and has testified as an expert historian in both state and federal courts, including in a interstate stream dispute before the United States Supreme Court.

Clive Strong was a deputy attorney general for the Office of the Idaho Attorney General from 1983 to 2017. Strong served as the Chief of the Natural Resources Division for 33 years. On his third day in the Office of the Attorney General, Strong was designated as the State's lead counsel in the Idaho Power Swan Falls cases, a role he held throughout his time with the Office of the Idaho Attorney General. Strong also served as lead counsel for the State in the Snake River Basin Adjudication from its initiation in 1987 through to its completion in 2014. He drafted portions of the Swan Falls Agreement and was the state's lead negotiator for the 2009 Swan Falls Reaffirmation Agreement. Strong is a nationally recognized expert in water law.

Made in United States
North Haven, CT
10 November 2026

22095622R00222